INTERNATIONAL SERIES OF MONOGRAPHS ON
PURE AND APPLIED BIOLOGY

Division: **MODERN TRENDS IN PHYSIOLOGICAL SCIENCES**

General Editors: P. Alexander and Z. M. Bacq

Volume 18

BIOCHEMICAL LESIONS AND LETHAL SYNTHESIS

OTHER TITLES IN THE DIVISION ON MODERN TRENDS IN PHYSIOLOGICAL SCIENCES

Vol. 1. FLORKIN—*Unity and Diversity in Biochemistry*
Vol. 2. BRACHET—*The Biochemistry of Development*
Vol. 3. GEREBTZOFF—*Cholinesterases*
Vol. 4. BROUHA—*Physiology in Industry*
Vol. 5. BACQ and ALEXANDER—*Fundamentals of Radiobiology*
Vol. 6. FLORKIN (Ed.)—*Aspects of the Origin of Life*
Vol. 7. HOLLAENDER (Ed.)—*Radiation Protection and Recovery*
Vol. 8. KAYSER—*The Physiology of Natural Hibernation*
Vol. 9. FRANCON—*Progress in Microscopy*
Vol. 10. CHARLIER—*Coronary Vasodilators*
Vol. 11. GROSS—*Oncogenic Viruses*
Vol. 12. MERCER—*Keratin and Keratinization*
Vol. 13. HEATH—*Organophosphorus Poisons*
Vol. 14. CHANTRENE—*The Biosynthesis of Proteins*
Vol. 15. RIVERA—*Cilia, Ciliated Epithelium and Ciliary Activity*
Vol. 16. ENSELME—*Unsaturated Fatty Acids in Atherosclerosis*
Vol. 17. BALABUKHA—*Chemical Protection of the Body against Ionizing Radiation*

OTHER DIVISIONS IN THE SERIES ON PURE AND APPLIED BIOLOGY

BIOCHEMISTRY

BOTANY

PLANT PHYSIOLOGY

ZOOLOGY

BIOCHEMICAL LESIONS
AND
LETHAL SYNTHESIS

SIR RUDOLPH A. PETERS
M.C., M.D., F.R.S.

Hon. Fellow of Gonville and Caius College, Cambridge
and of Trinity College, Oxford
Emeritus Professor of Biochemistry
University of Oxford

PERGAMON PRESS
OXFORD · LONDON · NEW YORK · PARIS
1963

PERGAMON PRESS LTD.
Headington Hill Hall, Oxford
4 & 5 Fitzroy Square, London, W.1

PERGAMON PRESS INC.
122 East 55th Street, New York 22, N.Y.

GAUTHIER-VILLARS ED.
55 Quai des Grands-Augustins, Paris 6e.

PERGAMON PRESS G.m.b.H.
Kaiserstrasse 75, Frankfurt am Main

Distributed in the Western Hemisphere by
THE MACMILLAN COMPANY · NEW YORK
pursuant to a special arrangement with
Pergamon Press Limited

Library of Congress Catalog Card Number 63-11698

Set in Monotype Imprint 11 on 12 pt and printed in Great Britain by
J. W. Arrowsmith Ltd, Bristol 3

CONTENTS

PREFACE

BIOCHEMISTS have the objective of describing in chemical terms and explaining the dynamic workings of the living cell; with physiologists they should know how the activities of cells are integrated in the wider whole of the animal or plant. This involves an understanding of the nature of the problems raised, and a consideration of the best means of reaching these objectives; in other words of the strategy of the attack. In this book I have tried to present one method of approach which can be fruitful in pure biochemistry and which certainly leads to practical considerations in medicine. It is not intended to suggest that other approaches are not also of value; some part of almost all the biochemical work in progress now is likely to be needed in the ultimate description of the living cell.

This can be regarded as a collection of short essays or lectures on subjects which have interested my colleagues and myself over a series of years. I have, however, done my best to bring the various subjects treated up-to-date so that they should stand in the right perspective. It has been impossible to do justice to all those who have worked over this rather large field, and there are, I am sure, quite unpardonable omissions. However, it was a choice between the wood and the trees. Too much detail would have obscured the main point which I have been trying to make. I have included references to various reviews in the hope that anyone who may wish to go more deeply into any part will find the references in the reviews to facts which are missing in my text.

RUDOLPH PETERS

ACKNOWLEDGEMENTS

I WISH first to thank all those colleagues who have taken part in the work described in this book; without them most of the research would have been impossible.

In the preparation of the text, I have been helped by numerous conversations with those around me and by the kind loan of a text in manuscript by Professor Gale. I would thank Mr. C. W. Carter and Mr. J. R. P. O'Brien for reading Chapter 2, Dr. V. P. Whittaker, Chapter 3, Dr. P. Alexander, Chapter 4 and the whole as editor, Miss Honor Fell, F.R.S., for parts of Chapters 4 and 5, Dr. P. F. V. Ward, Chapter 6, and Professor E. F. Gale, F.R.S., Chapter 7; also Dr. Jacobson for advice. I had the good fortune to spend 31 years as Professor in Oxford; after this I am indebted to Dr. I. de Burgh Daly, C.B.E., F.R.S., for the years spent at the A.R.C. Institute of Animal Physiology at Babrahem, and now to Professor F. G. Young, F.R.S., for the facilities he has provided in the Biochemical Laboratory, Cambridge. I am grateful to Dr. Aldous for reading Part 2 and to Mrs. M. Shorthouse for help with the proofs and with the index.

I wish to thank the Royal Institute of Public Health and Hygiene for permission to reprint the Harben lectures from the *Journal of State Medicine;* also the Faraday Society for the communication to their *Transactions*. The Waverley Press, Baltimore have kindly allowed the use of some blocks from my Dohme lectures in Johns Hopkins University. Original prints have kindly been sent by Dr. Porter, Dr. Whaley, Dr. Hagenace and Dr. Mercer, to whom I am grateful.

In addition to those mentioned, I wish to thank Editors and authors for the following figures:

Biochemical Journal: Figs. 2.2, 2.7, 2.8, 2.9, 2.10, 2.12, 2.13, 2.14, 3.6, 4.1, 4.2, 5.1, 6.7, 6.18, 6.19, 6.22, L5, L16.

Proc. Royal Society: Figs. 2.3, 2.4, 2.5, 6.3, L19.

Brit. Med. Bulletin: Figs. 2.11, 3.3, 6.4.

J. Physiology: Figs. 2.6, 6.4, L15.

Quart. J. Med.: Figs. 3.7, 3.8.

Biochem. Pharmacol.: Figs. 4.5, 4.6, 6.8.

Bull. J. Hopkins Hospital: Figs. 3.4, 6.5, 6.9, 6.10, 6.11.

Endocrinology: Fig. 6.14.

J. Comp. Pathology & Therapeutics: Fig. 6.15.

British J. Exp. Pathology: Figs. 16, a, b, c; L1, and L2.

Analytical Chemistry: Fig. 6.21.

J. Biol. Chemistry: Fig. on p. 137.

Quart. J. Exp. Physiology: Figs. 6.5, 8.1, 8.2, 8.3.

Gonville & Caius College, Cambridge: Portrait of W. B. Hardy.

J. Biochem. Biophys. Cytology: Figs. L12, L14.

Amer. J. Botany: Fig. L10.

Biochem. Biophys. Acta: Fig. L11.

Nature: Fig. L13.

J. Experimental Medicine: Fig. L9.

J. Anatomy: Fig. L17.

Birbeck and Barnicott, *Pigment Cell Biology*, p. 549, (1959) Line 18. Academic Press.

It is appropriate to mention that some of the work described was incorporated in the Dunham Lectures in 1946, in lectures in Liège, 1948, and in Italy 1948–9, also in the Dohme Lectures, in Baltimore, in 1954.

Lastly and most importantly, I wish to thank my family for their endurance and above all, my wife, for her wisdom and critical support.

Calcutta Medical Journal: Fig. 5.2.

PART 1

CHAPTER 1

Introductory

By the term "Biochemical lesion" is meant the initial biochemical change in tissue cells which precedes any damage visible with the light microscope. The term was introduced some 30 years ago to crystallize the idea that pathological disturbances in tissues were initiated by changes in their biochemistry. It is one of the more subtle ways in which the metabolism of the cell may meet interference; biochemical analysis of a tissue may provide evidence of changes, in tissue enzymes for instance, at a time when a histological abnormality is not detectable. Put crudely, it means substituting biochemical analysis for morbid anatomy, and taking disease at a point where it may still be reversible. The biochemical lesion is not necessarily due to change in the action of an enzyme, though this will often be the case owing to the extreme importance of enzymes in the tissue. If a succession of changes is essential to life, each of which is catalysed by a specific enzyme, that is to say if $A \rightarrow B \nrightarrow C \rightarrow D$ and there is a block between B and C, then the metabolic sequence cannot be carried out, and the block might be expected to produce serious damage to life, provided that there is no alternative pathway for the reactions giving energy. It has been pointed out that the biochemical lesion in cyanide poisoning is the blocking of the initial stage by which oxygen enters, through the enzyme cytochrome oxidase. This is a very direct instance, usually causing death in organisms depending upon a supply of oxygen. The essays in this book include other examples of greater subtlety, where there may be a partial damage giving time for some secondary changes to develop, and in some the effects may be reversible. In medicine we should concentrate our thoughts on the initial biochemical change rather than upon the appearance of the diseased tissue, wherever possible. Such an approach through biochemistry is the most fundamental one also to knowledge of the exact action of drugs. In the application of enzymes to disease, there is one thing to remember. Though it may be held that all enzymes are essential in a greater or less degree to life, this is not true in medicine. A biochemical lesion initiating a pathological condition is only likely to occur in an enzyme which for some reason or other is in short supply, and so becomes a bottleneck.

The original work upon the biochemical lesion in deficiency of thiamine (vitamin B_1) was put together for a lecture given at the Hospital for Nervous Diseases in London in 1936.[1] A perusal of this makes it clear, I think, that at that time we knew quite well that we had proved something revolutionary; namely that a new approach to pathology was now possible. I mention this because it was suggested to me recently that we were not clear then of the implication of our results. At that time, it is true, and over the years I was surprised that the point was not taken up more quickly by others, though it did need a rather fundamental adjustment of outlook. The idea seems so obvious to biochemists now.(2) Perhaps the position can be understood better by pointing out that even nine years later than our proof, an enzymic approach to the problem of finding an antidote to lewisite poisoning was stopped officially in all good faith by a committee which included many folk of considerable scientific weight. I do not mention this by way of criticism; but merely to show how little it was generally understood what had been proved.* It is therefore with great pleasure that one sees a progressive pathologist now feeling the need for a more biochemical approach. In his book on *New Pathways in Chemical Pathology*, Sir Roy Cameron(3) describes how his group were dissatisfied and often discouraged by the morphological outlook on cellular pathology, and consequently turned to recent biochemical and biophysical techniques for illumination in their difficulties, and have so come down to a study of intracellular biochemical lesions. A similar point of view has been put forward by Dawkins and Rees in their book *A Biochemical Approach to Pathology*.(4) In view of this evidence of a turn in the current of modern thought, it seemed advisable to collect in one book the various biochemical lesions which I have studied with colleagues, so that it was possible to appraise the value of this approach and to decide what advances are possible by tackling the problems in this way; it is essentially an attempt to work more consciously at "Co-ordinative biochemistry".

A short statement of the aims of this book may make the whole objective clearer. As will be indicated, when problems are tackled from this angle, conclusions may be reached from the biochemistry which are applicable to events *in vivo*; the method therefore constitutes a bridge between physiology and biochemistry; this bridge is more necessary than many think; essential as the more chemical studies on enzymes are, and without them there could be no progress, it is a truism that they must in the end be placed in the proper context of their reactions in the living cell. It could be argued that the moment has come now when attention must be more definitely turned in this direction. It is inevitable that only a small part of the chemical information obtained *in vitro* will be applicable owing to the conditions of pH, salt content and temperature at which various cells normally live.

* This is not surprising. It was many years before medicine as a whole appreciated that a deficiency could produce disease, so powerful was the concept of microbial attack.

A start is made in Part 1 with the biochemical lesion in thiamine deficiency. In this chapter it is shown how the attempt to unravel the biochemical events leading to convulsions of birds suffering from thiamine deficiency presented a proof that the initial change lay in a failure in the oxidation of pyruvate, which was due to a loss of the essential coenzyme cocarboxylase, or thiamine pyrophosphate, from the brain tissue of the pigeon. It was possible to show that addition of the vitamin even *in vitro* would restore the metabolism of pyruvate. This was the first time that an action of a vitamin *in vitro* had been discovered. It was also the first time that the oxidations connected with a vitamin and with enzymes had been proved to have a pathological effect. It was a true biochemical lesion, in the sense that no change of a histological character could be demonstrated by the light microscope in the tissue at a time when the convulsive signs appeared. If taken in time, the pathological condition could be completely reversed.

Following this the problem of the biochemical action of the tervalent arsenicals is considered, and the explanation in terms of enzymes is given for the difference between the actions *in vivo* of the mono- and di-substituted arsenicals. The work carried out in 1939 to 1942 was sufficiently practical to lead to the development of the antidote known as BAL (British anti-lewisite), whose chemical name is 2,3-dimercaptopropanol. The secret of the toxic action of a compound like lewisite, lies in its ability to form a ring with dithiols which is relatively stable; and it is this ring formation which makes the monosubstituted arsenical more toxic than the more potentially toxic disubstituted compound like the toxic arsenical smoke, diphenylchloroarsine. It is again shown that with these compounds it is the oxidation of pyruvate which is the first point of attack. This biochemical lesion is completely reversible by the antidote, though it is irreversible in its absence.

After this an account is given of the behaviour of mustard gas. In this case the effects are much more complex, and are not at present reversible, because they depend much upon the concentration. Mustard gas is an alkylating agent which operates in this fashion at room temperature and at neutral pH. After liberation of Cl^- ion, the mustard fragment can attach itself to a wide variety of chemical groups; some are more readily alkylated than others. Those attacked most easily are in the nucleic acid components of the cell, and this is why the mustards can become mutagenic. Their effect upon dividing cells has led to a wide development of synthetic nitrogen mustards with a variety of groups added to the N atom. Several of these compounds have proved to be anti-cancer agents in animals and in cultures of ascites cancer cells. Some of them have proved capable of prolonging life in leukaemia. The investigation of the biochemical lesion caused by mustard gas has given some practical results in medicine, as well as opening up a field of scientific research, Owing to the complicated nature of the reactions found, however, it seems to be an impossible task to get any antidote which could reverse the effects. The only thing which can be done is to find a "chaser", which has a sufficiently strong competitive action to combine with the "mustard" before it has combined with the tissue constituent which it might poison. The dominance of the sulphydryl groups in the reactions considered leads to a special consideration of these.

In a short chapter following that on "Mustard", an attempt has been made to bring together some of the known facts about toxicity and thiol groups.

In a later chapter a full account is given of the salient features in the action of carbon–fluorine compounds. In this the biochemical lesion is proved to start with a component of the citric acid cycle, the enzyme aconitase; but the effect is not due to the initial carbon–fluorine compound itself, such as fluoroacetate. This compound must first undergo an enzymic synthesis to fluorocitrate, which then blocks the enzyme, and this is called a "lethal synthesis". The proof that such a proccess can take place has opened the way to other possibilities. These have been explored in the U.S.A. with brilliant methods of synthesis; it appears that some azo and fluoropyrimidines and purines may even stop growth by substitution of a fraudulent nucleotide for the normal one. A most interesting example has been given where leukaemic cells have been induced to produce their own toxic substance from chemical fragments given to the cell. In regard to the carbon–fluorine compounds, the biochemical action is understood reasonably well; it differs from that of the "mustards", because the attack made by the monofluoro compounds is mainly due to a synthesis of fluorocitrate, which then inhibits one enzyme, aconitase. In principle this should be reversible by removal of the fluorocitrate. So far no compound has been found which will do this *in vivo*.

At this point a discussion of the biochemical lesion in thermal burns is introduced, a problem which involves the possible existence of a burn toxin. The presence of a burn toxin has been much debated for many years. On theoretical grounds, there is more than one area in a burn, firstly the part which is destroyed completely, "fried" in fact; and secondly the part where there is a partial damage. It is in the latter that the possible presence of a burn toxin must be sought. Low-temperature burns can be made artificially at a temperature at which it has not yet been possible to show damage to any enzyme. They are believed to be due to some alteration in the surface of the skin cells, possibly in the lipoproteins, which destroy the usual permeability, and allow the passage of some enzymes, especially a protease. If such a factor becomes loose in the body, it could be a potent source of producing toxic peptones, as well as inducing the change in the prickle cells which permits the skin to rise and form a blister. Whether in fact peptones are formed cannot be considered to be yet settled. Injections of these can give rise to many of the toxic symptoms of the toxaemia of burns; but it is still not entirely excluded that bacteria may be growing in the oedema fluid which may be responsible in part for the toxic effects described. This is another lesion which like mustard is not reversible, though it should be possible to mitigate the effects.

After the nine chapters on biochemical lesions and lethal synthesis, there is printed the Linacre Lecture for 1962, entitled "The Challenge of the Living Cell". This is essentially directed to the problem of integration in the cell. It may come as a surprise in relation to the rest of the book. In a way, it would be more logical to start with cell structure; but logic has been sacrificed to the development of biochemical research; the emphasis of which until lately has been upon enzymes and their isolation, rather than upon cell organization and surfaces. This has been so much the case, that much of membrane and surface chemistry is now being developed as biophysics. What is certain is that a full description and understanding of life will require an integration of structure and function, an integration which will bring to light more subtle ways in which biochemical lesions may commence.

In order not to confuse the general reader. I have attached to several of the chapters and to the Linacre Lecture a series of notes, amplifying and giving more detail than in the main text. An appendix at the end of the book contains four lectures given about 1930, which have been reprinted.

The enzymes have been named mostly by the trivial names which have been in use hitherto. For those who wish to bring their reading into line with the modern nomenclature proposed by the Enzyme Commission of I.U.B.[(5)] and now accepted by editors, I have included another appendix giving the new terminology.

References

1. Peters, R. A. (1936), The biochemical lesion in vitamin B_1 deficiency, *Lancet*, 1161.
2. Buffa, P. (1955), Lesione Biochimica, *Enciclopedia Medica Italiana*, **5**, 1737. Sansoni Ediziioni Scientifiche.
3. Cameron, G. R. (1956), *New Pathways in Chemical Pathology*, Williams & Wilkins, Baltimore.
4. Dawkins, M. J. R. and Rees, K. R. (1959), *A Biochemical Approach to Pathology.* Edward Arnold, London.
5. Report of the Commission on Enzymes of the International Union of Biochemistry, 1961. (President, M. Dixon). Pergamon Press.

CHAPTER 2

Biochemical Lesion in Thiamine Deficiency

IT WAS the work upon thiamine deficiency and convulsions which led to the first proposal of the term "Biochemical lesion" when I was working with the Roumanian biochemist N. Gavrilescu in 1931.* I had tried the term "Chemical lesion" before, but I was not satisfied that it expressed what I meant; for a full statement, see Peters, 1936[57]

THIAMINE

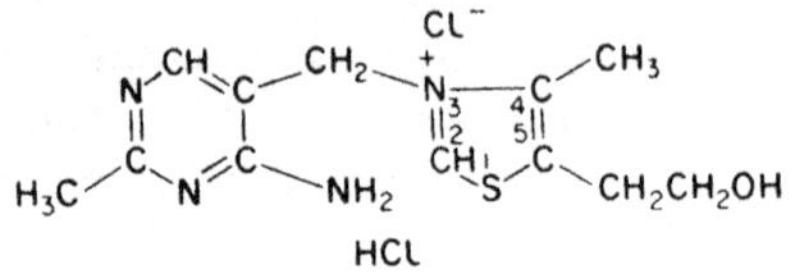

A total withdrawal of thiamine from the pigeon by feeding it upon polished rice, which has been thoroughly washed to free it from vitamin, has been known for a long time to induce signs† of "head retraction" and convulsions of a special type; the condition was called opisthotonus by McCarrison, and was well known to earlier workers on nutrition. Following upon the original finding of Eijkmann[12] (1889) that hens got nervous signs on feeding upon rice, this observation was rightly interpreted by Grijns[21] (1901) as being due to a deficiency in the diet rather than to a toxin; several workers used the pigeon instead of the hen; good pictures of various abnormalities produced were given by Schaumann[68] (1910). From 1922 onwards I used the pigeon prepared by feeding upon polished rice, the Patna variety, for our studies on the isolation of thiamine, which was then called the water-soluble vitamin B by E. V. McCollum; later it was called vitamin B_1 to indicate that it formed part of a complex of "B" vitamins. This work started in Hopkins's Laboratory, and was continued

* Professor N. Gavrilescu is now Professor in Bucharest, Roumania.

† I shall try throughout to use the term "sign" in the sense of those like Dr. Drysdale who taught me in medicine; "sign" to mean what is observed, and "symptom" what the patient observed.

when I went to Oxford in 1923. The isolation of thiamine from Baker's yeast occupied much of our time over 10 years (see Note 3, page 31). During all this period we used pigeons for our final tests. It was, in fact, the study of the pigeon test which led ultimately to an understanding of the biochemical lesion in the brain of the pigeon. When I first started to

(a)

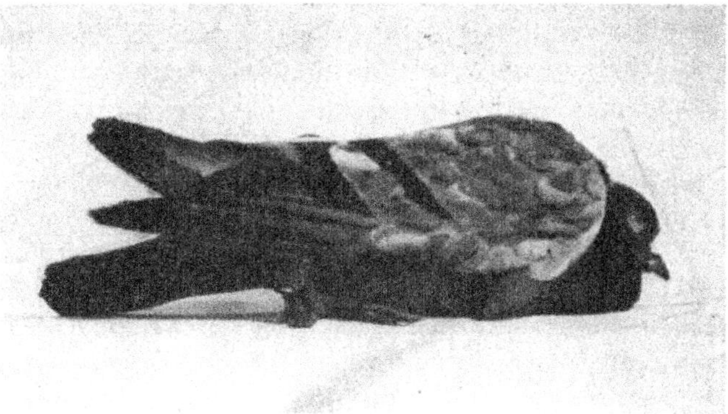

(b)

FIG. 2.1.

(a) Acute deficiency. Upper figure. Pigeon showing head retraction (opisthotonus) as the result of a diet of polished rice.
(b) Chronic deficiency. Lower figure. Pigeon showing chronic signs, with leg weakness, to be seen usually after about 30 days feeding upon a diet of rice containing traces of thiamine.
(Schaumann, 1910, *Archiv. für Schiffs-und Tropenhygiene* **14,** Beih. 8.)

use the so-called polyneuritic pigeon for the assay of vitamin B_1, the test was much out of fashion; it had been replaced by tests using as a criterion the growth of yeast or a comparison of the growth of young rats on diets with and without the vitamin; they grew more slowly on an avitaminous diet. The pathological phenomena occurring in the avitaminous pigeon are certainly confusing. I think that those using the "pigeon" test were deceived by the thought that the signs appearing in these birds upon the experimental diets were always due to the same cause. The appearance of head retraction of a certain type was always expected to have the same origin, whereas from the point of view of the physiologist, such a phenomenon would be very unlikely to happen. It would be somewhat like saying that the blocking of a nerve by the action of a local anaesthetic was exactly the same as severing it with a knife, because it produced the same final result of stopping the action of the nerve. We set ourselves to find out under what conditions the test was reliable and to eliminate what were in fact "pseudo" cures. When a batch of the pigeons which we were using was put to feed naturally upon a diet of polished rice, the signs of head retraction were variable in appearance. They would come on in a matter of 10 to 15 days, and sometimes later. We soon learnt to discard any pigeon in which the signs did not appear before the 30th day, because it was clear that other conditions were confusing the issue. In the birds getting "head retraction" before 30 days on the diet, it was found that warmth and a dose of water containing some glucose would often cure a bird for 2 to 3 days, while it was still eating the diet of polished rice. After this the convulsions and head retraction would reappear; when they did, glucose and water had no effect, and the birds could only be cured by giving the vitamin. Using birds prepared in this way, we were able to follow the vitamin in the various stages of isolation and finally to obtain from yeast the crystalline thiamine. I am stressing this because evidently an author of an excellent recent book[(79)] has not seen our publication in 1928 (Kinnersley, Peters and Reader[(32)]),* and has stated that the test is unreliable.

It was reflection upon the condition of the birds with acute signs which convinced me that deeper knowledge of the initiating factors must reveal some very important features of carbohydrate metabolism. In especial, the convulsive state induced by the deficiency drew my attention to the central nervous system and provided the stimulus to examine this for possible enzymic changes; until then attention had been focussed upon the

* Mr. H. W. Kinnersley was an analytical chemist who became a collaborator in the thiamine work over many years. He was a most agreeable colleague, was unassuming and very patient in the elaborate separations upon which we embarked. Miss V. B. Reader, a distinguished New Zealander, married Dr. E. Walker who died in 1941, and is now Mrs. Smith; after qualifying in medicine, she settled as an allergologist in Oxford and Bournemouth.

peripheral nervous system, which had become emphasized wrongly through an extensive use of the word "polyneuritic". The literature at that time was rather contradictory. Up till then there had been some attempts to connect the avitaminous condition with oxidations. Hess[24] had thought the condition to be like cyanide poisoning, because small doses of cyanide induced "head retraction" changes with convulsions. Marrian and Drummond[45,46] and ourselves had tried to find changes in oxidases such as the succinoxidase system in muscle; the attempts were unsuccessful. We started on the brain of the pigeon (Kinnersley and Peters, 1929[33]) by examining the metabolism of glucose *in vitro* using the brain "brei", which is another name for a mince of brain. In this way we found that there was a difference at the 3-carbon stage of lactate; the glycolytic path from glucose to lactate seemed to be normal (see Note 6 for scheme of glycolysis). Even in the intact brain there tended to be an increased amount of lactate in the brain of the pigeon in the avitaminous state, in opisthotonus. This may sound an overcautious statement; but the matter is better put this way, because there was a dynamic effect at work. To show up the difference at all we had to use a special guillotine which in one motion beheaded the animal and cut the brain in two halves so that these could be put immediately into liquid air. We always found more lactate in the avitaminous brain. If the birds were killed with iodoacetate (the inhibitor for muscle metabolism just introduced by Lundsgaard[44] (1930)), there was less lactate in the normal and avitaminous animal; even so there was always more in the avitaminous one (see Fig. H13). In the end it turned out that there was a defect not in the glycolytic stage glucose→lactate, but in the removal of lactate. This was shown to be true of the pigeon as a whole by R. B. Fisher[15]* (1931), and to depend upon the amount of exercise; there was more lactate present in the heart and blood after exercise in the avitaminous animal than in the normal. Similar facts were obtained by Inawashiro and Hayasaka[23] for the patient suffering from beri-beri, the disease due to deficiency of thiamine. At that time, these experiments were interpreted upon the idea, then prevalent, that lactate was an actual intermediate in metabolism. It would be thought now that the accumulation of lactate was due to a fault in the disposal of pyruvate.

Shortly after this we found (Gavrilescu, Meiklejohn, Passmore and Peters, 1932[16,17]) that there was a specific lowering of oxygen uptake *in vitro* with the brain brei from the pigeon suffering from thiamine deficiency (Fig. 2.2); the substrate could be either glucose or lactate. After many fruitless efforts, we succeeded in getting a restoration of oxygen uptake by the addition of a concentrate of vitamin B_1. It was in the first instance so small that we nearly missed it! To get this, it was essential that

* Now Professor of Biochemistry in Edinburgh.

the pigeon should be suffering from the acute opisthotonus signs. The figure (2.3) shows the essential facts about the vitamin effect. After this finding, the development of the research took time, but the crucial experiment had been done. Two problems then became especially important,

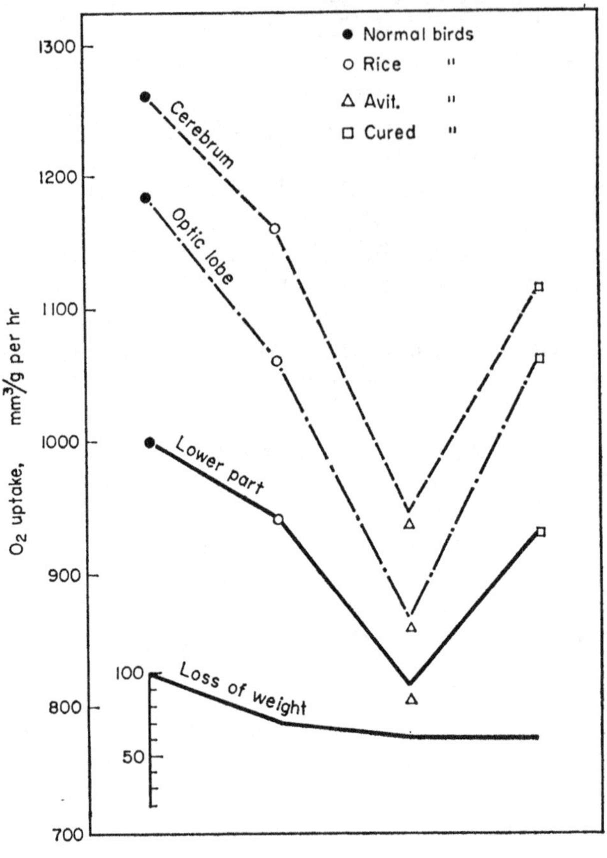

FIG. 2.2. Diagrammatic comparison of average oxygen uptake of different parts of the brain (pigeon) with the fall in weight at different states of avitaminosis. Ordinates: average oxygen uptake in $mm^3/g/hr$. Abscissae: periods in the feeding at which the average uptakes were observed. The interval between the first and second points represents 20–30 days; that between third and fourth points at 24–72 hr approximately. The diagram includes a curve for the average weight at the same periods, showing that average oxygen uptakes do not follow the fall in weight in the avitaminous periods.

Note that these effects only occur in presence of a substrate; in this case glucose. (Data of Gavrilescu and Peters, 1931.)

(a) to settle what was the essential substrate for the vitamin effect on the oxygen uptake, and (b) to determine whether the extra oxygen uptake produced by the addition of the vitamin ceased when the convulsions had disappeared.

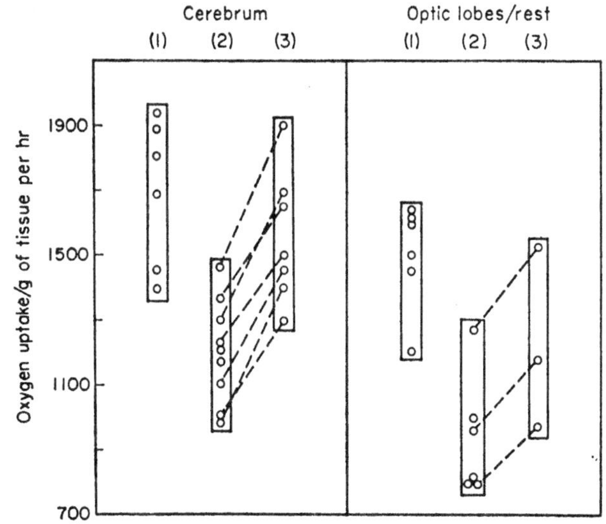

FIG. 2.3. Oxygen uptake mm³/g/hr in presence of 0·037 M lactate, for (1) normal, (2) avitaminous, and (3) avitaminous brain with addition of vitamin B_1 concentrates. Pigeon. From *Proc. R. Soc.* (1932), Ser. B, **110**, 439. (Data of Gavrilescu, Meiklejohn, Passmore and Peters.)

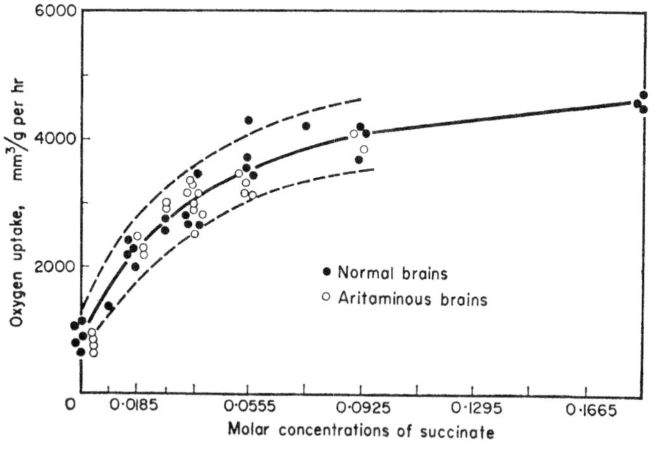

FIG. 2.4. Oxygen uptake for normal and avitaminous cerebrum in presence of varying concentration of succinate (see p. 29).

Regarding (a), we were able (Meiklejohn, Passmore and Peters[50]* to prove decisively that the "vitamin" effect upon the uptake of oxygen took place in lactate and not at all in succinate, even when a range of concentrations of the latter were used. In succinate the avitaminous brain behaved normally *in vitro*; the succinoxidase present in the brei was very active. (Fig. 2.4). (Note 1, p. 29).

Taking the next point (b), viz. the specificity of the extra oxygen uptake due to addition of the vitamin, we examined carefully the "vitamin" effect in the brains of normal pigeons on the one hand and on the other of

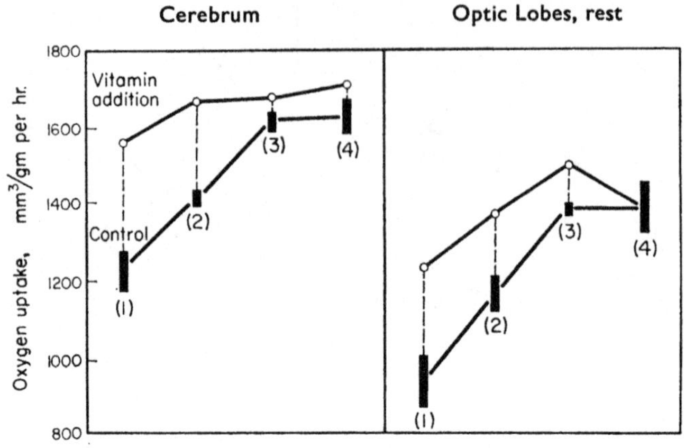

FIG. 2.5. Comparison of the average oxygen uptakes of avitaminous, dosed avitaminous and normal pigeon brain; substrate was lactate ± vitamin B_1: (1) Avitaminous; (2) Just out of symptoms; (3) Recovered well; (4) Normal.

The black rectangles represent the standard deviation of the mean of the observations. (Data of Meiklejohn, Passmore and Peters, 1932.)

* I owe a permanent debt of gratitude to Magdalen College, Oxford, for their generosity in giving Demyships to a brilliant group of young men who worked on these problems at that time; they were a happy set of collaborators. Dr. R. Passmore, who became a Lieutenant Colonel in World War II is now a Reader in Physiology in Edinburgh University; Dr. A. P. Meiklejohn, until his tragic death in a fishing accident in 1960, was Lecturer in Medical Nutrition at the same University; Dr. H. M. Sinclair is a Fellow of Magdalen and has been a Reader in Nutrition in Oxford University; Professor R. H. S. Thompson is Professor of Chemical Pathology at Guy's Hospital, London University; J. R. P. O'Brien is now Nuffield Reader in Clinical Biochemistry, Oxford. There were also others who took part; for instance, R. E. Johnson, a Rhodes Scholar, and now Professor of Physiology in University of Urbana, Illinois and Dr. G. K. McGowan, now at the University of Bristol. Mr. C. W. Carter is a colleague of many years standing. He joined my Department in 1925, and is still there, being now a Fellow of the Queen's College (Oxford) and Reader in Biochemistry. He has worked upon maintenance nutrition and upon dietary bradycardia in the pigeon, and upon some of the other B factors (see p. 290).

those acutely deficient, i.e. in opisthotonus, or of those which were on the way to recovery after dosing with the vitamin. If the vitamin effect is specific, we should expect to find this at a maximum in the acute stage, and gradually diminishing as the animal recovered; this was found. The respiration of the brain tissue increased during the cure, there being a diminution in the extra oxygen uptake due to the vitamin addition.

It should be stressed again that none of these effects can be seen properly until the acute condition of head retraction appears, and then they are seen best in the optic lobes and lower parts of the brain as distinct from the cerebrum. A study of birds showing leg weakness and other abnormal signs (Fig. 2.1b) was made by Rydin (see page 21). * Another point which was most carefully investigated at that time (Rydin[(66)]) was whether there was any difference in the capacity of the tissue to reduce methylene blue in presence of lactate; the results showed no difference, so that the vitamin action was definitely tied to oxygen uptake.

A few remarks may be made about the effect of thiamine on the bird with opisthotonus. The speed with which this condition can clear up is remarkable especially if the vitamin is injected into the subarachnoid space of the skull, under appropriate anaesthesia. These facts were known early in our work; the photos given in the illustration (Fig. 2.6) are from a more recent experiment. Within 15 min there is a big change and within 30 min the bird may have recovered sufficiently well to fly to the ceiling. In the acute stage, no pathological changes can be detected by microscopic observation of the brain. This was found originally by Dr. H. M. Carleton, the histologist in Oxford (in unpublished work), whom I asked to examine the brains. In any case it does not seem likely that any change detectable with the light microscope would clear up in 15 min, so that this is a good example of a pure biochemical lesion.

These facts about the acute and chronic stages of thiamine deficiency were well known to us at the time and formed the basis of our tests for thiamine during our isolation of this and also of the work upon the biochemical lesion. More recently (1941) the facts have been confirmed by Swank[(75,76)], who in careful work has also studied the pathology of the nervous tissues, and obtained correlations with the functional state.

The dramatic change in these birds upon injection of thiamine never fails to interest a student in the significance of biochemical detail. Years afterwards old students to whom I had shown this have told me how much impressed they were when they saw it. The only parallels known to me which are more dramatic, are the cure of "insulin" convulsions by injections of glucose or the cure of the tetany due to a lowered plasma calcium content by giving Ca salts.

* From Sweden, now a Research Director in an important pharmaceutical firm.

A B C D

FIG. 2.6. This shows the speed of the cure of thiamine deficiency in the pigeon by intracerebral injection of thiamine (under anaesthetic). At 10.08 a.m. Opisthotonus and rolling convulsions due to rice feeding. At 10.24 a.m. Cerebral injection was given of 1·5 mg thiamine Cl (HCl) + 0·45 mg glucose in 0·1 ml saline. B, 10.37 sitting. C, 10.42 beginning to stand. D, 10.52 cured.

(Peters, 1953, *Brit. Med. Bull.* **9,** 116.)

So the brain of the humble pigeon gave us the first proof of the action of any vitamin upon an enzyme system *in vitro*; it showed that this was related to the respiration by the increased uptake in oxygen specifically related to a dose of vitamin upon the avitaminous condition. In this way we also got a proof, in Oxford, of the essential importance of dynamic change in biochemical events at the level of pathological change. In spite of all the brilliant work upon muscle tissue up to that time, the exact connection between the biochemical change and the contraction had not yet been found. I do not think it is too much to claim that our work proved the theory of F. Gowland Hopkins of the essential importance of dynamic change of substances of small molecular weight in biochemistry—a thesis so wonderfully presented in his address to the British Association in 1913.

TABLE 2.1. DISAPPEARANCE OF PYRUVATE AND INCREASE IN O_2 UPTAKE BY THE ACTION OF VITAMIN B_1 WITH AVITAMINOUS PIGEON BRAIN RESPIRING *in vitro*. SUBSTRATE LACTATE. AVERAGE OF SIX EXPERIMENTS. FIGURES MG/G PYRUVATE PRESENT AT END OF 2-HR PERIOD OF RESPIRATION

L	L + V	Difference	Difference O_2 uptake mm^3/g
3·40	1·66	1·73	+ 1·102

Period of respiration = 2 hr. L = lactate. L+V = lactate+vitamin.
(From results of Peters and Thompson, 1934).

If we had stopped at this point we should still have laid the proof that the initial step in this pathological change was a biochemical one. In reality this was a very revolutionary fact; when taken to its logical conclusion, it meant that there should be less concentration by pathologists upon a histological picture obtained after the biochemical changes had run to secondary change, and that more attention should be given to the initial biochemical lesion which started the whole process. Fortunately we were able to go further and to decide that the biochemical lesion lay in the pyruvate oxidase system; both glucose and lactate had their metabolism blocked at the pyruvate stage in the absence of thiamine.* By that time it had been appreciated that pyruvate was the "hub" of metabolism, justifying the belief of C. Neuberg (1912[(52)]) in its importance. *In vitro* it was pyruvate which accumulated and which disappeared in the presence of the vitamin (Peters and Sinclair;[(59)] Peters and Thompson[(60)]). When pyruvate accumulated in the tissue, some leaked out into the blood, giving

* Both Sinclair and McGowan investigated the R.Q. of oxidation in the avitaminous brain, finding that this was lowered to a value indicating the metabolism of fat.

an increased concentration of pyruvate in the blood of animals (Thompson and Johnson, 1935[77]), and this was proved to happen also in cases of beri-beri in China by Platt and colleagues.[62–63] Prior to this work,

TABLE 2.2.

PYRUVIC ACID IN BLOOD

mg/100 g. B_1 deficiency as B.B.S.

	Normal	Avit.	Cured
Pigeon	3·9	11·3	5·3
Rat	4·2	9·4	—

Thompson and Johnson, 1935

BERI-BERI (human)

	as B.B.S.*	as Pyr. acid
Fulminating	10·1	2·72
Subacute	4·5	0·93
Cured	—	0·55

Platt and Lu, 1939

*Bisulphite binding substances

pyruvate had only been found in tissue metabolism studies as a result of adding tissue poisons, and it could perhaps have been argued that it was not a normal metabolite. That it was pyruvate in the pigeon's blood was confirmed chemically.[29] The presence of this compound in vitamin deficiency and its disappearance upon adding the vitamin gave stronger ground for the belief that it was a normal metabolite, accumulating in larger amounts in thiamine deficiency owing to a specific fault; thus proof was given to this important step in the Embden–Meyerhof scheme of glycolysis.

FURTHER DELINEATION OF THE POINT OF ACTION OF THIAMINE

We recognized quite early in our work that the oxidation of pyruvate was a complicated affair and that it seemed to involve several enzymic steps. In fact, the two Dutchmen, Westenbrink and Polak,* while working

* Professor H. G. K. Westenbrink is Professor of Biochemistry in Utrecht, and now Rector of the University.

in my laboratory in 1936, came to the conclusion that there was some preparatory step necessary for action of the vitamin.[78] So the stage was set for the exciting discovery by Lohmann and Schuster[40] in Meyerhof's laboratory that the cofactor for the fermentation of pyruvate in yeast, cocarboxylase, was thiamine pyrophosphate.

THIAMINE PYROPHOSPHATE (TPP)

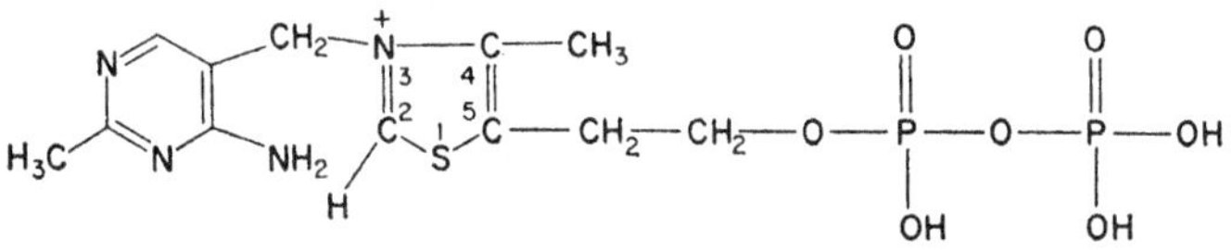

Cocarboxylase was a factor needed for the initial conversion of pyruvate into acetaldehyde in alcoholic fermentation. Hence the work on brain tissue and thiamine deficiency became linked to a wider field of observations. It brought into the picture the observations of Auhagen of the necessity of a cofactor for carboxylase in yeast, and it explained Simola's finding[73] that cocarboxylase activity decreased in the tissues of the avitaminous animal. That the same compound, thiamine pyrophosphate, should be responsible for the good fermentation of wine and for the rising of good bread, as well as protection against the disease of beri-beri opens wonderful vistas of Nature as a biochemist!

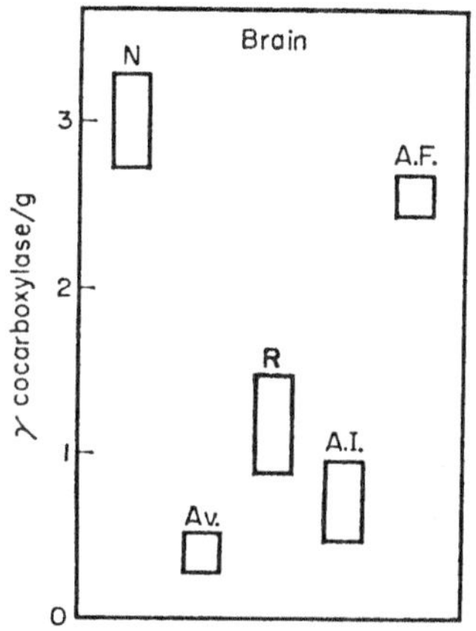

FIG. 2.7. Cocarboxylase in pigeon's brain. *N*, normal. *Av.*, avitaminous (with symptoms). *R*, rice-fed (no symptoms). *A.I.*, avitaminous after injection of vitamin B_1. *A.F.*, avitaminous fed vitamin B_1 for 3 days. Ordinates: mean ± 2ε. (Data of Ochoa and Peters, 1938.)

In view of this discovery, our next task was to prove in what form thiamine was active in brain. For instance the question arose whether the experiments *in vitro* involved a synthesis by the tissue of thiamine pyrophosphate from thiamine. Ochoa and Peters[(54)] studied the changes in cocarboxylase content of various tissues from normal and avitaminous pigeons, using an enzymic method of assay. We found that there was a larger diminution in the brain tissue than elsewhere, and that there seemed

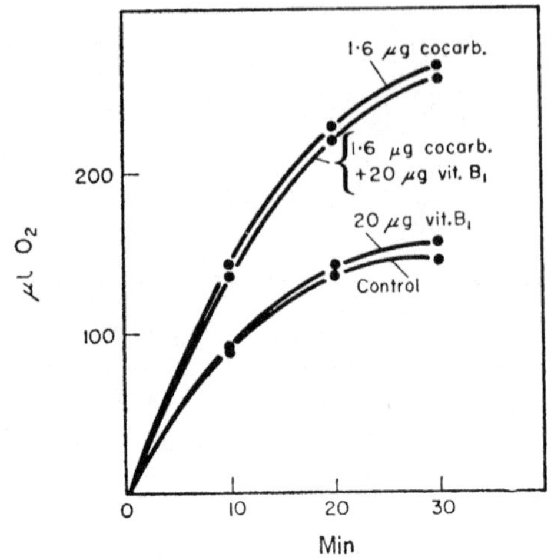

FIG. 2.8. Comparison of the effects of vitamin B_1 and cocarboxylase on the oxidation of pyruvate (0·018 M) in dispersions from avitaminous pigeon brain, in the presence of fumarate (0.004 M). (Data of Banga, Ochoa and Peters, 1939. Reproduced by permission of the Editor of the *Biochemical Journal*.)

to be definitely less when the acute signs appeared (Fig. 2.7). This was suggestive; but it did not prove that the pyrophosphate was the active form in the brain. A proof was reached finally by Banga, Ochoa and Peters (1939[(2)]) by a change of technique. Instead of a brain brei or mince, the tissue was dispersed in KCl solution (Fig. 2.8). This dispersion was much less stable than the mince, and it was necessary to reinforce it with a dicarboxylic acid such as fumarate and also an adenine nucleotide and magnesium to get proper activity; but nevertheless it gave us the required proofs; it was found that a dispersion from the avitaminous brain would only oxidize pyruvate when thiamine pyrophosphate was present: thiamine itself was inactive. So the biochemical lesion was traced to the lack of thiamine pyrophosphate and consequent failure of the pyruvate oxidase system to oxidize the crucial pyruvate stage in metabolism. Clearly in the earlier experiments when we got the "vitamin" effect, the brei had the necessary enzymes to carry out the phosphorylation of thiamine. Like other complex phosphates, thiamine pyrophosphate does not seem to get into the mitochondria in the brain brei, and penetrates as thiamine itself. Probably the blood brain barrier is impermeable to this *in vivo*. It is interesting to note (Fig. 2.9) that the dispersion of the brain from an avitaminous bird in our hands could detect as little as 0·01 μg of thiamine pyrophosphate, which emphasizes the activity of these effects.

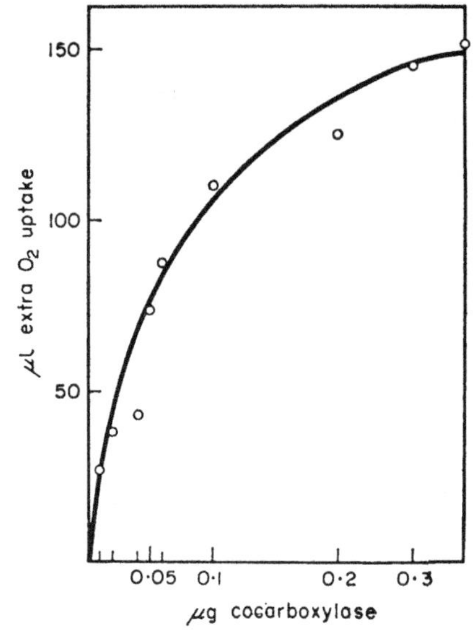

FIG. 2.9. Relation between the extra O_2 uptake and cocarboxylase concentrations in disperison from avitaminous pigeon brain. Substrate 0·027 M pyruvate + 0·005 M fumarate. O_2 uptake in 30 min. (Data of Banga, Ochoa and Peters, 1939.)

These researches have given a clinical dividend; changes in the α-keto acids, pyruvate and α-ketoglutarate are now used as a test for thiamine deficiency (in patients). It has been found best in practice to give the patient a loading dose of glucose as this brings out better any small changes in the values for pyruvate in the blood, see page 16 for references to the literature.

THE ORIGIN OF THE VARIOUS SIGNS IN THIAMINE DEFICIENCY AND THEIR RELATION TO POLYNEURITIS AND TO BERI-BERI

We may ask the question, how does the enzyme deficiency cause the convulsions? This matter is discussed more fully elsewhere. It might be thought that it is the increasing concentration of pyruvate in the tissue which causes the convulsions, but there is no evidence for this. When pyruvate is injected into the blood of the normal animal (the rabbit), it disappears with great speed (Lu and Needham[43]). The dysfunction appears to be due to the disruption of the usual energy supply in the cell required to keep it in equilibrium.

As was indicated on page 8 in the process of reaching thiamine deficiency in the brain in a pigeon feeding naturally upon polished rice, a complex adjustment is taking place; the curative experiments with sugar and water which have been mentioned already are best interpreted as showing that other tissues can sacrifice much of their thiamine to keep the brain running, and this can occur over a considerable time. Finally, in the pigeon, there is a critical lowering of the thiamine in the lower parts of the brain, where the "vitamin" effect is most marked, and convulsions occur. This indicates that thiamine is lost in the course of metabolism, and that these lower parts of the pigeon's brain (the optic lobes, etc.) are especially important to the bird, perhaps for the visual process.

If we realize that these adjustments are taking place during vitamin deficiencies, it becomes much easier to understand how the deficiency may not always run the same course in different animals.* There is a well known bradycardia due to thiamine deficiency in the rat which has been used as a test,[11] but this should not be confused with the dietary heart block† in pigeons (see page 291). In man, in beri-beri, clinical evidence suggests that in many cases the critical loss of thiamine occurs first in the heart, and dosing with thiamine rapidly relieves the cardiac condition (Williams *et al.*[80]). The course which the deficiency takes in pigs is also different.[81]

* As an example of the variation between animals, I can quote the goat, which is well known to one generation of Oxford medical students. We thought it would be a good idea to try and get a larger animal into a thiamine deficient condition, so that we could get samples of blood for experiment. We accordingly procured a goat, and fed this upon polished rice, which we had to supplement with roughage in the form of newspaper. It is interesting that the goat did not like *The Times*, and preferred a paper in which the betting news was prominent; probably the reason was because the paper on which *The Times* was printed was closer woven. However, we never succeeded in giving this animal any deficiency, and finally it succeeded in escaping and jumping upon a skylight above the classroom. Though the skylight consisted of wired glass, the goat to the consternation of those above was seen to disappear into the void of the classroom some 25 ft down where a class of medical students saw it descending as it were from the heavens. No one knows how the large pieces of glass missed anyone; but it was curious that the goat was the least moved of anyone! Someone who knew the mountains well said that this did not surprise him as goats quite frequently fought upon the mountain side and fell large distances without ill effects.

† Bradycardia means a heart which is beating more slowly than the normal. Heart block is a condition in which some of the heart beats reaching the auricle do not get through to the ventricle.

Turning to the rat, it may be asked whether the same facts about pyruvate and thiamine can be found in the rat, as in the pigeon. O'Brien and Peters[53] found a vitamin effect in brain brei from rats made deficient in thiamine; the animals were used for experiment at the point in the deficiency when handling induced spinning convulsions. These effects were observed in a medium containing lactate and pyrophosphate; they were definite, but they have not been repeated since. On the other hand, Banerji and Yudkin[4] found a deficiency in pyruvate oxidation in other parts of the rat. Upon the view which I have put forward of the selective nature of the deficiency, it may well be that the failure to demonstrate the effects in rat brain may be due to the way in which the loss of thiamine occurs from the different tissues on different diets. The problem of this type of pathological adjustment is most remarkable and has hardly been studied. It appears to involve competition in adsorptive equilibria between different organs (page 268).

In addition to the acute stage of thiamine deficiency in the pigeon which clears up rapidly upon dosing, there is the stage of chronic deficiency, (Fig. 2.1b) well known to those who have worked in this field; it occurs especially after a pigeon has been 30 days or more on the rice diet, without getting opisthotonus. A young bird may often develop genuine neuritic signs; for instance, it may not be able properly to support itself upon its legs: it usually sits down on the first joint. I have also seen several birds with optic disturbances at this stage; a bird may be blind in one eye, and so sensitive in the other that it crouches away from a hand shown to it some distance away. Even when birds in this condition are given thiamine, the leg weakness may take a month to clear up. The probable interpretation here is that we are dealing with a secondary change, occurring when a little thiamine is still present, which is sufficient to keep the brain just functioning. As has been mentioned this chronic state after 30 days upon the deficient diet has been investigated by Rydin,[67] Table 2.3, who found

TABLE 2.3. VITAMIN EFFECTS IN BIRDS WITH CHRONIC SIGNS, LEG WEAKNESS, ETC.

	% decrease in O_2 uptake over successive periods		
	¼–¾ hr	¾–1¼ hr	1¼–1¾ hr
Cerebrum	15	22	34
Optic lobes, and rest	8	12	11

(Substrate, lactate). (From data of H. Rydin, 1935).

vitamin effects *in vitro* mainly in the cerebrum, rather than in the optic lobes. This suggests that there also can be central effects in the chronic state, and we have known for some time of the possibility of haemorrhages in the brain of the rat from the work of Prickett.[64]* Much more nearly allied to this are the observations of Alexander *et al.*[1] upon the production of Wernicke's haemorrhagic polioencephalitis by giving pigeons this condition upon a diet of thiamine deficiency in which large supplies of other vitamins are also given; see Jolliffe *et al.*[31] There is now much evidence that Wernicke's encephalopathy in man can be cured by giving doses of thiamine and that excessive ingestion of alcohol is a predisposing cause.†

Based upon the behaviour of our birds, I would consider that any condition in the clinic which is not reversed rapidly by the giving of thiamine, must be complicated by some secondary factors. In man the deficiency of thiamine usually appears first in the heart, producing heart failure and the oedema of beri-beri; according to Professor Platt, this may occur in China especially after a hot restless night. Wernicke's encephalopathy, however, can occur, and Dr. Passmore (Davidson, Meiklejohn and Passmore[14]) has told me that the records of prisoners of war in World War II showed that in some cases thiamine deficiency led to this encephalopathy.

Other Work

The essential facts shown in the Oxford work on thiamine deficiency have been confirmed by Sherman and Elvehjem[71] upon chicks; they also found an increased cloacal excretion of pyruvate. In rats developing thiamine deficiency, Banerji and Harris[3] showed that there was an increasing excretion of bisulphite binding substances in the urine. A similar state of affairs was shown by Shils, Day and McCollum.[72] The administration of thiamine rapidly lowered the output of bisulphite-binding substances, which in their cases seemed to be mainly pyruvate. There have been also some studies on heart and other tissues. In rats, for instance, heart muscle from the auricle showed a fall of Q_{O2} in thiamine deficiency, whereas the ventricle showed little change (Muus *et al.*[51]). The normal muscle from the auricles gave twice the oxygen uptake of the ventricles, an interesting point. Differences were found by Olson *et al.*[56] in cardiac muscle from rats and *ducks*; both showed a decreased use of pyruvate in thiamine

* It has been pointed out to me recently by Professor J. Lehmann that F. Hofmeister thought in 1922 that vitamin B deficiency in the rat led to central disturbances in the nervous system, haemorrhages and cell neurosis in mid brain and cerebellum.

† In Wernicke's encephalopathy, there are lesions such as proliferation of blood vessels and necroses in the grey structures of the brain near the 3rd and 4th ventricles and aqueduct. The clinical signs are paralysis of the eye muscles, a reeling gait and disturbances of consciousness with a terminal somnolence.

deficiency, but the Q_{O_2} in pyruvate was only lowered in the duck. Some important work was done (Brin *et al.*[7]) in 1958 on the effect of thiamine deficiency on the oxidation of glucose by rat erythrocytes. It was found that the degradation of glucose by these took place mainly via the tranketolase system in the presence of methylene blue; in thiamine deficiency after 8 days, pentoses accumulated to an amount of 3 to 5 times the normal; addition of thiamine *in vitro* could change this. Here is a genuine example of another biochemical lesion; but it is not one which appears to be necessarily dangerous except as a warning signal, though it should be useful in diagnosis (see Note 5). In qualification of this, P. M. Dreyfus has found recently that in severe thiamine deficiency in rats the transketolase activity of some parts of the brain was greatly reduced—an observation which may well explain the haemorrhages[64] and our observations on the cerebrum and chronic deficiency.

SOME FURTHER BIOCHEMICAL POINTS (see also Note 2)

Though the essential nature of the biochemical lesion in thiamine deficiency has been described, a tidy understanding of the whole behaviour requires a more detailed knowledge of the pyruvate oxidase system. One of the difficulties of getting a complete analysis has been that the enzymes concerned are attached to particles which do not go into solution, so that even now the initial steps in pyruvate oxidation are not firmly understood, An analysis of the reactions carried out by a brei of pigeon brain during its oxidation of pyruvate was made by C. Long in 1937,[41] who described the chemical behaviour by three equations in the proportions given:

% of reaction present	
67	(1) $CH_3.CO{\cdot}COOH + \frac{1}{2}\,O_2 \rightarrow 3CO_2 + H_2O$
10·4	(2) $CH_3.CO{\cdot}COOH + \frac{1}{2}\,H_2O \rightarrow \frac{1}{2}CO_2 + \frac{1}{2}CH_2.CHOH{\cdot}COOH + \frac{1}{2}CH_3COOH$
19·6	(3) $CH_3.CO{\cdot}COOH + \frac{1}{2}O_2 \rightarrow CO_2 + CH_3COOH$

These three simultaneous equations are not of equal importance. It is likely that only (1) occurs in the uninjured brain, the reaction leading to the total combustion of the pyruvate to CO_2 and water.*

* These equations are now known to be only part of the story, because observations in Rome[5] show that in addition to the compounds mentioned, several other substances are formed from labelled pyruvate in varying small amounts. These include hexose-phosphate esters, phosphoglycerate, citrate, α-oxoglutarate glutamate, alanine, γ-aminobutyrate, in addition to some members of the citric acid cycle.

In 1936 Krebs with Johnson[38] enunciated the theory of the tricarboxylic or citric acid cycle, which was a masterly synthesis of previous work by Szent-Gyorgyi, and of Knoop and Martius.[35] According to this a 2-carbon fragment condensed with oxaloacetate to form a 6-carbon tri-COOH acid, which was then degraded by a regular series of reactions to oxaloacetate, when the cycle could start again with a fresh condensation of this with another 2-carbon fragment. At first it was not clear what compound was initially condensed; but this soon became delineated as the 2-carbon fragment arising from the decarboxylation of pyruvate. From the first suggestion of this cycle, we had to reckon with the strong possibility that, if proved, our oxidation of pyruvate in the brain brei to CO_2 and water might be going via the citric acid cycle; and this would carry as a corollary

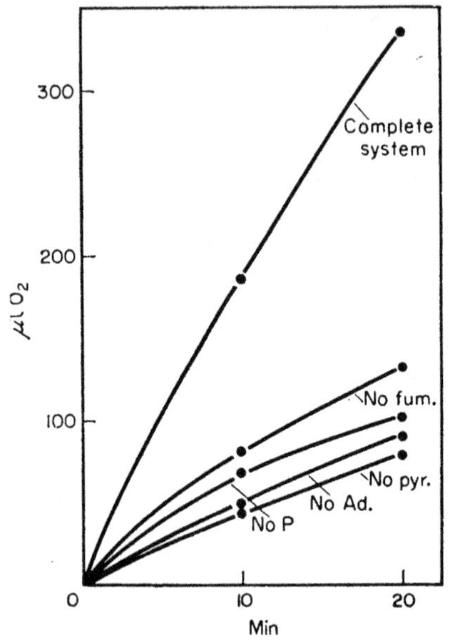

FIG. 2.10. Effect of phosphate (0·05 M), fumarate (0·005 M) and adenylic acid (0·00014 M) on the oxidation of pyruvate (0·0091 M) by dialysed dispersions from pigeons' brain. 1·5 ml enzyme (dialysed 2 hr) to 2 ml with additions including 0·1 mg Mg^{++} (as $MgCl_2$). The complete system contains enzyme + Mg^{++} + phosphate (P) + fumarate (fum.) + adenylic acid (Ad.) + pyruvate (pyr.). Sample with no phosphate buffered with $NaHCO_3$ to pH 7·3. Air, 38°. (Data of Banga, Ochoa and Peters, 1939.)

that the pyruvate system responsible for reaction (1) would include the enzymes of this cycle. We were unable to accept this at that time because of some contradictory facts in front of us. Firstly there was the fact that succinate oxidation had been proved in the brei to be completely insensitive to the vitamin effect in an avitaminous brain. Secondly Banga, Ochoa and Peters[2] found that the dicarboxylic acids were necessary for pyruvate oxidation by the dispersion, i.e. fumarate, malate or α-ketoglutarate, but that citrate had no catalytic action (Fig. 2.10), a fact which was not consistent with its place as an intermediate in the oxidation. The first difficulty is explicable if we think that the succinoxidase studied was in fact that from damaged mitochondria* (see page 28 for a full discussion of this). The second difficulty seems to be a matter of penetration, though one is always reluctant to fall back upon this; it is too easy a biochemical explanation! With the brain dispersion by using a block with fluorocitrate, it was possible to show that citrate accumulated coincidently with the decrease in the uptake of oxygen.[61] Fluorocitrate is a specific inhibitor of the enzyme aconitase in the citric acid cycle (see page 94) (Fig. 2.11). With the dispersion, addition of fluorocitrate leads to an accumulation of citrate. So the difficulty disappears if one assumes that citrate does not catalyse pyruvate oxidation in the dispersion because it does not penetrate the particles properly. In spite of other possible difficulties I have accepted provisionally the view that the citric acid cycle operates in the brain.

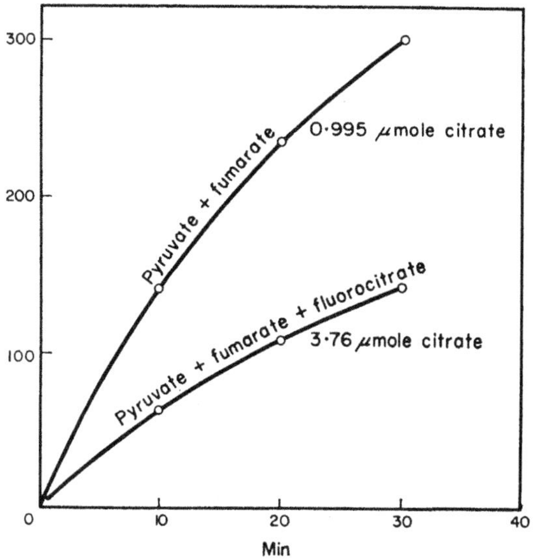

FIG. 2.11.

Thiamine pyrophosphate (TPP) is now known to be necessary for the metabolism of other α-keto acids in the animal, in particular α-ketoglutaric acid and sedoheptulose. One might expect that systems metabolizing

* A suggestion made to me by Dr. J. B. Chappell.

α-ketoglutarate would be also wrecked in a thiamine deficiency, especially because, in an isolated system, the thiamine pyrophosphate is said to be more dissociable in the α-ketoglutarate oxidase system (Green *et al.*(20)). However, this is not so; McGowan and Peters(49) could find no vitamin effect when α-ketoglutarate was the substrate, a fact confirmed by others. It is an interesting fact that brain tissue of the pigeon does not lose the thiamine pyrosphosphate from its α-ketoglutarate system at a time when the pyruvate system has become deficient. It is also to be noted that the

TABLE 2.4. COMPARISON OF CO_2 EVOLUTION BY YEAST AND O_2 UPTAKE FOR BRAIN WITH SEVERAL α-KETO ACIDS (μl)

α-keto acid	Yeast CO_2*	Brain O_2†	Mol oxidized
Pyruvate (C_3)	316	696	45·4
α-ketobutyrate (C_4)	396	171	45·7
α-ketovalerate (C_5)	273	18	—

* Evolution in 25 min. † per g/hr (3 hr).

(Data of Long and Peters, 1939). From *Biochem. J.* **33,** 759.

system in the animal differs from the yeast enzyme concerned with pyruvate, in so far as the yeast will deal with α-keto acids of increasingly long carbon chain, whereas the brain cannot metabolize chains of more than 4C atoms in length. The fact that Long and Peters(42) found that (Table 2.4) α-ketobutyrate was only degraded to propionic acid, and that there was no evidence of a further change, is consistent with the idea that any 3-C fragment formed by decarboxylation does not enter the citric acid cycle. The test for thiamine pyrophosphate which involves using yeast preparations has been much studied and improved by Westenbrink and his colleagues in Utrecht in a series of papers (see Jansen, (28).

THE EFFECT OF ANTIMETABOLITES AND OF ANTITHIAMINE FACTORS

Woolley and White(82) found that mice developed signs of thiamine deficiency when given the antimetabolite pyrithiamine

PYRITHIAMINE

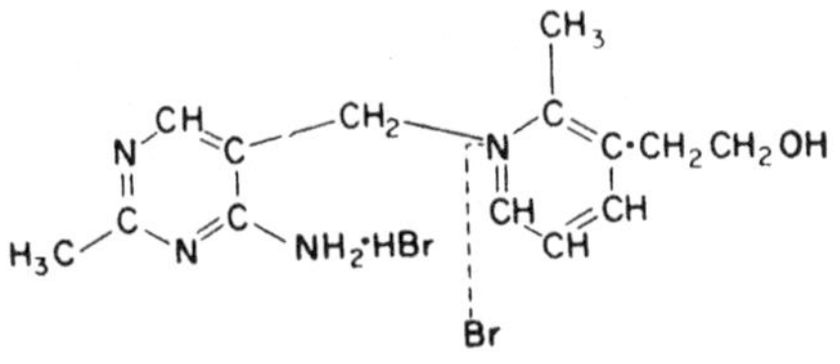

and that these were cured with an excess of thiamine in the diet. If the administration of an antimetabolite of thiamine were to induce the opisthotonus signs in pigeons on an ordinary diet, it would be an important support for the work which has been outlined so far, especially if there was an accompanying change in the content of thiamine PP in the brain tissue. This experiment has been done by Koedam.(36) 14 to 28 days of feeding on a normal diet with addition of pyrithiamine in pigeons gave a decrease in the brain of thiamine PP from 2·61 to 0·78 μg/g, together with the appearance of clinical signs. A large fall took place also in liver and especially in the heart. This seems to be valuable support both for the origin of the signs and also for the idea that the signs appear to be due to the utilization of thiamine in the brain, because the action of pyrithiamine is to stop the synthesis of thiamine PP from thiamine. It is to be noted that other antimetabolites of thiamine are known such as oxythiamine which have rather different actions (see also Note 4).

The Intimate Mechanism of Action of Thiamine Pyrophosphate

For years it has been clear that thiamine PP is the coenzyme for the conversion of pyruvate to a two-carbon fragment, which in yeast is converted to acetaldehyde, and in the higher animal to a two-carbon fragment which ultimately is attached to CoA to form the acetyl S CoA, the intermediate condensing with oxaloacetate to form citrate. In both the yeast and in the mammal, acetoin can be formed by the following

$$\underset{\text{Pyruvic acid}}{2CH_3{\cdot}CO{\cdot}COOH} \rightarrow \underset{\text{Acetoin}}{CH_3{\cdot}CO{\cdot}CHOH{\cdot}CH_3} + 2CO_2$$

reaction. It is also now clear that in the pyruvate oxidase system in the animal, the dithiol lipoate must be necessary, suggesting that this is an intermediate acceptor for the decarboxylated pyruvate. We have never been able to explain why in brain tissue, acetaldehyde is very toxic to the respiration, a fact which McGowan and I found long ago. Most important of all there seemed to be no reason why such a complicated structure as thiamine was necessary. One could never tell an inquiring student what the molecule actually did.

Now at last recent hypotheses and research has given us the answer. Breslow(6) postulated that there should be an intermediate in which a hydroxyethyl group would be attached to the C atom in position 2 of the thiazole ring. In support of this idea, Krampitz *et al.*(39) found that the phosphate-free 2-(α-hydroxyethyl)-thiamine would convert acetaldehyde to acetoin, a reaction induced by thiamine PP combined with apocarboxylase. These ideas have been proved for yeast by Holzer and colleagues in

Germany (1961),(25,26,69,70) who have established that decarboxylation of pyruvate occurs by the attachment of pyruvate to the 2 position of the thiazole ring to form active pyruvate-2-thiamine-pyrophosphate; this then becomes decarboxylated to α-hydroxyethyl-2-thiamine-pyrophosphate.

HOLZER COMPOUNDS

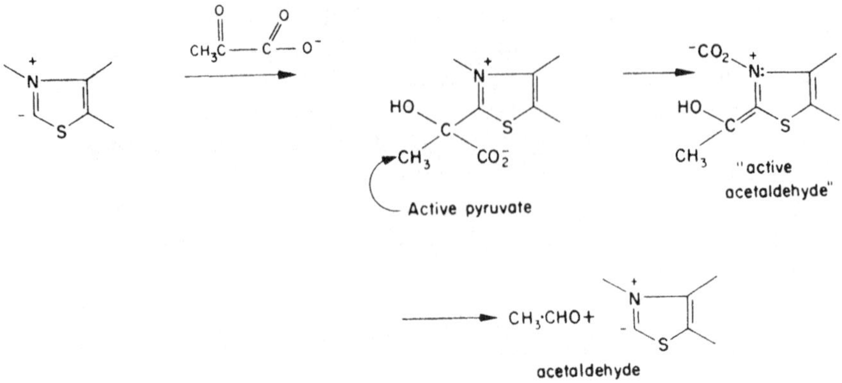

This may be regarded as an acetaldehyde compound. The acetaldehyde is liberated by apocarboxylase. Their crucial experiments were done by exposing brewer's yeast to labelled ^{14}C pyruvate for 5–10 sec, and then isolating the compounds in question. This appears at last to explain the presence of this unusual molecule; furthermore they have separated the same compounds from the pyruvate oxidase of pig muscle. So far as the brain is concerned it now remains only to demonstrate this in a brain brei. See Note 7.

FINAL REMARKS

I have tried in this short sketch to trace the course of researches which have led to a detailed knowledge of the biochemical action of thiamine. Much of this has followed naturally from a study of the biochemical lesion in the animal conducted in parallel with a study of the action of the vitamin *in vitro*. It has revealed a piece of biochemistry which is interesting in itself and also of significance in physiology and medicine.

NOTES ON CHAPTER 2

1. *Note on the Behaviour of Succinate*

It is a fact that the brain brei from the avitaminous pigeon showed no vitamin effect with succinate, whereas it showed this with the three-carbon acids lactate and pyruvate. This must be reckoned with in an assessment of the extent to which any citric acid cycle may be operating in brain tissue. It has been dismissed too lightly in some modern work. Figure 2.12 shows that the "succinate" increase was additive. The experimental facts were so definite that they must be explained, because according to the "cycle" theory, succinate must be an obligatory stage in the oxidation of pyruvate. When it became clear from the experiments with dispersions (ground brain preparations) that citrate could be formed in brain metabolism, we had a proof that in the dispersions some form of citrate cycle was in operation (Coxon, Liébecq and Peters;(8) Peters and Wakelin(61)). The question arose then whether a similar event was occurring in the brei, which is a much more stable preparation (presumably of mitochondria) with tightly bound reactions. Addition of fluorocitrate which causes such good accumulation of citrate in the dispersions has virtually no effect in causing this in a brei. If we wish, we can

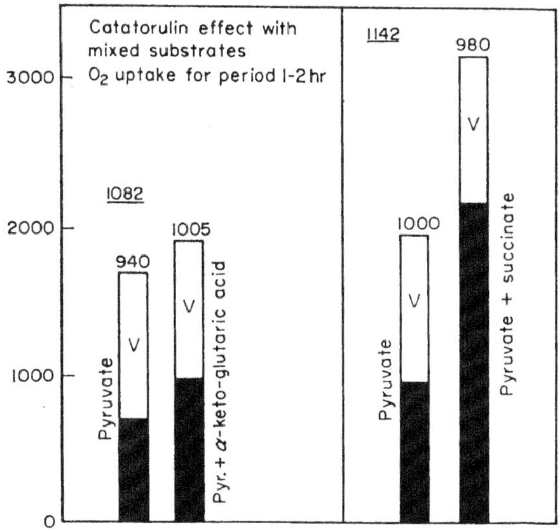

FIG. 2.12. Additive effect of succinate and the vitamin-catalysed pyruvate oxidation. (McGowan and Peters.)

attribute the inactivity to a failure of fluorocitrate to penetrate to the active centres in the brei. Certainly the brei does not require a sparking compound or in fact any other addition to make it capable of oxidizing the pyruvate molecule. Unavoidably we come to the explanation that in the brei there are two forms of the succinoxidase system, (i) a very active form which is insensitive to thiamine, and (ii) succinatedehydrogenase (succinate oxidoreductase) which is present with other enzyme of the citric acid cycle in the mitochondria, a conclusion receiving other support from other work of Coxon. The discrepant facts can then be explained by thinking that the succinoxidase system (i) is the enzyme system which has been liberated from damaged mitochondria. With this reservation that we have no direct proof for brain brei, we have accepted the citrate cycle for brain.

2. *Some Experiments on the Metabolism of Phosphoesters and Other Compounds with Brain Tissue from the Pigeon*

There are some experiments which have not been considered by those working on brain, which were done in Oxford in the early 1930's. Peters and Sinclair(59) investigated the effects of various additions of phosphoesters on the respiration of the brain brei, with the general results shown in Fig. 2.13. R. E. Johnson[29] found that when α-glycerophosphate was added to the pigeon brain brei, it was metabolized in the presence of pyrophosphate, methylglyoxal was formed and a substance giving a good 2,4-dinitrophenyl hydrazone, which he thought to be a sugar as it contain 6 C atoms. Evidently therefore synthetic reactions are taking place as well as those of degradation in this tissue. Methyl glyoxal has been found in the urine of vitamin B_1 deficient children by Geiger *et al.*[18] Its origin has never been clear.

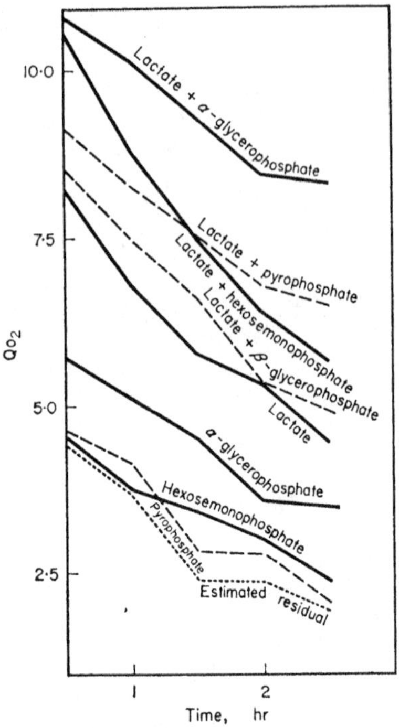

FIG. 2.13. Effect of various substrates upon respiration of minced brain. (Concentrations: lactate 0·033 M; α-glycerophosphate 0·024 M; pyrophosphate 0·019 M; hexosemonophosphate 0·013 M; β-glycerophosphate 0·023 M.)

3. *The Isolation of Thiamine*

The isolation of thiamine in crystalline form from rice polishings was first accomplished in 1925 by Jansen and Donath[27] in Java, where Eijkmann worked. This was the first vitamin to be crystallized as an entity, and it was a great achievement; it stopped at once views often mooted then that a vitamin was merely some peculiar state of a known biochemical compound. Until then in the last limit it was difficult to dispute this possibility. This vitamin must have been nearly pure; perhaps not quite, because it gave a Pauly reaction of ordinary type. The isolation from yeast took several years more to accomplish and we had the great pleasure of comparing our preparations in collaboration with Professor Jansen with his own.

At least 4 groups were working at it simultaneously; Windaus in Germany; Ohdake in Japan; Kinnersley, Peters and O'Brien in England; and Williams and Gurin in U.S.A.; the papers came out in this order, that of Japan being upon rice polishings, the rest on yeast. There was difficulty in arriving at the correct empirical formula. Windaus found the sulphur, and there was trouble over the last molecule of water; in fact the English workers were the first to publish the correct composition. I had been working upon the isolation for a long time, starting the work in 1922 in Hopkins' laboratory before I moved to Oxford; but R. R. Williams had started work earlier and has written a good history of the efforts given to getting pure thiamine. We made our isolation from Baker's yeast, and worked up in the laboratory by ourselves over 3 tons of yeast, because at that time we could not get help from commercial interests. We used absorption on Norite charcoal as one of our steps, but were not the first to use charcoal, as this had been used earlier by Vedder. We also used phosphotungstate in a new way as a selective precipitant for bases. The formula of the vitamin was $C_{12}H_{16}ON_2S$. There was some research required at that time to convince ourselves that the vitamin from yeast and that from rice polishings were the same, and we worked on this conjointly with Jansen and with van Veen. The structure of the compound which we had isolated eluded us. We knew that it had a distinctive smell. I had worked out also a formaldehyde azo test with a high specificity for the vitamin, which was useful in the isolation, and also a test based upon the use of the pigeon brain, which we called the catatorulin test. At that time also the concentrate which we had made in the course of the work proved of some value to others in the delineation of the other factors of the vitamin B complex. As far as I know, we were the

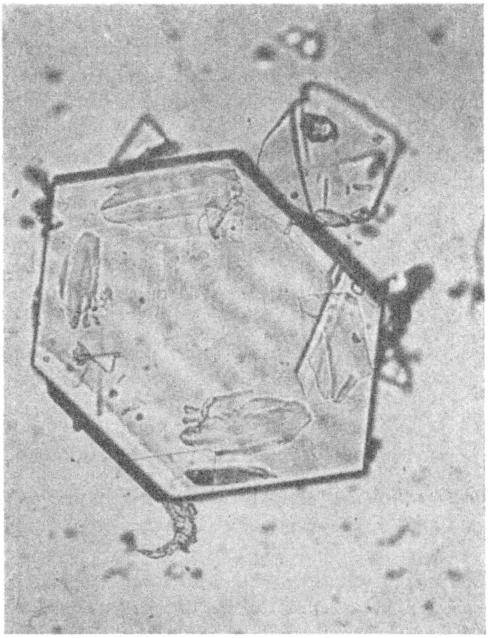

FIG. 2.14. Crystalline sulphate of thiamine with low melting point, 203°C. (Kinnersley, O'Brien & Peters, 1935).

first to prepare the nitrate and the sulphate of thiamine, the latter we found existed in two forms.[34] We also obtained in Oxford the first evidence of the conversion of thiamine by oxidation into an intense blue fluorescent substance, which is now known as thiochrome. In regard to the sulphur, it is an interesting fact that the sulphur in thiamine can get into a condition in which it is not readily detected by fusion with sodium and subsequent treatment with nitroprusside. But heating with nitric acid will always give the S, as sulphate.

It was very fitting that R. R. Williams and his co-workers should have been the first to get the structure of the molecule, as Williams had worked upon the vitamin for more years than anyone else. The important step in determining the structure was due to his long experience. He remembered that he had lost the activity of a batch of impure vitamin by treatment with sulphite. Treatment of the pure vitamin with sulphite split the molecule smoothly into its two parts, and so enabled the chemical structure to be determined, with its subsequent synthesis by his own group and others. It was this which paved the way for a really large-scale preparation of the vitamin, which can be now made very cheaply.

4. *Anti-thiamine Factors*

In addition to antimetabolites of thiamine, factors have been discovered in nature which can destroy thiamine.[19,74] In foxes a condition known as Chastek paralysis occurs when fresh fish (uncooked) is given to them. It was found to be cured by giving extra doses of thiamine (Green, Carlson and Evans[19]) and is due to an enzyme, thiaminase. Woolley and colleagues ran this factor to earth as an enzyme, called thiaminase, which can split thiamine. Another type of antithiamine effect was described by Evans and Jones.[13] They found that bracken contained a compound which was toxic to horses giving typical signs of thiamine deficiency. The poisoned animals could be restored to health by dosing daily with 50 to 100 mg of thiamine. Toxicity was due to two factors, a thermolabile and a thermostable one. In work upon the thermostable component from fern, Somogyi and Koller[37] have obtained a purified substance of acid character which forms a salt with brucine, and has a characteristic absorption at 285 μ. Unlike the enzyme from carp viscera, it does not split thiamine but it forms with it a stable salt like compound. If NaCl is added to test mixtures in increasing concentration the inactivation of the thiamine is reduced.

5. *Comments on Transketolase and Signs of Thiamine Deficiency*

Reactions in the pentose–phosphate cycle requiring *thiamine pyrophosphate*, and the enzyne *transketolase*.

(1) Ribose-5-phosphate + xylulose-5-phosphate
$\rightleftarrows$ sedoheptulose-7-phosphate + glyceraldehyde-3-phosphate

(2) Xylulose-5-phosphate + erythrose-4-phosphate
$\rightleftarrows$ fructose-6-phosphate + glyceraldehyde-3-phosphate.

Loss of thiaminepyrophosphate in the red blood corpuscles is the first event in thiamine deficiency; though most useful as a diagnostic warning, it does not seem to bring on the neurological signs; in support of this, Dreyfus[10] found that loss of transketolase activity in brain tissue did not drop as sharply as that in the blood. He got no signs of dangerous loss of transketolase activity until the thiamine pyrophosphate was reduced to 30 per cent of the normal value, agreeing with our figures for cocarboxylase. At 20 per cent of the normal the signs were severe with irreversible histopathological changes. Working with a micromethod of determining transketolase activity, he found that this was greater in the medullated structures and tracts of the spinal cord. The loss of activity of the enzyme in deficiency was greatest in the pons, a fact which would be in agreement with our feeling that the optic lobes and lower parts of the brain were most affected in the acute "head retraction" condition. Still, all points are not yet clear. He thinks that the haemorrhages found are secondary. Our work suggested that they came especially in the chronic condition in the pigeon. Could it be that the haemorrhages and the Wernicke condition require a small amount of thiamine to allow of the development of a secondary state?

A recent paper by Gubler(22) raises some new points which are difficult to reconcile with previous work. It may be that the criteria of deficiency used are different, as the author has mentioned anorexia as one of his signs. This is, of course, the earliest happening in a deficiency, but one has to take the acute stage as we have shown before vitamin effects in the tissue are demonstrable.

6. *The Glycolysis System*

The table gives the present knowledge of the glycolysis system. I am much indebted to Dr. M. Dixon, F.R.S., and Professor E. C. Webb for permission to use their table to appear in the second edition of their book on *Enzymes*, published by Longmans, Green. The enzymes are given their trivial names, but they are also numbered, and their present description can be found in the Report of the Commission on Enzymes of the International Union of Biochemistry.

TABLE 2.5. THE GLYCOLYSIS SYSTEM

XII.1.a α-glucose phosphorylase (2.4.1.1)

Glycogen + *n* H_2PO_4 = *n* glucose 1-phosphate

XII.1.b Phosphoglucomutase (2.7.5.1)

Glucose 1-phosphate + glucose 1,6-diphosphate
= glucose 1,6-diphosphate + glucose 6-phosphate

Alternatively, starting with glucose, XII.1.a–b are replaced by XII.1.c:

XII.1.c Hexokinase (2.7.1.1)

Glucose + ATP = glucose 6-phosphate + ADP

XII.1.d Glucosephosphate isomerase (5.3.1.9)

Glucose 6-phosphate = fructose 6-phosphate

XII.1.e Phosphofructokinase (2.7.1.11)

Fructose 6-phosphate + ATP = fructose 1,6-diphosphate + ADP

XII.1.f Aldolase (4.1.2.b)

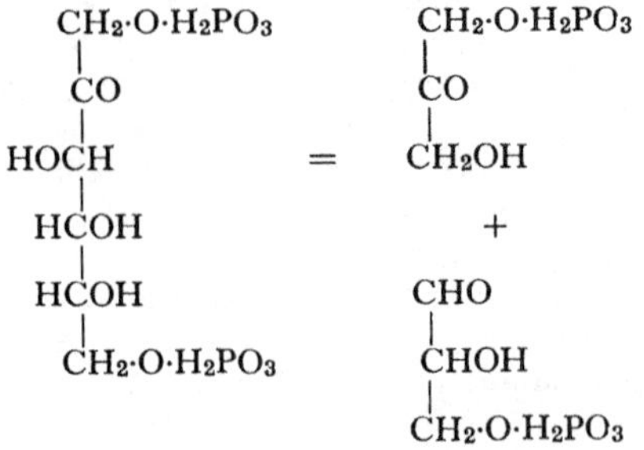

Fructose 1,6-diphosphate — dihydroxyacetone phosphate + D-glyceraldehyde 3-phosphate

XII.1.g Triosephosphate isomerase (5.3.1.1)

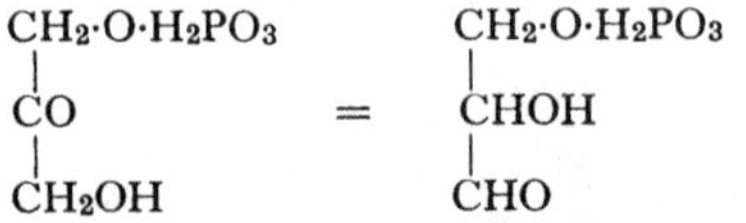

Dihydroxyacetone phosphate — D-glyceraldehyde 3-phosphate

XII.1.h Glyceraldehydephosphate dehydrogenase (1.2.1.12)

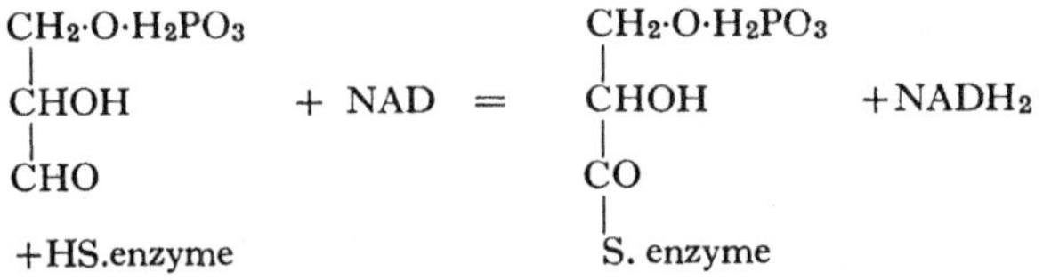

D-glyceraldehyde 3-phosphate + dehydrogenase — 3-phospho-D-glyceric acid—enzyme complex

XII.1.i Glyceraldehydephosphate dehydrogenase (1.2.1.12)

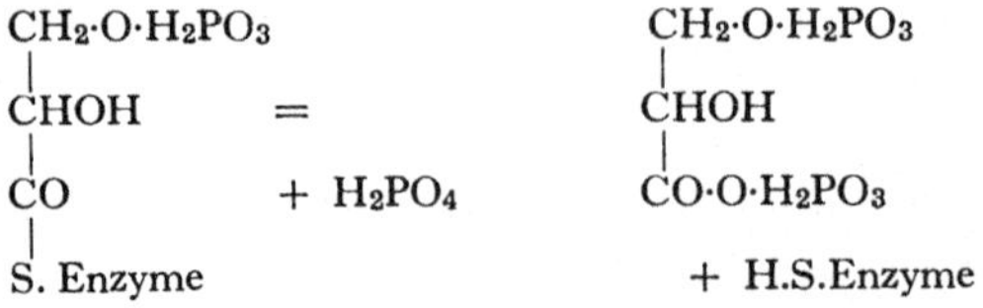

1,3-diphospho-D-glycerate

XII.1.j Phosphoglycerate kinase (2.7.2.3)

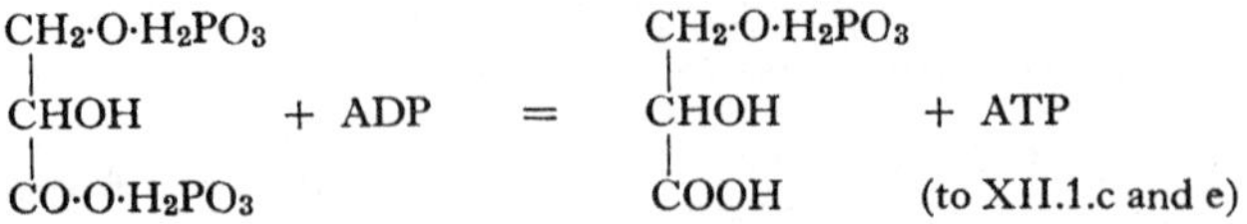

1,3-diphospho-D-glycerate — 3-phospho-D-glycerate

XII.1.k Phosphoglyceromutase (2.7.5.3)

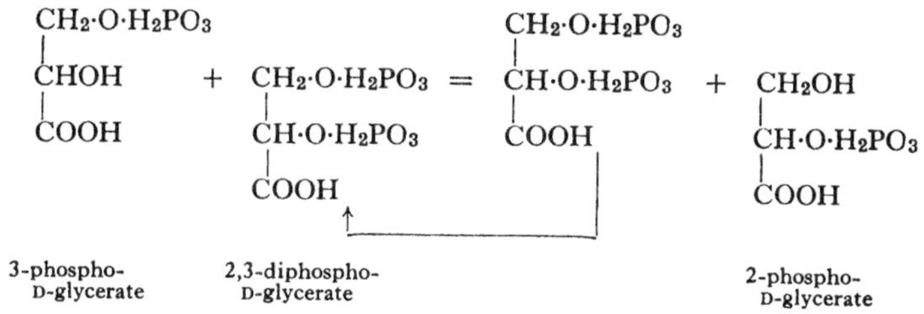

XII.1.l Phosphopyruvate hydratase (4.2.1.11)

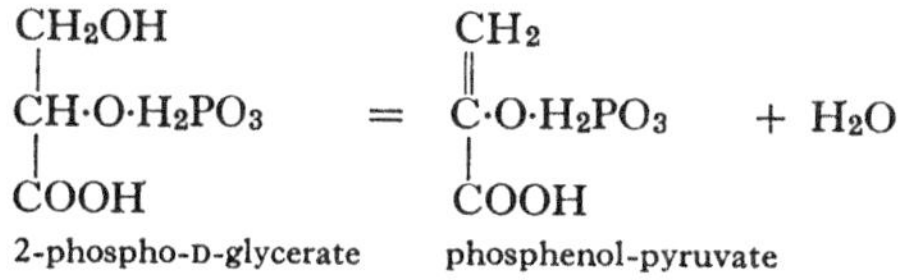

XII.1.m Pyruvate kinase (2.7.1.40)

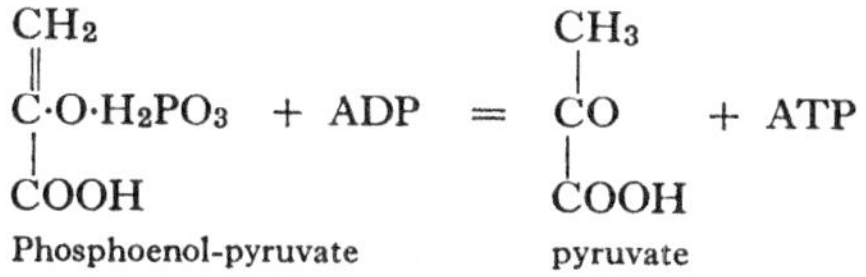

If O_2 is absent, the $NADH_2$ from XII.1.h is oxidized in muscle by XII.1.n and the end product is lactate:

XII.1.n Lactate dehydrogenase (1.1.1.27)

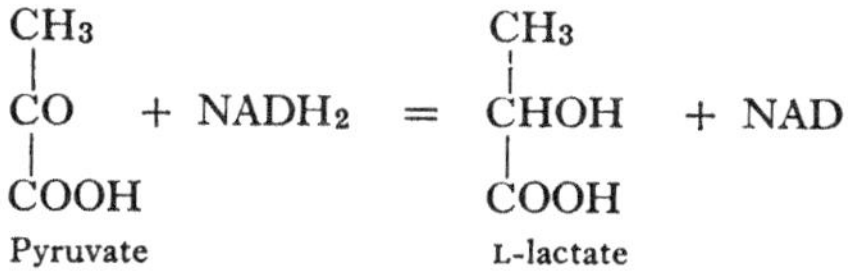

In yeast the end product is alcohol:

XII.1.o Pyruvate decarboxylase (4.1.1.1)

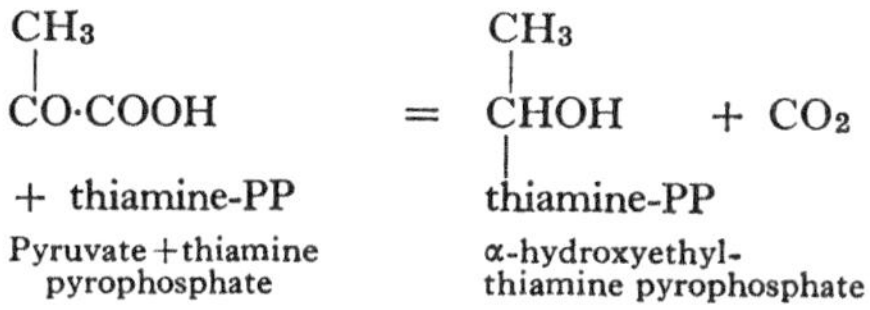

XII.1.p Pyruvate decarboxylase (4.1.1.1)

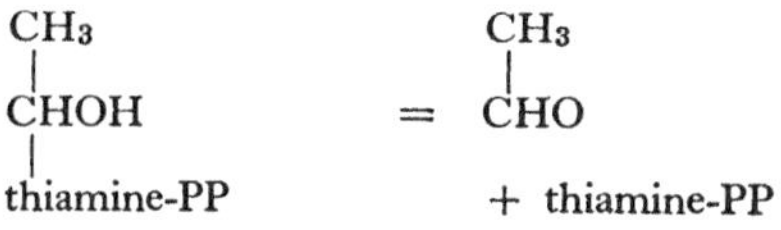

XII.1.q — Alcohol dehydrogenase (1.1.1.1)

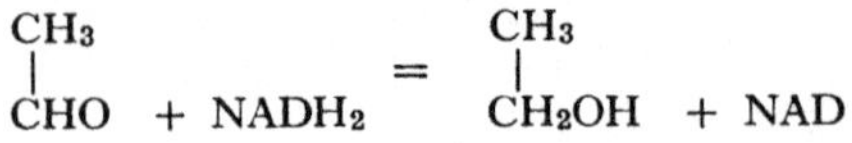

If O_2 is present, the $NADH_2$ is oxidized through the cytochrome system and is not available for reactions XII.1.n and q. Reaction XII.1.m is then followed by XII.1.r–v and the product is acetyl–CoA.

XII.1.r — Pyruvate dehydrogenase (1.2.4.1)

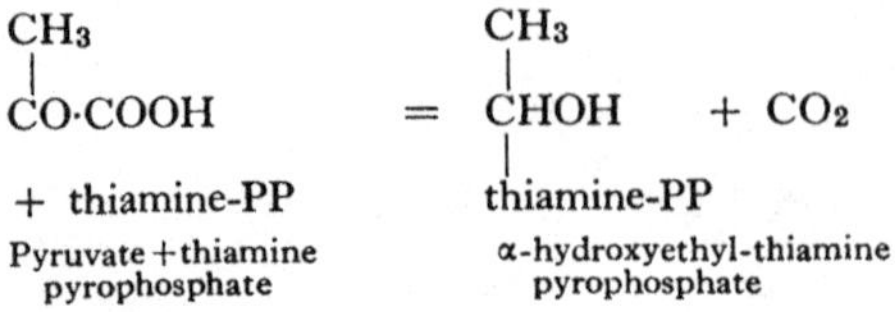

XII.1.s — Pyruvate dehydrogenase (1.2.4.1)

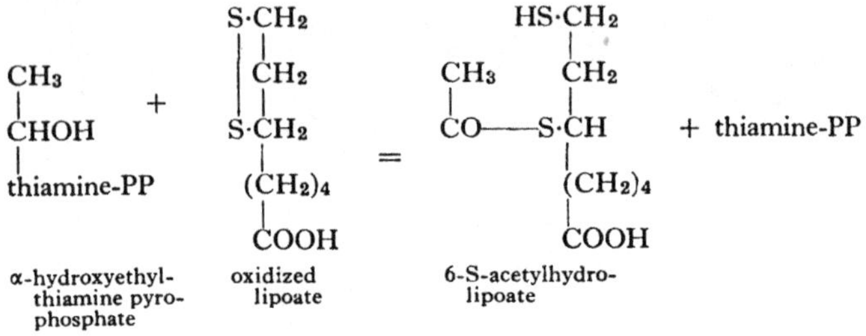

XII.1.t — Lipoate acetyltransferase (2.3.1.12)

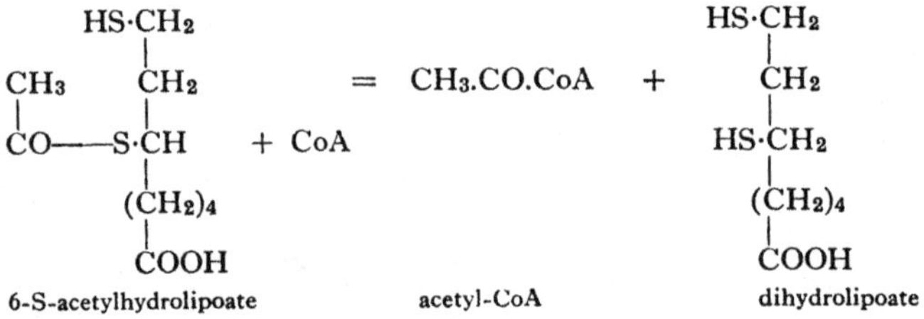

XII.1.u — Lipoamide dehydrogenase (1.6.4.3)

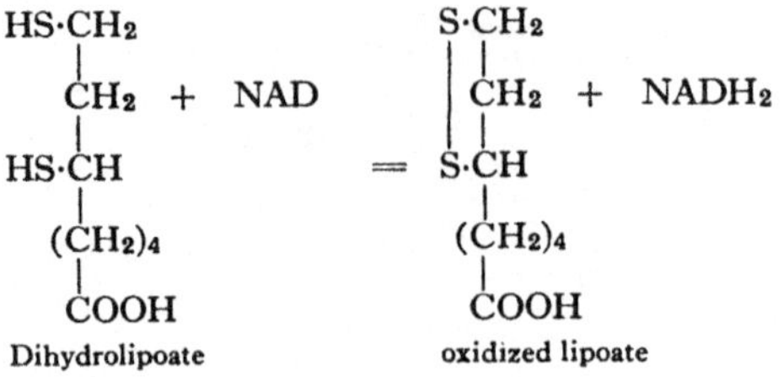

XII.1.v NADH$_2$ cytochrome *c* reductase (1.6.2.1) and cytochrome oxidase (1.9.3.1)

$$NADH_2 + \tfrac{1}{2}O_2 = NAD + H_2O$$

Sum of reactions XII.1.c–n:

$$glucose + 2ADP + 2H_3PO_4 = 2\ lactic\ acid + 2ATP + 2H_2O$$

Sum of reactions XII.1.c–m and o–q:

$$glucose + 2ADP + 2H_3PO_4 = 2\ alcohol + 2CO_2 + 2ATP + 2H_2O$$

Sum of reactions XII.1.c–m and r–v:

$$glucose + 2ADP + 2H_3PO_4 + 2O_2 + 2CoA = 2\ acetyl\text{–}CoA + 2CO_2 + 2ATP + 6H_2O$$

Note.—In adding the reactions it is necessary to remember that reactions XII.1.h onwards must be multiplied by 2, since 2 triosephosphate molecules are produced from one glucose molecule (reaction XII.1.f). For this reason XII.1.j produces enough ATP to supply XII.1.c and 1.e and the ATP produced in XII.1.m is an excess, available for other purposes. XII.1.v must be multiplied by 4, to allow for the NADH$_2$ produced in XII.1.h and 1.u.

When arsenate is added, XII.1.i–j are replaced by XII.1.w–x:

XII.1.w Glyceraldehydephosphate dehyrdrogenase (1.2.1.12)

$CH_2{\cdot}O{\cdot}H_2PO_3$ — CHOH — CO — S.Enzyme + H_3AsO_4 = $CH_2{\cdot}O{\cdot}H_2PO_3$ — CHOH — $CO{\cdot}O{\cdot}H_2AsO_3$ + HS.Enzyme

XII.1.x (Rapid, spontaneous)

$CH_2{\cdot}O{\cdot}H_2PO_3$ — CHOH — $CO{\cdot}O{\cdot}H_2AsO_3$ = $CH_2{\cdot}O{\cdot}H_2PO_3$ — CHOH — COOH + H_3AsO_4

The coupled phosphorylation of ADP is therefore abolished.

Note 7. (Page 28). It is interesting that TPP operates by an 'ylid' mechanism with the group 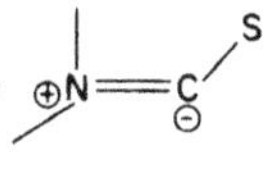 a reaction only recently characterized.

References

1. Alexander, L. (1940), *Amer. J. Path.* **16,** 61. (See also Alexander, L., Pijoan, M., Myerson, A. and Krane, H. M., 1938, *Trans. Amer. Neurol., A.* **64,** 135).
2. Banga, I., Ochoa, S. and Peters, R. A. (1939), *Biochem. J.* **33,** 1109–1121 and 1980.
3. Banerji, G. G. and Harris, L. J. (1939), *Biochem. J.* **33,** 1346.
4. Banerji, G. G. and Yudkin, J. (1942), *Biochem. J.* **36,** 530.

5. Beloff-Chain, A., Catanzaro, R., Chain, E., Masi, I., Pocchiari, F. and Rossi, C. (1955), *Proc. Roy. Soc. B.* **143,** 481 and subsequent papers.
6. Breslow, R. (1957), *J. Amer. Chem. Soc.* **79,** 1762.
7. Brin, M., Schohet, S. S. and Davidson, C. S. (1958), *J. Biol. Chem.* **230,** 319.
8. Coxon, R. V. and Peters, R. A. (1950), *Biochem. J.* **46,** 300.
9. Coxon, R. V., Liébecq, C. and Peters, R. A. (1949), *Biochem. J.* **45,** 320.
10. Dreyfus, P. M. and Victor, M. (1961), *Amer. J. Clin. Nutr.* **9,** 414.
11. Drury, A. N., Harris, L. J. and Maudsley, C. (1930), *Biochem. J.*, **24,** 1632.
12. Eijkman, C. (1890), *Genisk Tijkschrifit Ned. Ind.* **30,** 295.
13. Evans, W. C. and Jones, N. R. (1952), *Biochem. J.* **50,** xxviii.
14. Davidson, S., Meiklejohn, A. P. and Passmore, R. (1959), *Human Nutrition and Dietetics*. E. & S. Livingstone, Edinburgh, London.
15. Fisher, R. B. (1931), *Biochem. J.* **25,** 1410.
16. Gavrilescu, N. and Peters, R. A. (1931), *Biochem. J.* **25,** 1397 and 2150.
17. Gavrilescu, N., Meiklejohn, A. P., Passmore, R. and Peters, R. A. (1932), *Proc. Roy. Soc. B (London)* **110,** 431.
18. Geiger, A. and Rosenberg, A. (1933), *Klin. Noch.* **12,** 1258.
19. Green, R. G., Carlson, W. E. and Evans, C. A. (1941), *J. Nutr.* **21,** 243.
20. Green, D. E., Herbert, D. and Subrahmanyan, V. (1941), *J. Biol. Chem.* **138,** 327.
21. Grijns, G. (1901), *Geneesk. Tijdscht. Ned. Ind.* **43,** 3.
22. Gubber, C. J. (1961), *J. Biol. Chem.* **236,** 3112.
23. Hayasaka, E. (1929), *Tokohu, J. Exp. Med.* **14,** 85.
24. Hess, W. E. (1921), *Z. Physiol. Chem.* **117,** 284.
25. Holzer, H. and Beauchamp, K. (1961), *Biochem. Biophys. Acta* **46,** 225.
26. Holzer, H. (1961), *Angew. Chem.* **73,** 721–727.
27. Jansen, B. C. P. and Donath, W. F. (1926), *Proc. Sci. Amsterdam* **29,** 1390.
28. Jansen, B. C. P., *Vitamins and Hormones* vii. **84,** Acad. Press, N.Y. (Review).
29. Johnson, R. E. (1936), *Biochem. J.* **30,** 31 and 33.
30. Joiner, C. L., MacArdle, B. and Thompson, R. H. S. (1950), *Brain* **73,** 431.
31. Jolliffe, N., Wortis, H. I. and Fein, H. D. (1941), *Arch. Neur. Psych.* **46,** 569.
32. Kinnersley, H. W., Peters, R. A. and Reader, V. (1929), *Biochem. J.* **22,** 276.
33. Kinnersley, H. W. and Peters, R. A. (1929) (1930), *Biochem, J.* **23,** 1126; *J. Physiol.* **69**; *Proc. Physiol. Soc.* **xi.**
34. Kinnersley, H. W., O'Brien, J. R. P. and Peters, R. A. (1933), *Biochem. J.* **27,** 232.
35. Knoop, F. and Martius, C. (1936), *Z. Physiol. Chem.* **242,** 1.
36. Koedam, J. C. (1958), *Biochem. Biophys. Acta* **29,** 333.
37. Koller, A. and Somogyi, J. C. (1961), *Inter. Zeit.f. Vit.* **31,** 227.
38. Krebs, H. A. and Johnson, W. A. (1937), *Enzymologia* **4,** 148–156.
39. Krampitz, L. O., Greull, G., Miller, C. S., Bicking, J. B., Skeggs, H. R. and Sprague, J. M. (1958), *J. Amer. Chem. Soc.* **80,** 5893.
40. Lohmann, K. and, Schuster, P. (1937), *Biochem. Z.* **294,** 188.
41. Long, C. (1938), *Biochem. J.* **32,** 1711.
42. Long, C. and Peters, R. A. (1939), *Biochem. J.* **33,** 759.
43. Lu, G. and Needham, D. M. (1939), *Biochem. J.* **33,** 1544.
44. Lundsgaard, E. (1930), *Biochem. Z.* **217,** 162.
45. Marrian, G. F. (1928), *Biochem. J.* **22,** 843.
46. Marrian, G. F. and Drummond, J. C. (1926), *Biochem. J.* **20,** 1229.
47. Martius, C. (1937), *Z. Phyisol. Chem.* **247,** 104.

48. McGowan, G. K. (1937), *Biochem. J.* **31,** 1627.
49. McGowan, G. K. and Peters, R. A. (1937), *Biochem. J.* **31,** 1637.
50. Meiklejohn, A. P., Passmore, R. and Peters, R. A. (1932), *Proc. Roy. Soc. B.* **111,** 391.
51. Muus, J., Weiss, S. and Hastings, A. B. (1939), *J. Biol. Chem.* **129,** 303.
52. Neuberg, C. and Kerb, J. (1912), *Biochem. Z.* **47,** 405.
53. O'Brien, J. R. P. and Peters, R. A. (1935), *J. Physiol.* **85,** 454.
54. Ochoa, S. and Peters, R. A. (1938), *Biochem. J.* **32,** 1501.
55. Olivo, F., Rossi, C. S. and Siliprandi, N. (1962), *Biochem-Biophys Acta* **56,** 158.
56. Olson, R. E., Pearson, V. H., Miller, O. Neal and Stare, F. J. (1948), *J. Biol. C.* **175,** 489.
57. Peters, R. A. (1936), *Lancet* Lecture 1161, "The Biochemical Lesion in thiamine deficiency".
58. Peters, R. A. (1940), *Chem. and Ind.* **59,** 373. "Biochemistry of brain tissue".
59. Peters, R. A. and Sinclair, H. M. (1933) *Biochem. J.* **27,** 1677.
60. Peters, R. A. and Thompson, R. H. S. (1934), *Biochem. J.* **28,** 916.
61. Peters, R. A. and Wakelin, R. W. (1953), *J. Physiol.* **119,** 421.
62. Platt, B. S. and Lu, G. D. (1946), *Quart. J. Med.* **5,** 355.
63. Platt, B. S. and Lu, G. D. (1939) *Biochem,* J. **33,** 1525.
64. Prickett, C. O. (1934), *Amer. J. Physiol.* **107,** 459.
65. Rydin, H. (1935), Inuagural dissertation Uppsala. Almquist & Wiksills Boktrycherei—A.B.
66. Rydin. H. (1936), *Chem. Abstracts* **30,** 1101.
67. Rydin, H. (1935), *Biochem. J.* **29,** 860.
68. Schaumann, H. (1908), *Arch. Schiffs Tropen-Hyg.* **12,** Beihaft **5,** 37 and (1910) **14,** Beihaft, **3.**
69. Scriba, P., Schneider, S. and Holzer, H. (1961), *Biochem. Biophys. Acta* **54,** 115.
70. Scriba, P. and Holzer, H. (1961), *Biochem. Z.* **334,** 473.
71. Sherman, W. C. and Elvehjem, C. A. (1936), *Amer. J. Physiol.* **117,** 142.
72. Shils, M. E., Day, H. G. and McCollum, E. V. (1941), *J. Biol. Chem.* **139,** 145.
73. Simola, P. (1932), *Biochem. Z.* **254,** 299.
74. Spitzer, E. H., Coombes, A. I., Elevehjem, C. A. and Wisnicky, W. (1941), *Proc. Soc. Exp. Biol. Med.* **48,** 376.
75. Swank, R. L. (1940), *J. Exp. Med.* **71,** 683.
76. Swank, R. L. and Bessey, O. A. (1941), *J. Nutrit.* **24,** 77.
77. Thompson, R. H. S. and Johnson, R. E. (1935), *Biochem. J.* **29,** 694.
78. Westenbrink, H. G. K. and Polak, J. J. (1937), *Rec. Trav. Chim.* **56,** 315.
79. Williams, R. R., *Towards the Conquest of Beri-Beri,* p. 110. Harvard Univ. Press, Cambridge, Mass.
80. Williams, R. D., Mason, H. L., Wilder, R. M. and Smith, B. F. (1940), *Archiv. Int. Med.* **66,** 785.
81. Wintrobe, M. M. (1942), *Johns Hopkins Hosp. Bull.* **71,** 141.
82. Woolley, D. W. and White, A. G. C. (1943), *J. Biol. Chem.* **149,** 285.

CHAPTER 3

Arsenical Substances and Biochemical Lesions

ON pages 220–221 I have explained how I came to study toxic substances like arsenicals, and how inadequate I felt that a purely physical explanation was for the action of chloropicrin ($C.Cl_3.NO_2$) and phosgene ($CO.Cl_2$) upon the lung. In experiments in 1917 carried out in the laboratory which Sir Joseph Barcroft was directing, with the object of averting lung oedema due to gas poisoning, the effect of calcium salts was tried upon animals poisoned with chloropicrin. Calcium ions were thought then to improve permeability on the surface of cells so that the abnormal extravasation of fluid could be decreased. Thereby I got my first lesson in statistics. One batch of animals (goats) was divided into two and one of these was treated with an injection of calcium salts after the gassing. Those treated were improved, a result which stood up to a statistical test, applied by Professor Boycott who then kept us straight in these matters. We then went on to do a second batch. When the results of these two experiments were added together, there was no statistical difference between those treated and the controls! This is not to disparage the statistics; all that this could tell was that the chances were small that it was not a curative effect in our first experiment. Looking back upon this now with our present knowledge, it is extraordinary that anyone could have thought that by injections of Ca ions one could alter the concentrations of these at the surface of the lung. It was altogether too simple, when one thinks of the various tissue barriers to be traversed. Surface effects, however, are likely to be involved in the first instance; but it is wrong to think that the lung tissue was in general a passive filter; it has a metabolism of its own, which goes through the citric acid cycle, and therefore should be due to mitochondria. We have found lately that in poisoning by fluoroacetate or fluorodecanoic acid, there are large accumulations of citrate at death in the lung tissue (up to 40 times the normal) (page 93) and this proves how important the metabolism is. It makes it easier also to understand the mechanism of production of the oedema in gas poisoning, and even of rapid clinical oedemas.

The active form of arsenic is the tervalent one, formerly called trivalent. There are two broad classes of attack upon the tissues by tervalent arsenicals, and they are quite easily understood from a consideration of the chemistry. The classes are represented by diphenylchloroarsine on the one hand and by mapharsen on the other; the one is a disubstituted tervalent arsenical and the other is a monosubstituted. It so happens, as we know now, that the tissue constituents attacked by these two classes of arsenicals are different, because the second class can combine with two chemical groups at once. There will continue to be fundamental misunderstanding in this field until these two types of attack are understood. There are, of course, many modifications in detail.

Monosubstituted
R.As.O

Mapharside*

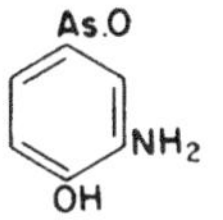

$CHCl{:}CH{\cdot}As{\cdot}Cl_2$
Lewisite

Disubstituted
R.R′.As·OH

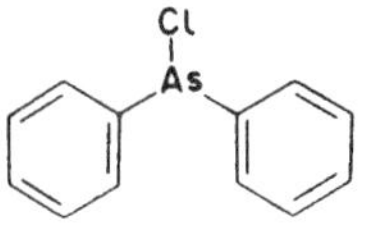

Diphenylchloroarsine

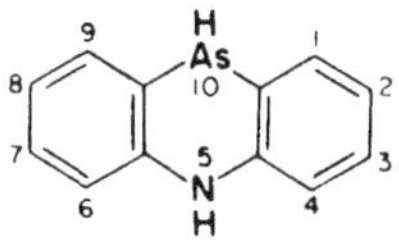

Phenarsazine

Diphenylchloroarsine Phenarsazine

FIG 3.1. Tervalent arsenicals (arsenoso)

To turn now to the groups with which arsenicals combine, we owe a bold guess as long ago as 1909 to Paul Ehrlich that As was combining with hydroxyl or sulphydryl groups(6) (see the passage on p.168). He made this guess at a time when chemical thought had not yet spread to entities inside tissues; there was a tendency even then to think that substances inside cells behaved in a mysterious way; the dynamic approach so well presented by Gowland Hopkins was barely understood. It is interesting to note that it was the isolation of the soluble —SH peptide, glutathione, by Hopkins in 1921(15) which gave the necessary precision and impetus to start —SH studies in tissues; in this respect his work was a turning-point in the biochemistry of arsenicals.

To digress a little, in following up the glutathione discovery, Hopkins with M. Dixon(16) envisaged two types of —SH compounds in the tissues, soluble ones like glutathione and "fixed" —SH compounds, in which the —SH group was very closely associated with some protein molecule in muscle tissue. The soluble —SH acted as a carrier of —H to the fixed —S.S.—. We found that a "fixed" layer containing —SH could be shown up readily in slices of skin (Peters and Walker(24,41)†) by the usual use of

* Also called arsenoxide, mapharsen, oxophenarsine.

† Dr. E. Walker was a Liverpool graduate who became a valued member of the staff of my Department in Oxford. He had a sudden illness and died at an early age in 1942.

nitroprusside and ammonia. Some 13 years later there came information slowly upon enzymes containing —SH groups, starting with the enzyme urease. Such larger protein molecules really form an intermediate group between GSH and the insoluble proteins containing the fixed —SH. It is now clear that the —SH found in proteins is that of the amino-acid cysteine in the molecule, though in the early 1920's this was difficult to prove.

No attempt is made in this chapter to give more than the outline of the arsenical researches. I have recently reviewed the arsenical work in 3 lectures;[22] there are also other good reviews; of the war work, Peters, Stocken and Thompson;[28] Waters and Stock;[43] of the pharmacology of BAL as studied in U.S.A.;[46] Stocken and Thompson.[37]

Our early investigations in this field were confined to effects upon skin slices and upon protozoa; there were very few intracellular enzymes defined in the early 1920's, and these showed little decrease in activity with arsenicals. An exception was the behaviour of some esterases[32] to diphenylchlorarsine (1920). The work of Voegtlin and colleagues was largely upon trypanosomes.[39,40] At that time none of us appear to have realized the difference between the action of the two classes of arsenicals. Certainly I did not appreciate this; both classes readily attacked the —SH group in glutathione as also the fixed —SH in skin tissue, so that the nitroprusside reaction was lost.

We soon showed that the most toxic compounds to the protozoa with which we were working was the diphenylchloroarsine, a compound highly irritant to the trachea and respiratory tract. As will be seen from the table, phenylarsenious oxide[20,42] is 10 times less toxic and this is not so irritant. It was shown by E. Walker that the —SH compound thioglycollate ($HS{\cdot}CH_2{\cdot}COOH$) could protect the protozoa from the action of diphenylchlorarsine; he also got a substantial reversal by adding a thiol compound immediately after the protozoa stopped moving as a result of poisoning with diphenylchloroarsine.

It should be realized that a reversal is much more important than a protection, because a protection may mean only that the protective agent has combined with the arsenical sufficiently strongly to prevent its combining with the —SH group in the tissues. The protection for instance might occur through some different chemical reaction from the reaction involved in the combination of the arsenical with the tissue component. For instance, one can protect ciliate protozoa against the action of mercuric chloride with ferrocyanide; but no one would imagine that there is a ferrocyanide group on a protein molecule; the mercury compound is almost certainly attacking the protozoon through some thiol group. I am stressing these points because so many of us have been led astray at times by this trap.

TABLE 3.1.* RELATIVE TOXICITY OF TERVALENT ARSENICAL AND ANTIMONY COMPOUNDS TO THE CILIATE PROTOZOA, COLPIDIUM (C) AND GLAUCOMA (G). DEATH CAUSED IN 3 MINUTES

Type	Aromatic	C	G	Aliphatic	C
			Disubstituted		
R_2 As OH	Diphenylchloroarsine	100†	100†	Dimethylchloroarsine	1—
	Diphenyl arsenious oxide	129	93		
	10-cyanophenoxy-arsine	145	117		
	10-chloro-5, 10-dihydrophenarsazine	57	42		
	Diphenylchlorostibine	101	—	Dimethylbromostibine	2
			Monosubstituted		
R. $As(OH)_2$	Phenylarsenious oxide	<5	4	Methyldichloroarsine	<1
	Phenyldichlorostibine	4	—	Ethyldichloroarsine	same

* This table is compiled from selected data of McCleland and Peters[20] as extended by E. Walker, where fuller details will be found.

† Concentrations (100) killing Colpidium were M/180,000 and Glaucoma M/420,000. Colpidium now called *Tetrahymena geleii*.

At the time when we were putting most of our effort into the action of diphenylchloroarsine, which seemed to us to be potentially the most toxic arsenical, Voegtlin and his colleagues were engaged in a study of —SH compounds and arsenoxide, which is also known as mapharsen, mapharside or oxophenarsine. This compound is essentially like phenylarsenoxide, because there is the possibility for two groups to combine with the arsenical. The U.S.A. workers found effects upon glutathione, and showed that trypanosomes could be protected *in vivo* by various —SH compounds, and that rats could be protected against the toxic action of arsenoxide by a previous injection of thioglycollate. It was an —SH effect because —S.S.— compounds had no action and no similar effects were found with amino-acids, lecithin, glucose or inorganic salts. They thought that there might be a combination with tissue —SH, and Rosenthal[33] proved this for coagulated egg albumin. It is to be noted that the protective effects were not large, and that a big excess of thiol compound was required. This dissociation of compounds of thiols and arsenicals at alkaline pH was proved in 1931 (Cohen, King and Strangeways[5]).

Mono Substituted Arsenicals

All this work was prior to the large extension of work on cell enzymes to a great extent promoted by the Embden–Meyerhof scheme of glycolysis and by the work inspired by Hopkins on oxidation enzymes and by the researches of Warburg. In the 1930's work on arsenic and enzymes was taken up again;[38] Krebs and Henseleit[17] used it in rather high concentration for blocking the metabolism of α-ketoglutarate. In 1936, Peters[21] found that in very small concentrations it blocked the vitamin-stimulated respiration of pyruvate in brain preparations from the pigeon. A similar effect was produced by the sulphone of mustard gas and led me to postulate that vesicants attacked the pyruvate oxidase system selectively. This was readily proved in 1939,[27] Table 3.2 and 3.3. Both arsenite and lewisite

Table 3.2. Comparison of the effects of tervalent arsenicals upon the pyruvate and succinate oxidase systems in brain (pigeon)

Preparation	Arsenical	Conc. mM	% inhibition	
			Pyruvate	Succinate
Brei	Arsenite	0·034	75	13
	Lewisite	0·017	89	5
Dispersion	Arsenite	0·04	55	5

The pyruvate oxidase system is also more easily poisoned than the glycerophosphate and lactate dehydrogenase. (Peters, Sinclair and Thompson, 1940)

Table 3.3. Effect of injected arsenite on the blood-pyruvate level of pigeons (expressed as mg pyruvic acid/100 g blood, as determined by 24-dinitrophenylhydrazone method. Duration of intoxication given in parentheses)

Normal	Acute	Chronic	Total dose As_2O_2 (mg)
1·28	4·02 (2 hr)	12·39 (4 days)	8·5
0·94	3·00 (2 hr)	14·15 (4 days)	8·5
0·94	3·12 (2 hr)	13·85 (43 days)	7
	6·30 (3 hr)		
Average 1·06	4·11	13·47	

Similar results were obtained for bisulphite-binding capacity (B.B.S.) of the blood in pigeons, and for pyruvate and B.B.S. with arsenite and lewisite in rats. (Data of Peters, Sinclair and Thompson, 1946)

at equal concentrations were far more toxic to the pyruvate oxidase system than they were to succinate oxidation, and also to the glycerophosphate and lactate dehydrogenase. It was not only an effect *in vitro*, because when animals were poisoned with an arsenical, a large increase took place in the concentration of pyruvate in the blood, somewhat similar to the condition in thiamine deficiency. This indicated the especial sensitivity of some component of the pyruvate oxidase system. It should be realized that with an increasing concentration of lewisite, other enzymes are attacked. In fact there is a marked selective action for arsenical inhibitors upon different enzymes. For instance, triosephosphate dehydrogenase, which is so sensitive to iodoacetate, is relatively insensitive to an arsenical such as lewisite. Vice versa, the pyruvate oxidase system is not so sensitive to iodoacetate, though it is more easily poisoned by iodoacetamide (see p. 76). The effect of an arsenical in increasing the amount of pyruvate in the blood was shown to occur also in man by Sexton and Gowdey[(34)] using mapharsen.

The marked effects of arsenicals upon the pyruvate oxidase system of brain shown *in vitro*, led to the use of brain brei and brain dispersions from the pigeon for the testing of possible antidotes and so ultimately it played an important part in the development of British anti-lewisite (BAL). The clue to the mode of action of the monosubstituted tervalent arsenicals was obtained by two members of our group. This period in World War II was the only time that we worked deliberately as a team with regular consultation between the members. I am sure that those who were having their hair cut in Oxford did not realize that they were contributing to the war effort! The hair was used by Stocken* and Thompson[(35)] as a source of the keratin required for making the reduced "kerateine". When they made compounds of lewiste and of arsenite with reduced keratin, known as kerateine, analysis showed that one As group was combining in a rather stable combination with two —SH groups especially with lewisite. The compound with arsenite was less stable than that with lewisite. This led them to the belief that this was due to ring formation, and so to the testing of dithiols with vicinal —SH groups as antidotes. As one which would form a 5-membered ring with As, likely to be the most stable combination, 2,3 dimercaptopropanol (Fig. 3.2). was made. When tested, it was found to free the component in the pyruvate oxidase system from its arsenical inhibition, so that it was again active in pyruvate oxidation. It is known now as BAL (British Anti-lewisite) or dimercaprol.

It will be understood that these synthetic efforts by Stocken in the mercaptan field were not without their repercussions upon the physical atmosphere

* Dr. L. A. Stocken joined the Oxford Department in 1937 and is now a Senior Research Officer, and a Fellow of St. Catherine's College, Oxford.

of the laboratory. This was rather difficult at the time because the work was secret and because the students studying biochemistry wondered what was happening. One of my staff with studied evasion pointed out that onion substitutes became very necessary in war-time.

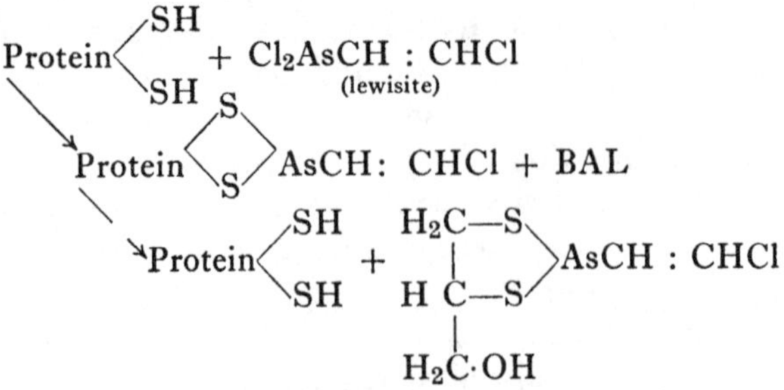

Action of BAL (2·3 dimercaptopropanol) as proposed originally.
See lipoate, p. 50.

FIG. 3.2.

The dithiol was active *in vivo* not only as a protective agent but also as an antidote, when added after poisoning by lewisite was established (Table 3.4). It could even reverse commencing pathological damage upon the skin.*

TABLE 3.4. EFFECT OF THIOLS UPON RATS WITH A SKIN CONTAMINATION OF A LETHAL DOSE OF LEWISITE (30–38 MG APPLIED PER KILO BODY WT.)

	No. of animals used	Survivors	
		No.	%
Untreated	13	0	0
Treated with monothiol			
2-mercaptoethanol applied after 30 min	6	0	0
Treated with dithiol			
BAL applied after 30 min	21	21	100
BAL applied after 60 min	18	17	95

(Data of Stocken and Thompson)

Figure 3.3 illustrates the effect of BAL upon the vesicant action of lewisite, as compared with an anti-gas ointment. The application of anti-gas

* It must be emphasized that in our experiments a *mono*thiol did not protect against or reverse lewisite poisoning. This statement is made because in some reviews, it has been implied that monothiols have some effect, a statement also at variance with theory.

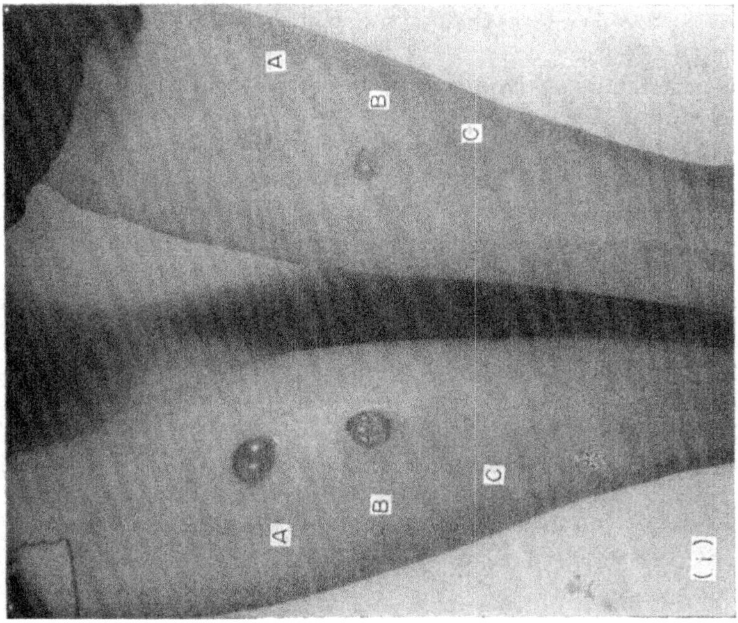

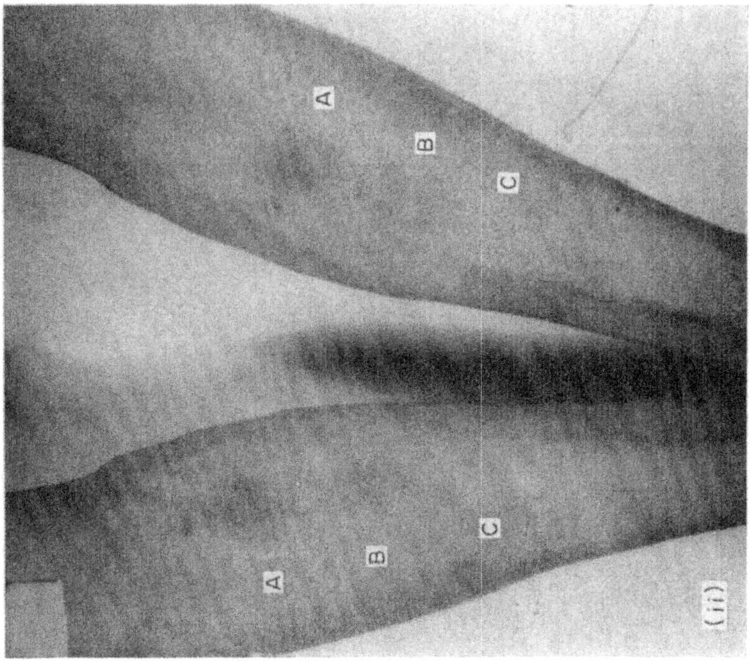

FIG. 3.3.

Area	*Right arm*	*Left arm*
A	Treated with anti-gas ointment 2 hr after contamination, when erythema was well established.	Treated with BAL 10 per cent 2 hr after contamination.
B	No treatment (control)	No treatment (control).
C	Treated with anti-gas ointment 2 min after application of drop of lewisite.	Treated with BAL 10 per cent 2 min after application of drop.
	(i) Photo taken at time immediately before treatment.	(ii) Photo taken 24 hr after test.

(Photographs by permission of Ministry of Supply.)

ointment at 2 min after contamination was successful. Treatment with BAL at 2 hr, after the beginning of erythema was successful for BAL. The control (left arm) shows a smaller blister at 24 hr than the right arm, which is probably due to a slight spread of the BAL on the skin.

Fig. 3.4. Convulsions induced in a pigeon by intracerebral injection (under ether) of 100 μg lewisite. (Peters and Hastings, 1952).

The analogy *in vitro/in vivo* was nearly complete, but it had not been proved that lewisite injected into the brain of the pigeon would cause convulsions; this was finally done and shown to happen after an injection of 100 μg.[26] So (Fig 3.4) lewisite wrecks the pyruvate oxidase system *in vivo*.

This is not the place to consider in detail the practical value of BAL in medicine as a therapeutic agent; but it has been useful in cases of arsenical and mercurial poisoning; it is another example of a substance which has been found in the laboratory and applied to the clinic as a last step, of which we now have several instances. Though the first made, it is the least toxic; even the 1,3-dimercaptopropanol is more toxic (Stocken[37]).

I only learnt later that BAL was used in hairdressing establishments for producing permanent waves. I always wondered how they managed to deal with the mercaptan-like smell.

LIPOIC ACID

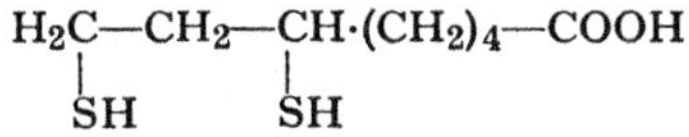

Reduced

When the BAL work was released for publication in 1945–6, it seemed to be clear that a dithiol component must be present in the pyruvate oxidase system, and that this must form a ring, rather larger than the 5-membered ring formed with BAL; otherwise it was not easy to see why the ring with BAL should be more stable than the compound with the enzyme. This was the conclusion of V. P. Whittaker.(45)* (Fig. 2.5) However, it was not clear how two—SH groups could coexist on a protein in this

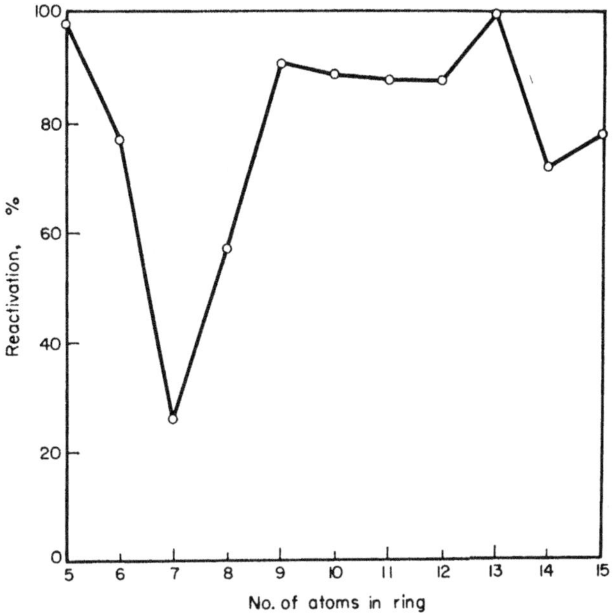

FIG. 3.5. After poisoning by lewisite α-ω-dithiols were added with increased number of carbon atoms. Reactivation with α-ω-butane dithiols is much less efficient. (Data of V. P. Whittaker.)

vicinal position to one another. Ten years after the work on BAL was completed, the answer to this puzzle came in a most interesting set of observations upon the components of α-keto acid oxidase systems. Various researches in the U.S.A. converged upon the discovery that these systems

* Dr. V. P. Whittaker is now on the staff of the A.R.C. Institute of Animal Physiology (Babraham), Biochemistry Department. He was a University Demonstrator in Oxford until 1951, when he went to Cincinnati for 4 years.

had a component now known as lipoic acid, or thioctic acid, and the work of Gunsalus and of Reed with their colleagues was especially important.[12,13,30] Lipoic acid is an octanoic acid with the 6,8 groups having an —SH group when in the positions substituted with S in the reduced form. The bacterial systems were shown to be blocked by arsenite, and the arsenical block was released by BAL. This is a very satisfactory result, because it shows why α-keto acid systems are specially sensitive; also, since arsenicals will form with reduced lipoate a 6-membered ring, it explains why the As was slowly transferred from lipoate to BAL to make the 5-membered ring, with its extra stability. Until a year ago it could be argued that the presence of lipoate in brain tissue had not yet been proved; but in the last year this possible objection to a complete story has disappeared; E. M. Gal and D. E. Razevska (1960) have proved the presence of lipoate in brain. This appears to settle in principle the mode of action, or the biochemical lesion produced by mono-substituted tervalent arsenical substances. It must be emphasized again that a monothiol does not reverse the effects of monosubstituted arsenicals.

Disubstituted Arsenicals

There remained the problem of the toxicity of the disubstituted arsenicals of which diphenylchloroarsine is the best example, the compound used as a highly irritant arsenical smoke. Even so, its systemic toxicity has never been considered to be pronounced, even though to protozoa as has been mentioned, it is much more toxic than the monosubstituted compounds. In theory, from its chemical structure the monosubstituted compound could only combine with one —SH group at a time; it could not therefore make more than a monothiol attack upon some individual enzyme. This paradox caused me some trouble until I found with Dr. Lotspeich* an instance of an enzyme which was very sensitive to diphenylcloroarsine and the analogous phenarsazine and insensitive to

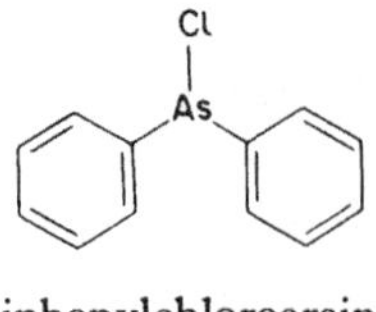

Diphenylchloroarsine

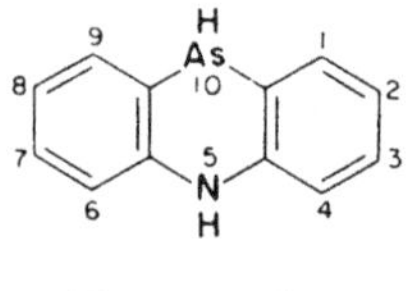

Phenarsazine

* Dr. W. D. Lotspeich is now Professor of Physiology in the University of Rochester,. N.Y.

the action of lewisite.[19] This was the isocitrate dehydrogenase. Figure 3.6 shows the effect of small amounts of diphenylchloroarsine upon isocitrate dehydrogenase; lewisite in much larger concentration did not have any inhibitory effect upon the enzyme. With this enzyme, therefore, the action of the two arsenical compounds is much that which occurs with the ciliate protozoa. Furthermore, the effect of the disubstituted compound is readily reversed by glutathione or by another monothiol compound; in this way the enzyme experiment is a good example *in vitro* of the effect of a monothiol in reversing the action of DA upon the ciliates in our older experiment.

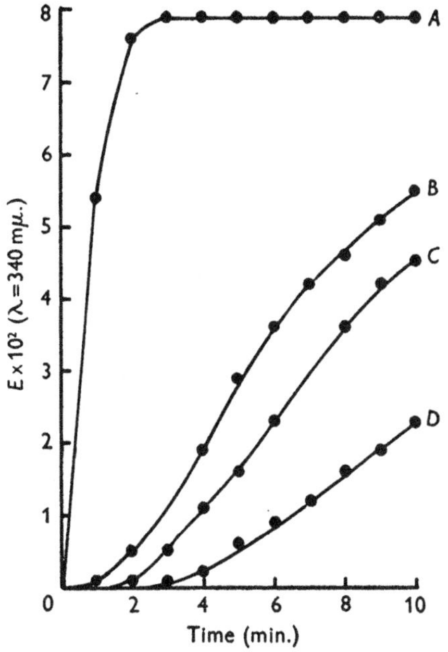

FIG. 3.6. Inhibition of isocitric dehydrogenase by diphenylchloroarsine and reactivation by monothiols. Curve A shows the rapid activity of the control in reducing coenzyme II (NADP). Curve D the activity is much inhibited by DA (0·063 mM). Curves B and C the activity is reversed by glutathione (1·3 mM) or by monothioethylene-glycol (0·63 mM). Note lewisite 0·65 mM had no inhibitory effect. (Lotspeich and Peters, 1951.)

Based upon these experiments with the isocitrate dehydrogenase, which is one of the enzymes of the citric acid cycle, we can now form a reasonable hypothesis for the very different behaviour of these two classes of arsenical upon protozoa and upon the higher animal. Diphenylchloroarsine and the phenarsazine compounds of the same class are potentially the most toxic; but in the higher animal their toxicity is at once counteracted by the circulating glutathione. On the other hand, with a compound like lewisite, any compounds formed with a monothiol are very unstable, and the arsenical is handed on eventually to lipoate where it blocks the oxidation of α-keto acids, the most important of these being the initial stage of pyruvate decarboxylation. In some experiments in which Mr. Wakelin and I poisoned brain particles with DA,[23] we found that 50 per cent of this poisoning was reversed with a monothiol, but that the remainder required a dithiol. I interpret this to mean that there is a conversion *in vivo* of some of the diphenylchloroarsine to phenyl arsenious oxide which could then react as lewisite with a dithiol component. Such an action may well explain the puzzling losses of sensation (skin anaesthesia) found among some patients exposed to arsenical smokes during World War I.

The remarks made about arsenicals by Albert[1] in chapter 10 of his interesting book call for comment in regard to the reversal of the pyruvate oxidase system by monothiols after poisoning with arsenicals. Gordon and Quastel[11] claimed that the inhibition of pyruvate oxidation by mapharside in ground brain tissue from the rat could be reversed by glutathione. As this was discrepant with our own results we reinvestigated it, and confirmed that the residual respiration is poisoned by mapharside and that this was reversed by glutathione. But when pyruvate was added, the extra oxygen uptake so induced is powerfully poisoned by mapharside, and this extra toxicity is not reversed by glutathione. Hence, our former results are confirmed that the true pyruvate oxidase system when poisoned is not reversed by a monothiol. We do not know what the enzymes or substrates are which take part in the "residual" oxidation, but part of this may indeed be the reaction by which brain tissue when damaged converts pyruvate direct to acetate, a reaction in which only one atom of oxygen is involved.

TABLE 3.5. MAPHARSIDE: EFFECT OF GLUTATHIONE ON TOXICITY TO BRAIN BREI (RAT): O_2 μl 30–90′

Expt. ————>	(1)	(2)
Substrate ————>	Nil	Pyruvate
No addition	700	2040
+GSH (50 μ mol.)	761	1911
+Maph. (3 μ mol.)	90	75
+ Maph. +GSH	670	645

(Peters and Wakelin, 1954)

The behaviour with isocitrate dehydrogenase is explained by thinking that diphenylchloroarsine makes a monothiol attack. It is still not clear how this arsenical brings about irritation of the nerve endings in the eye and the trachea. By extrapolation from these results, one would infer that there is an —SH group in the nerve ending, and that a stimulus is caused when this combines with the arsenical compound. The mechanism of this effect would be worth investigating because of the possibility of relating it to a definite chemical reaction. Clearly it should be slowly reversible to account for the fact that the irritation disappears (see p. 78).

Some Other Considerations

When the lewisite–BAL compound is injected, it is non-toxic; but when a similar equimolecular compound of mapharside and BAL is injected, it is more rapidly lethal than mapharside alone (Peters and Stocken[25]). We think that this is because the compound is more lipid-soluble so that the arsenic is carried more quickly to the active site, where it dissociates to liberate the arsenical fragment. Support is given to this view by the fact that injection of BAL together with the mapharside–BAL prevents the toxicity. The presence of the —NH_2 in the aromatic ring appears to make the compound with the dithiol more dissociable. This experiment enables us to understand why an injection of large amounts of a monothiol together with mapharside in the case of the experiments of Voegtlin gave a temporary effect. Furthermore, as might be expected from the dissociation, it takes more mapharside than lewisite to poison pyruvate oxidation in a brain preparation.

Tissue cultures. Working with lewisite and BAL, Fell and Allsopp[7] confirmed that the dithiol stopped the toxic action of the lewisite in cultures from the chick. In bone marrow cultures, Harrison and Randoll[14] found another case in which mapharside–BAL was toxic to the cultures; but if the concentration of BAL was increased, there was a transference of the As to the medium in which the culture was suspended. On the other hand with the protozoon *Colpidium* (work unpublished), I could not reduce the toxicity of mapharside–BAL by adding more BAL. This is difficult to explain, but it may be similar to a recently published experiment in which oxidized BAL was found to be toxic to a virus.[29]

Clinical Trials

Extensive clinical trials were made upon patients suffering from arsenical exfoliative dermatitis both in this country and in the U.S.A. It was found that the compound was too irritant to be applied regularly to the skin, and that it was best administered in a mixture containing BAL 5 or 10 per cent benzoyl benzoate 20 per cent in arachis oil. It can be injected into the buttocks, but there is often an abscess at the site of injection. Nevertheless, the results are dramatic; the toxic condition is rapidly cured and within 14 days of commencing treatment the skin may be virtually normal. These clinical results are sensational. The results in the treatment of lead poisoning are disappointing; there seems to be even a tendency for the

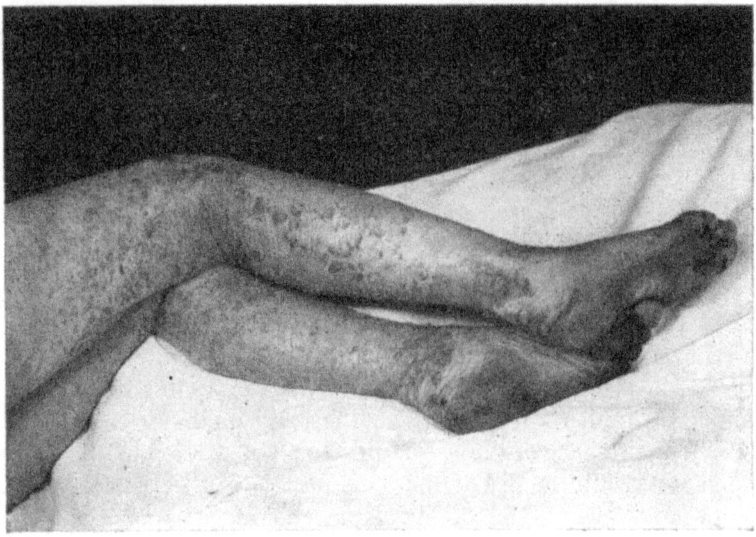

(a)

(b)

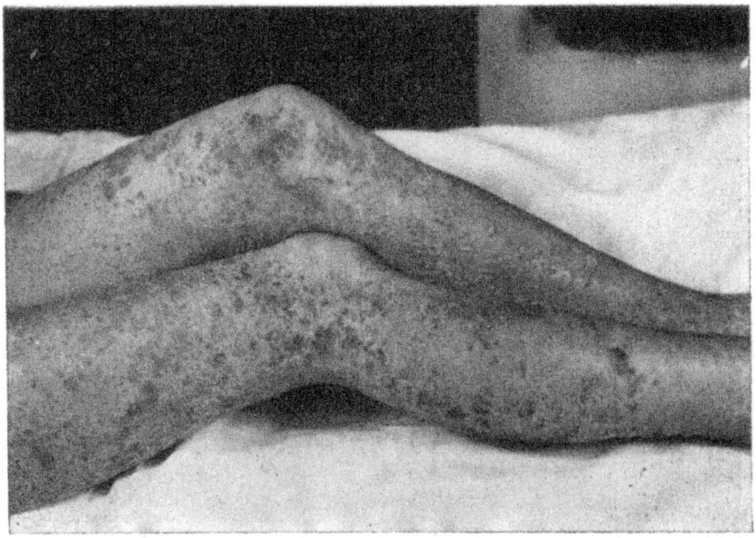

Fig. 3.7. Effect of BAL upon a case of exfoliative arsenical dermatitis. Condition before treatment. (Carleton, Peters and Thompson, 1948.)

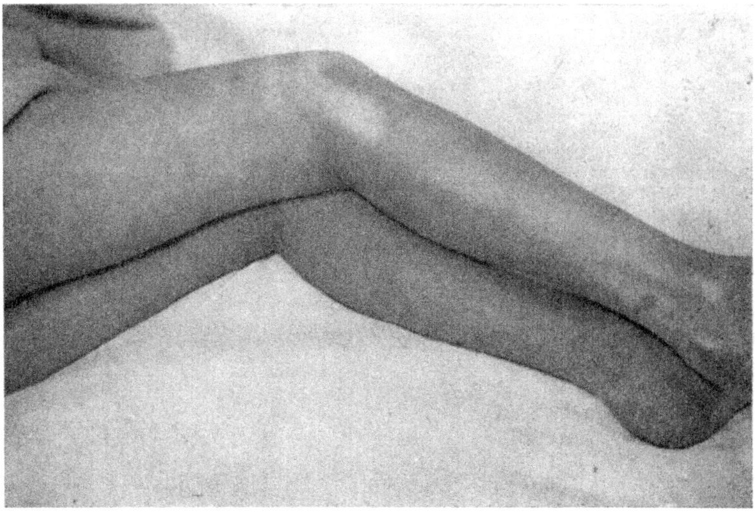

(a)

(b)

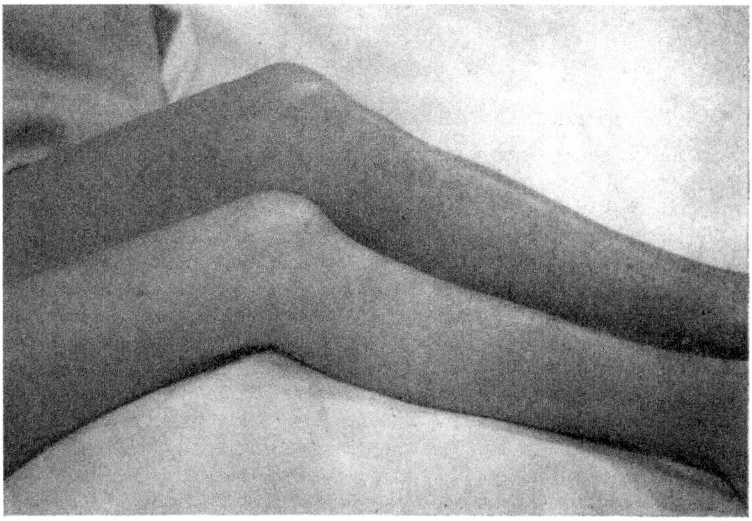

FIG. 3.8. Effect of BAL upon a case of exfoliative arsenical dermatitis. Condition 14 days after commencement of therapy. (Carleton, Peters and Thompson, 1948.)

BAL to redistribute the lead to the disadvantage of the patient. On the other hand, very striking effects have also been obtained against poisoning by mercury compounds especially in Baltimore by Longcope and Luetscher.[18]

Summary

In the case of the arsenicals, we have another instance in which a parallel study of events *in vivo* and *in vitro* of a biochemical lesion has yielded a good scientific as well as a practical dividend. It is interesting to note the close parallel also between the effects of thiamine deficiency (Chapter 2) and of the toxic action of the monosubstituted As compounds; both lead to the failure in function of the initial reactions of the decarboxylation and dehydrogenation of pyruvate. In the one case it is the loss of thiamine PP, which means that there is no acceptor for the acetyl group first formed, as has been shown in very recent work. The arsenical in its turn blocks the thiol group of lipoate which accepts the acetyl from the thiamine PP before it is handed on finally to become acetyl.S.CoA. The connection between biochemistry, pharmacology and pathology stands out here in clear relief.

It can be claimed to be the first instance in which a therapeutic substance has been discovered by a logical theoretical approach i.e. one based on chemistry. Another case has been found recently, viz. an antidote to cholinesterase inhibition (p. 162).

APPENDIX

Arsenical Substances and Dithiol Compounds

The short sketch given so far has been confined to the bare bones of the work on the biochemical lesion as we know it today. It does not do justice to the large mass of good work which was conducted both on the effect of arsenicals upon the whole animal and also upon enzymes on the one hand, and upon various thiols to find out which was the most suitable dithiol, (see also Webb and van Heyningen,[44] Barron *et al.*[3] Work done in Oxford was confirmed and amplified in the U.S.A. For instance, several other dithiols were tried by Stocken and Thompson and found to be more toxic, and a large number of other dithiols were synthesized in the U.S.A. and tried on larger animals including cats and dogs. One of these a 2,3-dimercaptopropyl ethyl ether was not so potent as an antidote, 7 times more had to be used, but it was stated to be less toxic (Riker[31]). All the others were more toxic. It is interesting to note the greater toxicity of the 1,3 dimercaptopropanol, which would form a 6-membered ring similar to that of lipoate, rather than the 5-membered ring with BAL. It is generally agreed that these dithiols including BAL are irritant, and not more than a few millilitres can be applied by the skin route. There is no doubt that the slow release of BAL into the circulation from an intramuscular injection tends to keep up the concentration without leading to too much irritation. The source of the irritation seems to be complex. On the one hand BAL was proved to act as a powerful chelating agent for various metals. Cytochrome oxidase was not inhibited, but it oxidized the BAL rapidly. On the other hand, there was a strong inhibition of polyphenol oxidase (copper protein), carbonic anhydrase (zinc), catalase (iron), peroxidase (iron), glyoxalase. There is no doubt, therefore, of its action upon metals which may be part of enzymes. There is also another way in which the compound can be active; it can reduce —S.S.— bonds, and therefore upset the —SH equilibria in cells.

Much work has been done upon other metals. It has been mentioned already that the effects upon mercury poisoning were excellent, and these clinical results have been confirmed in Belgium (Bacq and Roskam, 1951[(2)]) in 2 cases of poisoning by mercuric oxycyanide. It is interesting that urinary excretion was re-established after 3–5 days of anuria. A most careful study of cadmium poisoning was made by Gilman and his colleagues. They found that BAL could exist in two complexes with cadmium. Unless great care was taken with the doses, the effect of BAL was to remove cadmium from some tissue where it might be doing much damage and to deposit it in the kidney, where its effect was disastrous, and where it led to a rapid rise in the non-protein nitrogen in blood.[(10)] This behaviour is not a contradiction of the use of BAL, which is one of the only compounds available in these cases of cadmium poisoning. All that it shows is the necessity for using it with previous knowledge of these facts. BAL should not be used in lead poisoning.

References

1. Albert, A. (1960), *Selective Toxicity*, 2nd ed. Methuen, London, and John Wiley, New York.
2. Bacq, Z. M. and Roskam, J. (1951), Congrès Francais de Medicine, 28th Session Rapports **1,** 212–247.
3. Barron, E. S. G., Miller, Z. B., Bartlett, G. R., Meyer, J. and Singer, T. P. (1948), *Biochem. J.* **42,** 69.
4. Carleton, A. B., Peters, R. A. and Thompson, R. H. S. (1948), *Quart. J. Med.*, New Series **17,** 49.
5. Cohen, A., King, H. and Strangeways, W. I. (1931), *J. Chem. Soc.* 3043.
6. Ehrlich, P. (1909), *Ber. Dtsch. Chem. Ges.* **42,** 17.
7. Fell, H. B. and Allsopp, C. B. (1946), *Brit. J. Exp. Path.* **27,** 310.
8. Gal, E. M. and Razevska, D. E. (1960), *Archiv. Biochem. Biophys.* **89,** 253.
9. Gilman, A., Allen, R. P., Philips, F. S. and St. John, E. (1946), *J. Clin. Invest.* **25,** 549.
10. Gilman, A., Philips, F. S., Allen, R. P. and Koelle, E. S. (1946), Supplement, *J. Pharmacol.* **87,** 85.
11. Gordon, J. J. and Quastel, J. H. (1947), *Nature, Lond.* **159,** 97.
12. Gunsalus, I. C. (1952) (July), Communication to 2nd Int. Congress of Biochemistry, Paris. Also (1954) in *Mechanism of Enzyme Action* page 545, Ed. W. D. McElroyd and B. Glass, Baltimore.
13. Gunsalus, I. C., Struglia, L. and O'Kane, D. J. (1952), *J. Biol. Chem.* **194,** 859.
14. Harrison, K. and Randoll, F. W. (1948), *Quart. J. Exp. Physiol.* **34,** 141.
15. Hopkins, F. G. (1921), *Biochem. J.* **15,** 286.
16. Hopkins, F. G. and Dixon, M. (1922), *J. Biol. Chem.* **54,** 527.
17. Krebs, H. A. and Henseleit, K. (1932), *Z. Physiol. Chem.* **210,** 33.
18. Longcope, W. T. and Luetscher, J. A. jun. (1946), *J. Clin. Invest.* **25,** 557.
19. Lotspeich, W. D. and Peters, R. A. (1951), *Biochem. J.* **49,** 704.

20. McCleland, N. and Peters, R. A. (1919), *J. Physiol.* **53,** Proc. XV.
21. Peters, R. A. (1936), *Curr. Sci.* **5,** 207.
22. Peters, R. A. (1952), Harben Lectures, *Roy. Inst. Publ. Hlth. & Hyg.* **15,** 89, 122 and 151 (Review).
23. Peters, R. A. (1955), *Bull. Johns Hopkins Hospital* **97,** 1. (Dohme Lecture No.1).
24. Peters, R. A. and Walker, E. Unpublished Reports to Government 1924–25.
25. Peters. R. A. and Stocken, L. A. (1947), *Biochem. J.* **41.** 53.
26. Peters, R. A. and Hastings, A. B. (1952), see Ref. 23.
27. Peters, R. A., Sinclair, H. M. and Thompson, R. H. S. (1946), *Biochem. J.* **40,** 516.
28. Peters, R. A., Stocken, L. A. and Thompson, R. H. S. (1945), *Nature, Lond.* **156,** 616 (Review).
29. Philipson, L. (1962), *Biochem. Biophys. Acta* **56,** 375.
30. Reed, L. J. (1953), *Physiol. Rev.* **33,** 544.
31. Riker, N. F. (1946), In Supplement to *J. Pharmacol. Exp. Ther.* **87,** 66.
32. Rona, P. and Gyorgy, P. (1920), *Biochem. Z.* **111,** 115.
33. Rosenthal, S. M. (1932), *Publ. Health Rept. Wash.* **47,** 241.
34. Sexton, G. B. and Gowdey, C. W. (1947), *Archiv. Derm. & Syphil.* **56,** 634.
35. Stocken, L. A. and Thompson, R. H. S. (1946), *Biochem. J.* **40,** 529.
36. Stocken, L. A. and Thompson, R. H. S. (1946), *Biochem. J.* **40,** 535.
37. Stocken, L. A. and Thompson, R. H. S. (1949), *Physiol. Rev.* **29,** 168 (Review).
38. Szent-Gyorgyi, A. (1930) *Biochem. J.* **24,** 1723.
39. Voegtlin, C., Dyer, H. A. and Leonard, C. S. (1923), *U.S. Publ. Health Rept.* **38,** 1882.
40. Voegtlin, C. (1925), *Physiol. Rev.* **5,** 63.
41. Walker (1925), *Biochem. J.* **19,** 1082, 1085.
42. Walker, E. (1928), *Biochem. J.* **22,** 292.
43. Waters, L. L. and Stock, C. (1945), *Science,* **102,** 601.
44. Webb, E. C. and van Heyningen, R. (1948), *Biochem. J.* **42,** 74.
45. Whittaker, V. P. (1947), *Biochem. J.* **41,** 56, and 1944 Thesis (Oxford).
46. *Note*: A special number of *J. Pharmacol. Exp. Ther.* Supplement **87,** 60 is devoted to papers on BAL from the U.S.A.

CHAPTER 4

The Biochemical Lesion in Mustard Gas Poisoning

RESEARCH upon the biochemical lesion induced by mustard gas has provided another instance like that of arsenic, where good has come out of the attempt to find an antidote which would make the hostile use of mustard gas of no avail. Though no antidote was found, the work led to most interesting developments in the genetic field, and to an intensive study since 1947 of new agents with a somewhat similar chemical activity which have proved their value in the treatment of leukaemia. It is interesting to note that long ago Berenblum(13) used sulphur mustard gas as an anti-carcinogenic substance.

Since a description of the more recent work takes us into the field of the metabolism of nucleic acids, a short account is needed of some of the main points established in recent work. Though development is proceeding at a very rapid rate, there appear already to be some generalizations which are likely to stand.

There are several excellent reviews on the nucleic acids; if a reader wishes to get a balanced picture of the whole subject without too much detail, important as this may be, he would be well advised to read the book by Davidson(21) (1961). For other good reviews see the references(24,33). There are now known to be two main classes of nucleic acid, deoxyribonucleic acid (DNA) and ribonucleic acid (RNA). Though at one time the protamines or their equivalent were thought to be essential for genetic behaviour, it is now considered that the deoxyribonucleic acids can behave genetically alone. This view is based in the first instance on the finding by Avery, Macleod and McCarthy(8) (1944) that one type of pneumococcus can be transformed into another by addition of a transforming principle containing DNA. The other is that the nucleic acids from viruses are able by themselves to direct the chemical change in the host, which leads to the formation of infective virus. This can happen when the nucleic acid is either DNA or RNA; but the emphasis here seems to be upon DNA for the initial genetic change. The DNA in the nucleus is in the form of duplex chains of nucleic acid; it is now thought that the theory of Crick and Watson(19) for the duplication of these strands of nucleic acid fits the facts closely, down to the detailed arrangement of the purines and pyrimidines which make up the nucleic acid. Since DNA is to be found in the nucleus and since it is localized in the chromosomes, it has become an accepted idea that DNA is concerned with the first stages in the regulation of cellular synthesis. The DNA in the nucleus forms ultimately RNA which then carries the desired information to the cytoplasm for the synthesis of the proteins upon templates connected with the RNA of the ribosomes. It is held that the RNA in the ribosomes is different from that which is present in the cytoplasm and that the nuclear RNA acts as a messenger. The individual amino-acids are carried to the templates on a form of RNA known as transfer RNA (s-RNA). In addition to these forms, the literature contains references to other small molecular RNA. Though the detail in this picture will undergo changes with the progress of research, the main outlines now seem to be correct. That the protein part of the nucleoprotein must also matter would seem obvious. Looking at the facts known already, however, there are evidently two ways in which there can be interference with the behaviour of the chromosomal substance or with its synthesis; either there can be direct replacement of a group upon the strand of the nucleic acid, or at least some interference with its chemical structure; or again there can be some interference with the coiling of these strands. We shall see that mustard gas can interfere in both of these ways.

MUSTARD GAS DICHLORODIETHYL SULPHIDE (MG)

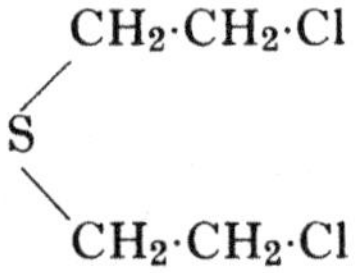

Mustard gas was discovered in 1887 by Meyer, and some of its properties were known then. No one thought that it could be made on the large scale until this was demonstrated in World War I when it was used in warfare in 1916. Its effects were so remarkable that intensive research into the pathology of its action was begun; in this chapter no attempt will be made to describe the various physiological or pathological effects of this substance. I shall confine myself as far as possible to the initial biochemical change, the biochemical lesion.

In the early 1920's an attempt was made to understand the biochemical changes. One of the first theories of its action was that it penetrated into the interior of a cell in virtue of its lipoid solubility, where it liberated acid within the cell, thereby upsetting the pH of the cell and so its metabolism.* This was based upon the fact, then well known, that "mustard" is converted in an aqueous solution into a mixture of acid (HCl) and thiodiglycol;

* Fell and Allsopp (1941) found that sulphur mustard gas penetrated very rapidly as such into the interior of cells of a tissue culture.

the latter is quite inert. The "acid liberation" theory of vesication was examined in Cambridge by Peters and Walker (1923) who were not able to support it.[39] They found that on the one hand dichlorodimethylsulphide liberated acid much quicker than did "mustard" and had little irritant effect on the skin. On the other hand the sulphone of "mustard"

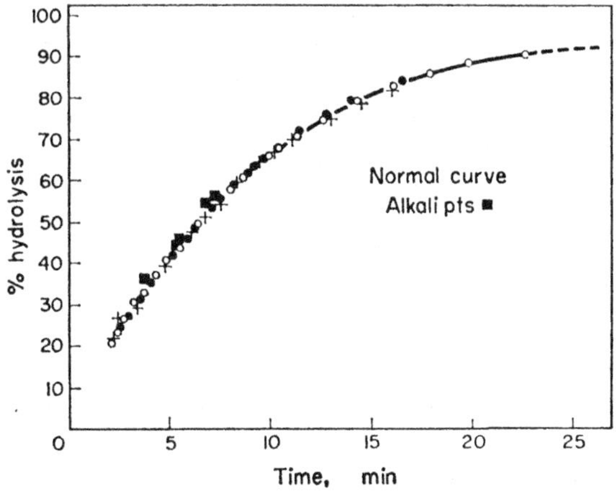

FIG. 4.1. Formation of acid by 5 per cent alcoholic solution. Ordinates per cent acid liberated. Abscissae time. Temp. 24.5 ±0·10. Points ● ○ + represent three different experiments. Points marked ■ obtained in the presence of alkali.

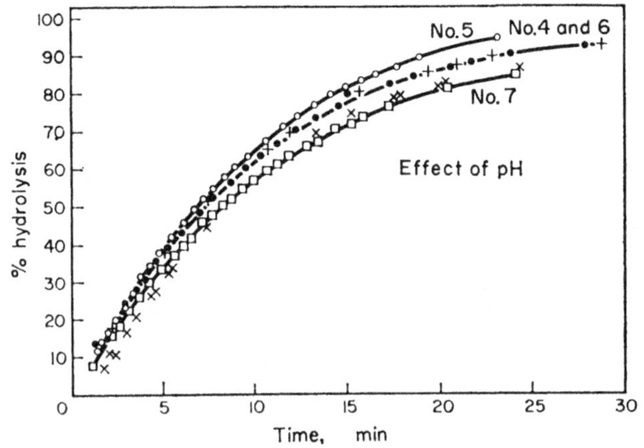

FIG. 4.2. Effect of pH on formation of acid. Reaction of titration end point varied temp. 24·6° ± 0·1. Curves: No. 4, pH 5·3 (control). No. 5, pH 9·0. No. 6, pH 7·1. No. 7, pH 9·0 exposed specimen of DES. × = pH 7·0. (In presence of phosphate M/1000.)

liberated acid far more slowly than "mustard" itself, and yet was almost as vesicant. During the course of this research some facts were discovered which became of importance later. The experiments on the rate of liberation of acid were done at neutral pH; but they found that there was little effect of pH upon the reaction over rather a wide range, pH 4–10. The presence of chloride markedly slowed the rate of liberation of acid; even the amounts of chloride in the tissue cell could slow the rate. Another fact of interest emerged in the relation of "mustard" to —SH groups. Whereas the —SH groups in skin were readily attacked by diphenylchloroarsine as well as by the sulphone of "mustard", there was much less effect of "mustard" itself, which suggested that an attack upon —SH groups was not the main mode of action. These facts lay dormant until 1939, when the main points were confirmed, and when they could be interpreted.

During the 1930's, a development of physicochemical theory upon electronic lines by Ingold and Hughes[(32)] had led to better understanding of the reactions involved. Their ideas were applied to mustard gas by Ogston and Philpot *et al.*[(35,36)] "Mustard" is an electrophilic reagent in their terms, that is to say that it is a reagent which acquires electrons (or a share in electrons) belonging to some other atom or ion. The opposite to this is a nucleophilic reagent which is one which transfers electrons to a foreign nucleus. The fact that the liberation of acid in water is not much affected by change in pH is consistent with a nucleophilic substitution reaction known as S_n1. "Mustard", in short, is an alkylating reagent which can alkylate at room temperature and at neutral pH. It first ionizes according to Fig. 4.3, this being an intrinsic property of the molecule, to form a

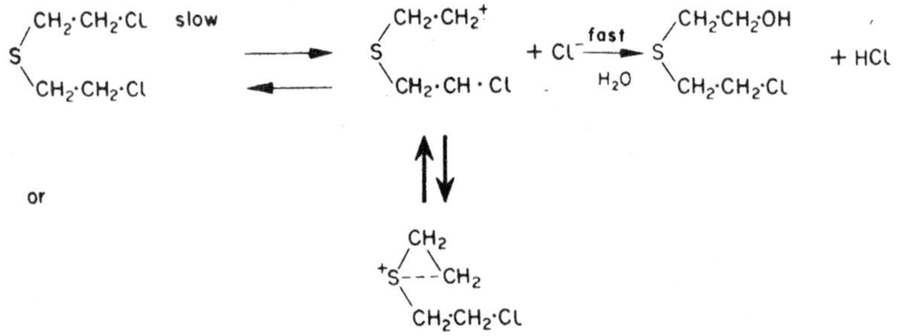

FIG. 4.3. Reactions of mustard gas with water

charged ion, which can be written also in ring form. Following this the activated ion will combine either with water or a variety of other groups. Figure 4.3 shows the course of the first reaction of mustard gas with water to form the half hydrolysed product, known as β-chloro-β′-hydroxydiethyl sulphide. The second, $—CH_2·CH_2·Cl$, reacts similarly to form the innocuous thiodiglycol, $S(CH_2·CH_2·OH)_2$. If Cl^- is present in the reacting medium, the apparent rate of reaction will be slowed because the activated ion will recombine with Cl^-. These reactions can be followed by the changes observed in the rate of liberation of acid; and the recombination explains the effect of chloride in slowing the liberation of acid. The nitrogen mustards which have the general formula $R·N(CH_2·CH_2·Cl)_2$ react similarly to the sulphur mustard gas.

In the activated form, mustard gas has the capacity of combining with a large number of other groups, which are nucleophilic (electron-rich); these can be organic acids, —OH compounds, sulphydryl, amino or phosphate groups. In 1940 Ogston and Philpot with Holiday worked out a method of determining the groups which the activated mustard would select first, when a number of possible groups were provided. This was called the competition factor, a list of some of which is given in Table 4.1.

TABLE 4.1. COMPETITION FACTORS

Thiophosphate	7×10^4	BAL (2,3 dimercaptopropanol)	$7·8 \times 10^2$
Thiophosphonates	$1·0 \times 10^4 - 10^5$	Amino-	$1·8 \times 10$
Thiosulphate	$2·7 \times 10^4$	Carbonyl-	$1·0 \times 10$
Diethyldithiocarbamate	$3·4 \times 10^4$	Phosphate	$7·5 \times 10$
Cysteine	$1·2 \times 10^3$		

Selected from A. G. Ogston (1948) Biochem. Soc. Symposia No. 2, p. 2.

Independently, Bartlett and Linstead(10) developed a similar idea in the U.S.A. Later in careful work Ross(43) worked out the similar competition factors for the carbonium ion of the nitrogen mustards,(44) though there was variation in detail. For organic acids the form $R·COO^-$ is more reactive than R·COOH. So the more mustard there is available, the more groups of different kinds will be alkylated; hence it is clear that the problem of finding antidotes to mustard gas is the problem of finding a series of compounds which will simultaneously take to pieces a number of alkylated derivatives of different types. For instance, those with —COOH may be easily hydrolysed by alkali. If compounds are formed with —SH, they become thioethers; $Cl·CH_2CH_2·S·(CH_2)_2S·R$. A method of taking these to pieces with silver salts was found, but not under physiological conditions (Peters, Ford Moore and Wakelin(41)).* In regard to combination

* It is interesting to note that this silver reaction was used by Paul(37) for the thioether group in cytochrome C.

with —SH groups, this has been found to depend upon the ionization of the group, as it does for $—COO^-$ groups; the form which is not ionized is less reactive.* In very elaborate studies made by Wormall and his colleagues[9] during World War II, using radioactively labelled mustard, it was proved that combination took place with a large number of groups, and this was confirmed and extended later by Elmore, Gulland and co-workers in careful work.[25] Canadian research showed that the nucleic acids combined readily with "mustard" and that guanylic acid was the most easily attacked;[47] Brookes and Lawley[17] have studied the nature of this reaction in detail. This was in line with earlier observations of Berenblum and Schoental in 1941, who had found a precipitation of nucleoproteins with "mustard".[14] The amount of work in this field is so large that it would be impossible to do justice to it without a long review; in fact such reviews exist, as for instance that by Phillips (1949). For the chemistry see especially Ross.[43]

Sulphone and Sulphoxide

The mustard gas sulphone is toxic, but it combines in a different way, and can attack —SH groups rather specifically. There is evidence that divinylsulphone is the toxic intermediate. The sulphoxide itself is relatively stable. It is non-toxic under conditions in which the other compounds

Sulphone

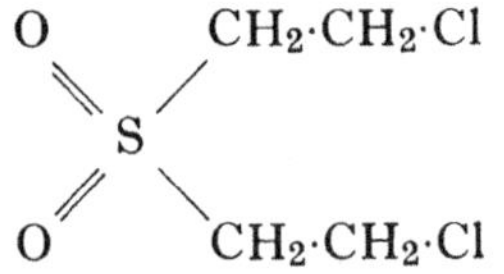

Sulphoxide

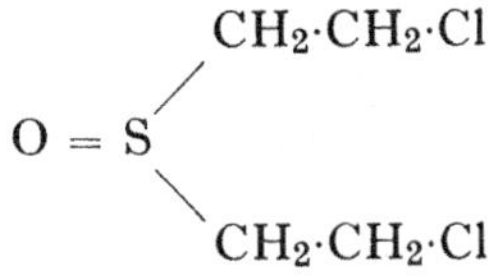

are toxic; but there is evidence (Lutwak-Mann) that it can become toxic after injection *in vivo*; this is likely to be due to a conversion to one of the toxic forms.

Vesication

The exact mechanism by which sulphur mustard gas produces the damage in skin leading to vesication as well as erythema cannot yet be settled.[38] Dixon, Needham [22–23] and their colleagues found that the isolated hexokinase was very powerfully poisoned by mustard gas and attributed the damage in skin to this, owing to the key role played by hexokinase in the initiation of the metabolism of carbohydrates. I felt unable to accept this attractive view for the reason that the hexokinase in

* Mustard gas can also form elaborate sulphonium compounds: Stein *et al.*[44]

skin appears to be poisoned slowly by sulphur mustard gas, whereas the damage can be shown to be very rapid to some skin constituent, Cullumbine (1947); Cullumbine and Rydon[20] found that leucotaxine was set loose. Beloff and Peters[11] (1944) after finding that thermal burns of the skin in rats liberated a proteinase from the cells of the skin, obtained some evidence that treatment with mustard gas led to a similar result. In consequence of this, the writer has concluded that the initial attack *in vivo*, namely that made by the smallest doses of mustard gas producing damage in the skin appears to be an alkylation of some constituent in the surface of the skin cell; this could be an enzyme, but it could equally well be something else. Very small amounts of sulphur mustard gas can cause damage; as little as 6 $\mu g/cm^2$ causes vesication in 50 per cent of white subjects, even 1/100 of this in sensitized people, and in the rat 1$\mu g/cm^2$ damages the corneal epithelium permanently; 0·01 $\mu g/cm^2$ in the rat again will cause suppression of the division of the epithelium of the cornea. We may ask at this juncture whether there are any compounds known to be especially sensitive. The answer to this is that some viruses are more sensitive to sulphur mustard gas than even the enzymes hexokinase and chicken pepsin, and these are those which contain predominantly deoxyribosenucleic acid (DNA) (Herriott[30]). Hence it is a reasonable hypothesis that mustard gas produces changes in the permeability of the skin cells by attacking some nucleoprotein in the surface of these cells. A similar explanation could apply to other tissues where the damage induced by mustard gas is especially severe. So far no one seems to have considered the possibility of the existence of nucleoprotein in the cell surfaces of the animal, though nucleic acids have been reported to be extruded by bacteria.

A valiant attempt was made during the war to protect against the action of mustard gas by the use of a chaser which had a very high competition factor for the activated ion of mustard gas, and which might capture and inactivate unreacted mustard gas; for this purpose a number of thiophosphonates were made and a limited success was reached.[31]

The complicated action of sulphur mustard gas *in vivo* was illustrated well in the work of Fell and Allsopp[26] upon cultures of tissue fibroblasts; and their main points were later confirmed for one of the nitrogen mustards. With a high concentration of sulphur mustard gas, there was immediate fixation, which was followed by a rapid entry of the compound into the cells and its accumulation in the fat. With lower concentrations there was a large distortion of the outline of the cells, the cutting off of blebs and much abnormality in the nucleus. In its effect on mouse skin the sulphur mustard resembled the action of X-rays, but it was not the same as that of the carcinogenic agents like benzpyrene, in that there was interference with mitosis but not with differentiation, a fact confirmed for nitrogen mustard again by Gillette and Bodenstein.[28a] A similar action was not found for the products of hydrolysis, thiodiglycol and HCl, though HCl alone would fix the cells. Observations of this kind required much patience; they are consistent with the view that low concentrations of mustard gas attack surfaces.

The observation on nucleoproteins paved the way for a much more fundamental discovery in 1946 by Auerbach and Robson; suitable applications of sulphur mustard could cause mutations in the fly *Drosophila*. Hence the work upon mustard gas has led to a very large development since 1946 in the genetic field, and particularly in applications to the therapy of cancer. For several reasons sulphur mustard has not been the main compound used. Its volatility is objectionable, and it rapidly sensitizes those working with it. Furthermore it does not lend itself readily to substitution and to the synthesis of a wide variety of new compounds. Hence attention has been turned to the nitrogen mustards, upon which an intensive study of the simpler compounds was also made during 1940 to 1945. The fundamental reaction of these compounds is the same in so far as they are also alkylating substances at low temperatures and neutral pH; but their structure can be varied almost infinitely. Several substituted nitrogen mustards have been tried in leukaemia, and some have had success in prolonging life. Though not strictly upon the biochemical lesion of sulphur mustard, this extension of the field is of such interest that it seems relevant to my main subject to mention briefly some of the points of interest.

As a result of research since 1950, especially under Prof. Haddow at the Chester Beatty Institute and at the Sloan-Kettering Institute, New York, we know now that there are other compounds in addition to the "mustards" which have similar properties. These are epoxides, ethyleneimines and methane-sulphonates. The active agents are all electrophilic, and react with nucleophilic groups which are rich in electrons. In the case of the sulphur and nitrogen mustards and the substituted mustards the reaction follows equation

$$R_2N \cdot CH_2CH_2Cl \rightarrow R_2N \cdot CH_2CH_2^+ + Cl^-$$

For some of the other compounds, namely the epoxides, ethyleneimines and mesyloxy compounds, the electrophilic character arises from the displacement of one nucleophilic group by another in a bimolecular process, according to the equation

$$R \cdot O \cdot SO_2CH_3 + R'COO^- \rightarrow R \cdot OOCR' + OSO_2CH_3^-$$

The possibility has been raised that the reaction is with free radicals, but this has not received support.[1,2]

Assuming that the most damaging reaction is the one with nucleic acid, with the emphasis upon DNA, we can ask what types of reaction are known, in terms of the actual groups contained in various substances; a list of these, which can be called also the radiometric alkylating reagents is given

Epoxide	CH_2-CH- (ring closed through O)
Ethyleneimine	CH_2-CH_2 (ring closed through N)
Sulphur mustard	$R\cdot SCH_2CH_2Cl$
Nitrogen mustard	$RN\cdot(CH_2CH_2Cl)_2$.
Mesyloxy	$CH_3\cdot SO_2OR^1$ —where R^1 is $(CH_2)_n$).
	(Example is myleran), $CH_3\cdot SO_2\cdot O\cdot(CH_2)_4\cdot O\cdot SO_2\cdot CH_3$

FIG. 4.4. Radiomimetic reactive groups.

in Fig. 4.4. The observations of Elmore, Gulland *et al.*[25] on the reaction of mustard with DNA and RNA *in vitro* were extended to all these agents by Alexander[3]. The very large number of compounds made by organic chemists by attaching various groups to the above residues have been tested against various tumours and also for their effects *in vitro* on nucleic acids and on proteins. A very complete account of these as well as of the philosophy which has guided their synthesis will be found in a paper by Bergel[15]. If we follow up the thread of the effects upon nucleic acid, it was shown by Alexander that some radiomimetic alkylating substances esterified the phosphate groups in DNA very easily. When the mustard made from phenylalanine by Bergel and Stock was tested, Alexander, Cousins and Stacey[6] found that with minimal amounts of the mustard, the only reaction detected with DNA was the esterification of the phosphate groups; there was virtually no reaction with amino groups. It seems to be probable that the primary reaction with the phenylalanine mustard is esterification of the phosphate groups, and that then there follows a transfer of the group to the N7 atom of the guanine residue; this is in agreement with the fact that there is no spectral change associated with the initial step. On the other hand, the matter is complicated, as the reaction appears to be different with the sulphur mustard. In Canada, Walker and Watson,[46] who have confirmed the above results for the phenylalanine mustard, support the work of Brookes and Lawley,[17] according to whom the first reaction when small amounts of sulphur mustard are used is a direct alkylation of the N7 atom of guanine. Hence it seems that the attack made by individual alkylating agents may be different in the actual group attacked; but that it comes down to an interference by combination with the nucleic acid. It is interesting to mention again that some 16 years ago, Young and Campbell[47] thought that the purine guanine was the one with which sulphur mustard most readily combined; it was found[18] also about then that a nitrogen mustard induced a change in the absorption spectrum of DNA, which was interpreted as a reaction with purine or pyrimidine groups.

The whole matter is full of surprising chemical details, which extend even into the different response of growing mouse leukaemia cells in tissue culture as shown by Alexander and Mikulski[3] (1960). The HN_2 mustard, $CH_3N(CH_2CH_2Cl)_2$, immediately arrests division; but with dimethylmyleran, $CH_3SO_2O{\cdot}CH(CH_3){\cdot}(CH_2)_2{\cdot}CH(CH_3)\ O{\cdot}SO_2{\cdot}CH_3$, at least one division occurs before irradiation. It is to be noted that the cells do not stop dividing because the DNA is no longer synthesized; the synthesis of this goes on normally for a day or two after treatment with a corresponding increase in the size of the cells. The relation to various stages of division has been summarized recently by Berenbaum[12] (Fig. 4.5) in work on production of antibody to TAB vaccine.

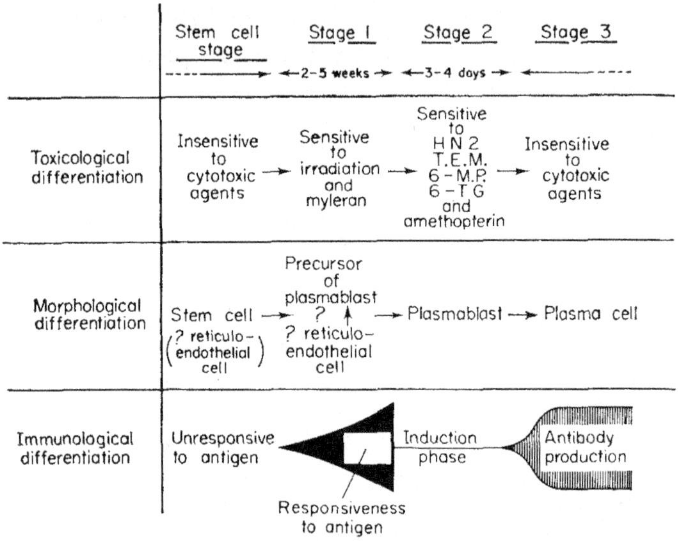

FIG. 4.5. Stages in the functional and morphological differentiation of immunologically competent cells and their varying susceptibility to cytotoxic agents. (Berenbaum)

Cross-linking

Another interesting question is whether a so-called bifunctional compound is more active biologically than a monofunctional one. We used to discuss this in 1940–41 at our informal laboratory tea parties; was it a fact that vesication was more readily produced by a two-armed mustard than by a one-armed one; the one with two arms might carry out some other reaction? This matter has been taken up very actively with a wealth of experimental work by Alexander and others at the Chester Beatty Institute, in terms of the actual reactions with nucleic acid. They have developed the cross-linking hypothesis, according to which the most active of the mustards and other compounds in this series owe their activity to their possession of polyfunctional groups. If these alkylating substances combine with more than one strand of the nucleic acid (intermolecular binding) or even with more than one point in a long molecule (intramolecular binding), then they might be expected to induce more disturbance than those which only alkylate one grouping. Experiment showed that on treatment of nucleic acid with N mustard there is a marked change in viscosity, which is more pronounced the more concentrated the solution; this might be expected if this induced more inter-molecular combination. The immediate fall in viscosity of DNA (according to Alexander and colleagues) is not due to the degradation of the molecule, but to an alteration of their shape by intermolecular linking; on standing degradation occurs. The experiments upon this were done with the heads of herring sperm and Fig. 4.6 is from the data of such an experiment. It will be

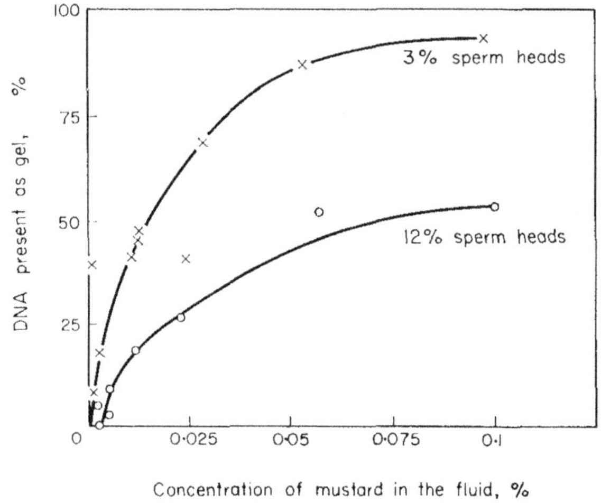

FIG. 4.6. Cross-linking of DNA in the heads of herring spermatozoa following treatment with different amounts of the nitrogen mustard derived from phenylalanine ($(ClCH_2CH_2)_2N \cdot C_6H_4CH_2CH(NH_2)COOH$). (Alexander *et al.*)

noticed that there is considerable gel formation with some of these. Rather similar results were got when nucleoprotein fibres from the spleens of rats were used. In all this work, it is surprising how well these viscosity experiments fit with the effects seen by Fell and Allsopp on the nuclei in cultures of tissue fibroblasts. The cross-linking hypothesis in general follows the biological activity, as assessed by the capacity to inhibit the growth of tumours, and so by quantitative experiments provides a reason for the generally greater activity of the bifunctional as distinct from the monofunctional alkylating agents. For instance, it is pointed out that "With the mustard, melphalan, the threshold gel dose

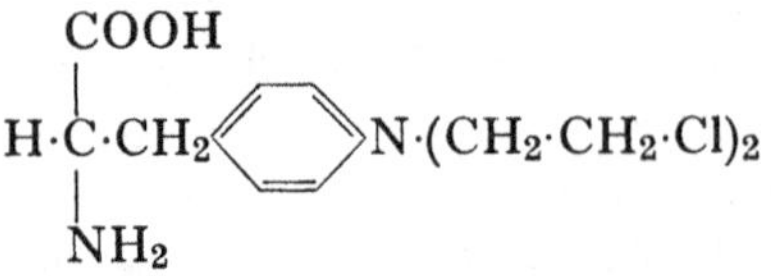

is approximately 0·15 mg/g of nucleoprotein and this reaction with the nucleoprotein *in vitro* was not affected by the presence of extra protein in the solution. The same mustard inhibits the growth of the Walker tumour in the rat at a dose of 1 mg/kg of animal, a dose at which a significant amount of cross linking would be expected to be produced."* When the alkylation has taken place, the reacted DNA is unstable and the viscosity gradually falls, due to a slow hydrolysis with a consequent break in the DNA chain.

The Definition of Radiomimetic and Mutagenic

There are three possible actions of the substances which are "alkylators": they may imitate the action of radiation, being called radiomimetic; or they may be mutagenic, that is capable of inducing mutations as was shown originally for sulphur mustard; or finally they may be "biologically active" in the sense that they can be cytotoxic for grafted tumours like the Walker tumour. It might be expected upon *a priori* grounds that a substance could be mutagenic, without being specially cytotoxic, because one might imagine that a less severe action might change some group on the DNA without producing the terrific damage caused by cross-linking. Hence it is easy to see why the one-armed or monofunctional mustards should be as mutagenic as the bifunctional (Stevens and Mylroie[(45)]). Dr. Alexander, whom I have consulted on this matter, says that there is such variation that it is difficult to draw hard and fast lines. For instance, the one-armed myleran ($C_2H_5 \cdot O \cdot SO_2 \cdot CH_3$) is a moderate carcinogenetic agent and one of the most effective mutagens known though it is not a cross-linking agent. He thinks that the term radiomimetic should be confined to those compounds which are (1) mutagenic in different types of cell and not, like caffeine, confined to plants only, (2) carcinogenic, (3) much more toxic to dividing cells, causing delayed death after one or two division cycles in all the daughter cells.

* Quotation from P. Alexander, Mrs. A. Swarcbort and K. A. Stacey, 1959.

An Appraisal of the Present Position

Important as the modern observations are, they do not yet tell us exactly how the biochemical lesion inducing vesication with a mustard agent starts. Arguing from this work, however, it would seem likely that some phosphate group either in a phospholipid or possibly some guanine group in the RNA possibly near the surface of the cell becomes esterified; it may even be in one of the internal surfaces, where one may say that the presence of RNA is proved by its existence in ribosomes attached to the endoplasmic reticulum. In most of the modern work the emphasis has been upon the DNA, for which an entry into the cell is necessary, as far as the nucleus. It seems unlikely that the mustard does not attack some other constituent of the cell on its way to the nucleus. The finding of E. Walker and myself (1923) that the attack upon the —SH groups in the skin was much less pronounced with mustard than with its sulphone or the arsenicals is now well confirmed. At the same time, it has been shown that some —SH groups are among the most sensitive, witness the thiophosphonates. The explanation of the difference is reasonable; that this is because at the cell pH of 7·0 the —SH groups of proteins are little ionized. In work upon the effect of sulphur mustard and its sulphones on the pyruvate oxidase system in the 1940's, it was found by Wakelin and myself[(40)], that the sulphones were more toxic to the total system than was sulphur mustard or the nitrogen mustard. On the other hand there was a greater toxicity of the mustard to the pyruvate system than to the succinoxidase. The toxic action was not only upon the pyruvate dehydrogenase but also upon something else in the pyruvate oxidase system. One may speculate that the attack is upon the coenzyme A or the lipoate needed for the operation of the citric acid cycle both at the pyruvate and the α-ketoglutarate stage. More will be said upon these matters in the special section devoted to the "thiols".

NOTES ON CHAPTER 4

1. *Mustard Gas and Enzymes*

The course of events *in vivo* is puzzling and discussed well by D. M. Needham (1948). Forty enzymes tested were virtually untouched by mustard gas; these included cytochrome oxidase, glyoxalase, trypsin and even the —SH enzyme triosephosphate dehydrogenase. Enzymes sensitive to mustard gas were hexokinase, creatine phosphokinase, glycerolphosphokinase, adenosine triphosphatase, pyrophosphatase, pyruvate phosphokinase, adenylic deaminase and swine pepsin.

Whereas creatinephosphokinase was the most sensitive enzyme *in vitro* it was not poisoned *in vivo* in the skin of the rat. The slow poisoning of hexokinase may be due to a secondary effect following on change in permeability, like the leakage of proteinase after painting rat skin *in vivo* (Beloff, Peters and Wakelin, 1945).

2. *Some Other Points*

While the action of the bifunctional mustards as cross linking agents is very clear, there are some other facts known about them which should not be forgotten. I found with Wakelin using a kerateine preparation similar to that used by Stocken and Thompson for the arsenicals, that after a treatment of this with sulphur mustard, there were evidently compounds present in the preparation of the type, kerateine $-S-CH_2CH_2{\cdot}S{\cdot}CH_2CH_2Cl$. Their stability suggested that they might have become also sulphone-like in structure. These would retain the vesicant action of the mustard for long periods and may well be a factor in some of the delayed actions of mustard gas which are known to take place. Whether they could last long enough to be responsible for the mustard gas keratitis, which occurs years afterwards, is unknown; but they might be even influencing some of the modern experimental results. We also were surprised by the long periods during which sulphur mustard could remain stabilized in the presence of dried kerateine as if it were adsorbed. It took a very thorough treatment with acetone to remove this adsorbed mustard gas.

References

1. Alexander, P. (1954), *Advances in Cancer Research* **2,** 1 (Review).
2. Alexander, P. (1952), *Nature* **169,** 226.
3. Alexander, P. and Mikulski, Z. B. (1960), *Biochem. Pharmacol.* **5,** 275.
4. Alexander, P. and Stacey, K. A. (1958), *Proc. 4th Int. Congress of Biochemistry Vienna.* Symposium No. IX. Pergamon Press.
5. Alexander, P., Swarcbort, A. and Stacey, K. A. (1959), *Biochem. Pharmacol.* **2,** 145.
6. Alexander, P., Cousins, S. F. and Stacey, K. A. (1957), Ciba Found. Symp. Drug resistance in micro-organisms, p. 294.
7. Auerbach, C. and Robson, J. M. (1946), *Nature* **158,** 432.
8. Avery, O. T., Macleod, C. M. and McCarty, M. (1944), *J. Exp. Med.* **79,** 137.
9. Banks, T. E., Boursnell, J. C., Francis, G. E., Hopwood, F. L. and Wormall, A. (1946), *Biochem. J.* **40,** 734 and subsequent papers.
10. Bartlett, P. D. and Linstead, R. P. (personally communicated); see also Bartlett, P. D. and Swain, C. G. (1945), O.S.R.D. Reports, U.S.A.
11. Beloff, A. and Peters, R. A., with R. W. Wakelin (see Peters, *Nature* 1947).
12. Berenbaum, M. C. (1962), *Biochem. Pharmacol.* **11,** 29.
13. Berenblum, L., Kendall, L. P. and Orr, J. W. (1936), *Biochem. J.* **30,** 709.
14. Berenblum, I. and Schoental, R. (1947), *Nature, Lond.* **159,** 727.
15. Bergel, F. (1958), *Ann. New York Acad. Sci.* **68,** 1238.
16. Bergel, F. and Stock, J. A. (1954), *J. Chem. Soc.* 2409.
17. Brookes, P. and Lawley, P. D. (1960), *Biochem. J.* **77,** 478, and **80,** 496 (1961).
18. Chanutin, A. and Gjessing, E. C. (1946), *Cancer Research* **6,** 599.
19. Crick, F. H. C. and Watson, J. D. (1954), *Proc. Roy. Soc. A* **223,** 80.
20. Cullumbine, H. and Rydon, H. N. (1946), *Brit. J. Exp. Path.* **27,** 33.
21. Davidson, J. N. (1960), Monograph on Biochemistry 4th edition. Methuen (Review).
22. Dixon, M. and Needham, D. M. (1946), *Nature, Lond.* **158,** 432 (Review).
23. Dixon, M. (1948), Biochem. Soc. Symp. No. 2 (Review).
24. Drysdale, R. B. and Peacocke, A. R. (1961), *Biol. Rev.* **36,** 537 (Review).

25. ELMORE, D. T., GULLAND, J. M., JORDAN, D. O. and TAYLOR, H. F. W. (1948), *Biochem. J.* **42,** 308.
26. FELL, H. B. and ALLSOPP, C. B. (1948), *Cancer Research* **8,** 145.
27. FORD-MOORE, A. H., PETERS, R. A. and WAKELIN, R. W. (1949), *J. Chem. Soc.* 1754–57.
28. FRIEDENWALD, T. S. and WOOD, A. C. (1948), *Bull. Johns Hopkins. Hosp.* **82,** 2. Papers from Wilmer Institute.
28a. GILLETTE, R. and BODENSTEIN, D. (1946), *J. Exp. Zool.* **103,** 1.
29. GILMAN, A. and PHILIPS, F. S. (1946), *Science* **103,** 409 (Review).
30. HERRIOTT, R. M. (1948), *J. Gen. Physiol.* **32,** 221.
31. HOLIDAY, E. R., PHILPOT, J. ST. L. and STOCKEN, L. A. (1950), *Biochem. J.* **47,** 637.
32. INGOLD, C. K. and HUGHES, E. D. See (1935) *J. Chem. Soc.* 244, 255, and other papers, also (1937) p. 1183 (also with J. L. Gleave).
33. LOFTFIELD, R. B. (1957), *Progr. Biophys. & Biophys. Chem.* **8,** 347 (Review).
34. NEEDHAM, D. M. (1948), Biochem. Soc. Symposia No. 2.
35. OGSTON, A. G. (1948) Biochem. Soc. Symposia No. 2.
36. OGSTON, A. G., HOLIDAY, E., PHILPOT, J. ST. L. and STOCKEN, L. A. (1948), *Trans. Faraday Soc.* **44,** 45.
37. PAUL, G. (1950) *Acta Chem. Skand.* **4,** 239 and (1951) **5,** 379.
38. PETERS, R. A. (1947), *Nature, Lond.* **159,** 149 (Review).
39. PETERS, R. A. and WALKER, E. (1923), *Biochem. J.* **17,** 260.
40. PETERS, R. A. and WAKELIN, R. W. (1947), *Biochem. J.* **41,** 555.
41. PETERS, R. A., FORD-MOORE, A. H. and WAKELIN, R. W. (1949), *J. Chem. Soc.* 1754.
42. PHILLIPS, F. S. (1950), *J. Pharmacol.* **99,** Pt. II, 281 (Review).
43. ROSS N. C. J. (1953) *Advances in Cancer Research* **1,** 397 (Review).
44. STEIN, W. H., MOORE, S. and BERGMANN, M. (1946), *J. Org. Chem.* **11,** 664 and succeeding papers.
45. STEVENS, C. M. and MYLROIE, A. (1952), *Biochem. Biophys. Acta* **8,** 325.
46. WALKER, I. G. and WATSON, W. J. (1961), *Canad. J. Biochem. Physiol.* **39,** 365.
47. YOUNG, E. G. and CAMPBELL, R. B. (1947) *Canad. J. Res.* **25B,** 37.

CHAPTER 5

Inhibitors of Enzymes Containing Thiol Groups and Toxicity

THE relation of the —SH groups in enzymes to biochemical lesions is so cogent and also presents so many facets that it requires some special consideration separately. As has been stated already, the original impulse was given to the study of —SH substances unquestionably by the work of Hopkins on glutathione in 1921. We soon applied the nitroprusside reaction of —SH to skin sections, and Voegtlin and his colleagues also used it in their studies of the therapeutic arsenicals. In our early work in the 1920's (page 42), we found definite relations between toxicity to the —SH groups and toxicity *in vivo*, except as already mentioned in the case of sulphur mustard. At the same time with the limited enzymes then at our disposal, we could get no evidence for the "enzyme" theory of vesication; there were few intracellular enzymes then known, a situation which has changed in the intervening years. For instance, in work by Hellerman *et al.*[20] (1933) the enzyme urease was proved to behave as an enzyme containing —SH groups; and later Hellerman[21] measured the —SH groups in urease determining which were essential for activity. From 1936 onwards there was a definite attempt to relate the —SH groups to the activity of some enzymes. M. Dixon[12] (1937) studied the effect of iodoacetate ($I\,CH_2{\cdot}COO^-$), upon a series of enzymes, finding that the alcohol dehydrogenase was inactivated at the low concentration of M/3000. With M/100 iodoacetate, he found inhibition of the triosephosphate dehydrogenase, the α-glycerophosphate and lactate dehydrogenases, and aldehyde mutase. Dickens had shown previously that iodoacetate was an alkylating agent for —SH groups in cysteine,[10] and study of this type of reaction[1] was made by Schubert[38]

$$ICH_2{\cdot}COOH + HS{-}CH_2{\cdot}CHNH_2COOH \rightarrow HOOC{\cdot}CH_2S{\cdot}CH_2{\cdot}CHNH_2{\cdot}COOH + HI \quad (1)$$

In 1936 I found that the pyruvate oxidase system in brain was very sensitive to mustard gas sulphone and also to arsenicals; this was the part of the system involving the thiamine component.[29] We had found previously that iodoacetate (0·54 mM) diminished the oxygen uptake of the pyruvate oxidase system and that this affected the thiamine component of the respiration;[32] the amount of pyruvate accumulating was increased. As these observations suggested that there was an —SH enzyme involved, I took up the matter again with Wakelin in 1939. Using the technique which Hopkins *et al.*[22,23] had used to demonstrate the presence of an —SH group in succinate oxidation, we found that treatment with —S.S.— groups (cystine ester) inactivated the pyruvate dehydrogenase; the activity was restored at any rate in part by the addition of a compound containing —SH groups (cysteine ethyl ester), this being on an early step in the oxidation.[28] Comparative studies of the relative effect of the double bond compound, maleate, upon the oxygen uptake of the brain system in the presence of the two substrates pyruvate and succinate respectively showed that at a low concentration maleate was much more inhibitory to pyruvate.* As the double bond compound attacks —SH groups, these observations were consistent with the presence of an essential —SH group in this system, showing that on the whole this —SH group was not particularly sensitive to iodoacetate; that is to say that the sensitivity was intermediate between that of triosephosphate dehydrogenase and of succinodehydrogenase to which it is relatively insensitive. On the other hand we found that the pyruvate oxidation was remarkably sensitive to iodoacetamide ($ICH_2{\cdot}CONH_2$); the amidation made all the difference to its activity to an important —SH group in the enzyme system. These large variations in the inhibitory or toxic action of substances which appeared to be attacking —SH groups convinced me at the time that there were many varieties of —SH groups so far as the reactions with —SH reagents were concerned. We do not yet understand all the points involved in these reactions. One has been unravelled in the work

* Other groups than —SH can react slowly with iodoacetate; it has been found that ribonuclease is alkylated by iodoacetate and it contains no —SH groups (see ref. 17). It is interesting to note that iodoacetamide does not react with ribonuclease, though it so readily inhibits a component of the pyruvate oxidase. The reaction with ribonuclease is complex, being probably on imidazole groups and methionine. The reaction with methionine was studied and takes place when iodoacetate is allowed to react for more than 20 min at pH 4–8, best at pH 2·5. Methionine carboxymethylsulphonium iodide is formed.

$$\begin{array}{l} CH_3{\cdot}\overset{+}{S}{-}CH_2{\cdot}CH_2{\cdot}CHNH_2{\cdot}COOH \\ \quad\;\; | \\ \quad\;\; CH_2 \\ \quad\;\; | \\ \quad\;\; COOH \end{array}$$

TABLE 5.1. COMPARISON OF THE CONCENTRATION INHIBITING THE PYRUVATE OXIDASE SYSTEM UNDER STANDARD CONDITIONS (Pigeon brain, brei)

Class	Substance	μM
Arsenicals	Lewisite	16
	Arsenite	34
Non-arsenicals	Iodoacetamide	60*
	Divinylsulphone	67
	Mustard gas sulphone	110
	Allyl iso-thio-cyanate	242
	2.2.2. Trichlor-triethylamine	750
	Mustard gas	1000

Data of Peters and Wakelin, 1947.[30]

*Iodoacetate has approx 20 per cent of the toxicity of iodoacetamide.

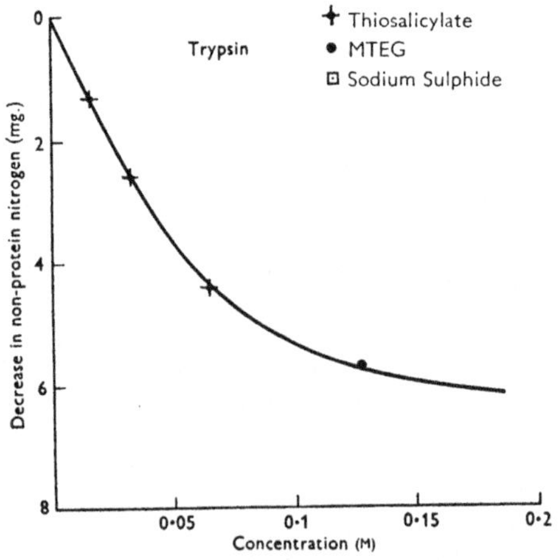

FIG. 5.1. Inactivation of chymotrypsin by monothiols. Effect of previous exposure to varying concentrations of thiol compounds on the digestion of casein substrate by crystalline chymotrypsin. Addition of 1 mg enzyme preparation to 6 ml casein solution, pH 7·6, 38°. Digestion time 45 min. Results expressed as decrease in mg of NPN formed. (Peters and Wakelin, 1948.)

on alkylating agents, namely whether the —SH is ionized at the pH of the cell or not* (page 64).

In the life of the cell, the —S.S.—, —SH equilibria must be very important, and —SH compounds and —S.S.— must live in separate compartments. It has been long known that insulin has necessary —S.S.— groups, in 1948[31] we found that this was so for chymotrypsin and trypsin, which was quickly inactivated by substances like monothioethylene glycol; the degree of inactivation for 4 monothiols depended only on the molar concentration. The modern protein work of Sanger and others proved the presence of—S.S.—very clearly in insulin.

When Belgium was isolated (1940–45), Bacq and his colleagues investigated independently the effects of many of these reagents, for which he has proposed the general term "substances thioloprives", an illuminating generalization. Like ourselves they arrived at the conclusion that the attack was upon —SH groups, and this was pursued actively with many of the toxic agents,[1,2] including chloropicrin ($C{\cdot}(NO_2)Cl_3$) and bromopicrin.[9] I think that it is still true, as I wrote in 1948,

> The varied views which can still be taken are shown from the fact that M. Dixon regards the lachrymators, for whose action see Eqn., as the test agents *par excellence* for —SH
>
> $$\underset{\text{chloroacetophenone}}{C_6H_5{\cdot}CO{\cdot}CH_2Cl} + HS{\cdot}R = C_6H_5{\cdot}CO{\cdot}CH_2SR + HCl$$
> $$\text{SH-compound}$$
>
> enzymes (like hexokinase), whereas Guzman Barron considers that the supreme agents are arsenicals.[5,6] Bacq thinks that both vesicants and lachrymators owe their power to their capacity for altering —SH groups whether it be by oxidation or by combination; whereas in an American statement (Gilman and Phillips[14]) in 1946, based largely upon experiments upon the sulphur and nitrogen mustards gases, there was not much stress laid upon enzymes or —SH groups.

The Belgian workers contributed several facts of interest; for instance, chloropicrin attacks both the free and the masked —SH groups in native albumin without changing the molecular weight. Studies of —SH groups in proteins are now in progress (cf. Cecil and McPhee[8]), and we hope that there will soon be enough theory to enable us to predict what type of —SH group in what enzyme will be susceptible to attack by one of these toxic agents. One of the most widely used substances for the detection of —SH groups is now the *p*-chloromercuribenzoate introduced by

* Bremer and Greenberg (1961) have described a transmethylating system in mammalian tissues, mainly microsomes; this catalyses the transfer of "methyl" from S-adenosylmethionine to BAL, mercaptoethanol and other "non-physiological" sulphur compounds. Cysteine and glutathione do not accept the methyl group. Again this emphasizes the wide difference in behaviour of —SH groups.

Hellerman.[19] Another compound often used is iodosobenzoate.[3] As Dixon pointed out, it is reasonable that substances like acrolein should be lachrymators (CH_2:CH·CHO), because the compound would combine with —SH substances. To consider briefly a physiological matter, if it is an —SH group in the eye with which the lachrymator combines to produce the irritation, where is this —SH group in the eye? Is it the nerve itself or is it some sensitive nerve ending? Since the lachrymatory effect wears off, the combination would not be irreversible, unless there can be some rapid regeneration of the group in question (see Dixon[13] for discussion of this).

It would be interesting to know the state of the —SH group in the isocitrate dehydrogenase, which we studied and which seems to be a monothiol enzyme. The effect of arsenicals has been described already (page 44). Table 5.2 gives the other observations which we made. Unlike the pyruvate

TABLE 5.2. INHIBITORS ON ISOCITRATE DEHYDROGENASE*

Inhibitor		Conc. mM	% inhib.
Iodine compounds:	Iodoacetate	1·08	0
	Iodoacetamide	0·54	0
	Iodosobenzoate	0·81	50
Metallic compounds:	$CuSO_4$	0·052	50
	p-chloromercuribenzoate	0·02	88
	phenyl mercuric nitrate	0·004	77
Other —SH inhibitors:	Na maleate	2·08	0
	chloroacetophenone	0·65	0
	CH_2.N.(CH_2.CH_2.Cl)$_2$	1·28	0
	Alloxan	4·0	0
Arsenical:	Diphenylchloroarsine	0·063	100
	Lewisite	0·655	0

Lotspeich and Peters, 1951.

* The inhibitor acted on the enzyme for 15 min before addition of the isocitrate. The enzyme was a dialysed phosphate extract of acetone dried powder from pig heart.

system, iodoacetate and iodoacetamide had no toxicity, but iodosobenzoate was inhibitory. *p*-chloromercuribenzoate and Cu were toxic.[27] On the other hand, the lachrymators and sodium maleate had no effect.

SIGNIFICANCE OF —S.S.—, —SH EQUILIBRIUM *in vivo*

The equilibrium in question may be of great importance *in vivo*. Within 30 to 60 min after the injection into dogs and rabbits of oxidized thiosulphate, tetrathionate ($Na_2S_4O_6$), there is complete anuria, and a specific degeneration is found in the tubule of the kidney.[15,16] At the same time there is evidence that the tetrathionate is reduced to thiosulphate ($Na_2S_2O_3$); it readily oxidizes thiol groups to the disulphide form *in vitro*. Attempts were made to discover whether the succinoxidase system was inactivated *in vitro* by tetrathionate, and it was found to be sensitive to this, though succinate protected it to some extent. On the other hand it could not be discovered that there was any inactivation of this enzyme *in vivo* after administering doses known to cause severe damage to the kidney. In explanation of this discrepancy between the *in vivo* and *in vitro* events, one can say that it is probable that the tetrathionate does not pass easily into the cells, and that its action is to reduce the amount of glutathione in the —SH form necessary for a proper functioning of the oxidation-reduction equilibria; even so, the problem is crying out for more work to clarify the precise nature of the biochemical lesion involved. As a practical measure, the anuria is stated to be relieved by the intravenous injection of 30 per cent NaCl with the idea of reducing the swelling of the cells of the proximal tubules.

It is a truism to say that we have always to be careful to note whether events found *in vitro* apply *in vivo*. We had another good instance of this during our work on mustard gas.[30] We found that the addition of diethyldithiocarbamate (or of diethanoldithiocarbamate) to mustard gas potentiated the toxic action of this on the pyruvate oxidase system in brain some 20 times *in vitro*. The matter was followed up and it was found eventually that the potentiation was due to the formation of a compound between the "mustard" and the diethyldithiocarbamate, which had the formula I;

$$(C_2H_5)_2{\cdot}N{\cdot}\overset{\overset{\large S}{/\!/}}{C}{-}S(CH_2)_2S{\cdot}(CH_2)_2Cl$$

I

$$(C_2H_5)_2N\overset{\overset{\large S}{/\!/}}{C}{-}S(CH_2)_2S{\cdot}(CH_2)_2\overset{\overset{\large S}{\backslash\!\backslash}}{C}{\cdot}N(C_2H_5)_2$$

II

When this was made, it proved to be more toxic. On the other hand the disubstituted thiocarbamate II was non-toxic to the enzyme. The potentiation was therefore due to the easier combination of the pyruvate oxidase system with the remaining —Cl group. In support of this the one-armed

Cl compounds could not be potentiated. In practice the observation was unimportant *in vivo*, because when I was injected into the living animal it was not toxic, which was interpreted to mean that it did not penetrate to the active centre conerned *in vivo*.

Cell Division and Thiol Agents

Finally, we may consider some of the results with a few of the agents in question upon dividing cells and upon cells in tissue culture. The modern position upon the nucleic acids suggests that any genetic changes are associated especially with the DNA and with substitution by alkylation or incorporation in the nucleic acid chains. We may ask whether there is any indirect evidence that —SH groups may also be involved. It is convenient to start this discussion with the detailed study made by Hughes and Fell[24] (1949) with sulphur mustard gas upon cultures of normal chick osteoblasts using a cinematographic method of recording and phase contrast microscopy. Although from the arguments developed in Chapter 4 it is clear that much of the attack made by the sulphur mustard is on guanine groups, it is not certain that there may not be some effects upon —SH groups which happen to be ionized. According to their description, the process of division starts with the *prophase*, when

> the cytoplasmic processes are largely withdrawn, the nucleoli vanish, and faintly grey, diffuse chromosomes materialize throughout the nuclear area. At the same time the nuclear membrane disappears, but cytoplasmic inclusions remain outside the nuclear area until the late *anaphase*. The chromosomes contract, become increasingly distinct and assume a radial orientation in the plane of the coverslip; they probably lie in this plane as the mechanical result of the flatness of the cell. Whether the spindle has already begun to form is not yet known and since it is therefore uncertain whether this stage should be regarded as late prophase or early metaphase, we have termed it the *radial* stage. As the cell becomes more nearly spherical, the chromosomes rotate from the plane of the coverslip to one at right angles to it, presumably under the influence of the spindle elements; but in normal cells the spindle is indistinct with phase contrast microscopy and the details of its formation cannot be seen. During metaphase the chromosomes move to and fro in the equatorial region of the spindle with unsynchronised linear motion. Without warning the chromatids suddenly separate and pass quickly to opposite poles (the start of *anaphase*). About 3 min after the beginning of anaphase, the granules and mitochondria of the surrounding cytoplasm bulge into the interzonal region and constriction into daughter cells begins. The contrast of the chromosomes falls during *telophase* so that usually nothing is clearly visible in the nuclear area until 20 min after anaphase, by which time the nuclear membrane and nucleoli are present and the daughter cells are flattening.

This is quoted at length, because it is important to bear in mind the complications which have to be understood ultimately in biochemical terms. With the sulphur mustard even at as low a concentration as

12·5 μg/ml distorted mitotic figures could be seen. Among these changes, the mitoses were characterized by the delayed arrival of certain chromosomes at the equatorial plate or their failure to reach it, by lag at anaphase and telephase, and by the formation of one or two micronuclei derived from the lagging chromosomes in addition to two daughter nuclei of nearly normal size. Under the influence of this sulphur mustard prophase and metaphase were much delayed and prolonged, up to 4 times and sometimes the cells did not enter metaphase. On the other hand the anaphase proceeds at a normal rate. They record that the effects are much like those of irradiation and are like those seen upon pollen grains.

The complicated nature of the changes induced by the sulphur mustard may have an —SH component, though it will not be the main attack. Turning now to some other results by one of these authors which are likely to be essentially an effect upon the—SH groups, Hughes[25] (1949) studied the action of iodoacetamide upon cell division in the same tissue. He found that there were two main effects on the cells of the chick in mitosis, an inhibition of the mitotic spindle formation in the sense of a disorganization, and also an inhibition of the reconstruction of the nuclei in telophase. The effects with the —SH inhibitor are therefore different, though they are effects upon the process of division. It may be recalled that even in 1922, Shearer[39] already described an increase in —SH substances in the sea urchin egg after fertilization, and that Rapkine concluded in 1931 that the —SH content of the cell increased before division.[35,36]

There are other observations. Beatty[7] (1951) studied, on amphibian eggs, the effect of the lachrymator, chloroacetophenone, a selective inhibitor for another class of —SH groups, In his view this led to what he called a "blocked —SH syndrome", the chief points in which are

(1) A delayed early cleavage.
(2) Inhibition of cell division, leading to the production of giant cells with giant nuclei and polynucleate cells. The giant cells are definitely heteroploid and rarely divide.
(3) Changes to be seen later in gastrulation.

The point about these effects of selective —SH inhibitors is that they show that somewhere in the whole process —SH groups are concerned; this seems now clearer still since the work of Mazia (see page 203). The purely —SH inhibitors as compared with the sulphur mustard seem to produce much more inhibited cell division, perhaps because the supply of energy is in some way reduced. This could happen if they attacked part of an essential pyruvate oxidase system necessary for supplying the energy for the division.

The Pyruvate Oxidase System and Toxicity

From our present knowledge the pyruvate oxidase system including the enzymes of the citric acid cycle will contain many component enzymes which can be attacked by —SH reagents. The decarboxylation of pyruvate and of α-ketoglutarate needs the dithiol lipoate, the succinoxidase system has an —SH group, the isocitrate dehydrogenase has a monothiol group, the aconitase probably has one, as it is sensitive to *p*-chlormercuribenzoate. Coenzyme A is an —SH substance. Several of these enzymes have their —SH groups in a different station of activation. It can be assumed that if any one of these is attacked, it will cause the stoppage of the oxidation of pyruvate with its consequent cell dysfunction. Hence the total pyruvate oxidase system should be a useful test for vesicants, because if the —SH inhibitor does not attack one enzyme it may well attack another. This explains why we found that the pyruvate oxidase system was a useful guide to vesicants. High pressure oxygen also attacks it, as Dickens and others found in oxygen poisoning(11) (Table 5.3). Toxicity to hexokinase also follows vesicant action.(12)

Table 5.3. Comparison of toxicity to brain pyruvate oxidase system and vesicant action (dichlorodiethyl-sulphone = 100

Substance	Oxidase toxicity	Vesicant action
Iodoacetamide	186	Unknown(1)
Divinylsulphone	198	2·5
Dichlordiethylsulphone	100	100
	37	vesicant(2)
Iodoacetic acid	19·8	10–20
Trichlorotriethylamine	9	nil
Phenyl-2-chloroethylsulphide		
Dichlorodiethylsulphoxide	nil	nil
Thiodiglycol		
Dihydroxydiethylsulphone	nil	nil

1. 1·5 mg produced vesicles on arm of R.W.
2. Readily produces vesicles on arm of R.A.P.

Data of Peters and Wakelin, 1949.(34)

Epidemic Dropsy and Sanguinarine

Related to this field of thiols and the pyruvate oxidase system there is some work on epidemic dropsy, which is a disease found in India, occurring in persons using for their cooking some varieties of mustard oil. The biochemical lesion in epidemic dropsy has been considered to be related to thiamine deficiency and Wilson and Ghosh(41) found that there was a raised concentration of pyruvate in the blood, as in thiamine deficiency. It is a serious nutritional disease, and in its more severe manifestations

TABLE 5.4

BLOOD CONTENT OF B.B.S. IN EPIDEMIC DROPSY.

	mg/g
Normal	3·38
Sick	8·33

Data of Wilson and Ghosh.

INCREASE OF PYRUVIC ACID IN BLOOD OF RATS POISONED WITH SANGUINARINE (CHLORIDE)

	mg/g
Control	1·52
Poisoned	3·18

Data of Sarkar.

may induce angiomatous swellings in the skin. It can therefore cause a condition which leads to the proliferation of the blood vessels of the skin (Fig. 5.2). Lal and others (see 37, for refs.), traced the disease to the use of a mustard oil which was contaminated with an oil from the seeds of a poppy, known as *Argemone mexicana*. In searching for the toxic principle, Sarkar and Rahman found that the nitrogenous bases present could be precipitated as crystalline picrates. At this stage Sarkar came to work out the detail of the compounds in my laboratory in Oxford, and after purification, he got a substance with the formula $C_{20}H_{15}O_2N$, which he succeeded in identifying as a rare alkaloid known as dihydrosanguinarine.

SANGUINARINE

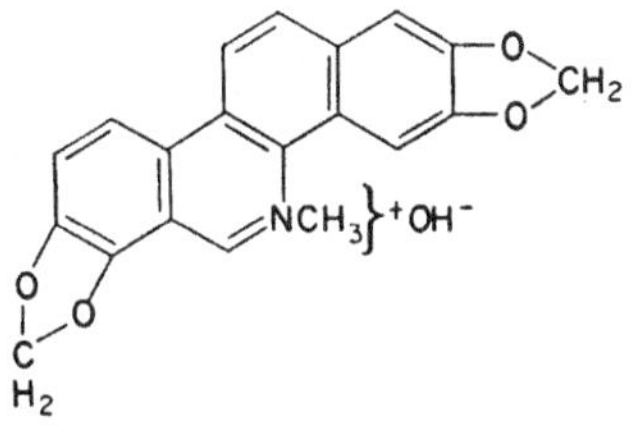

DIHYDROSANGUINARINE
(Dihydroanhydrosanguinarine)

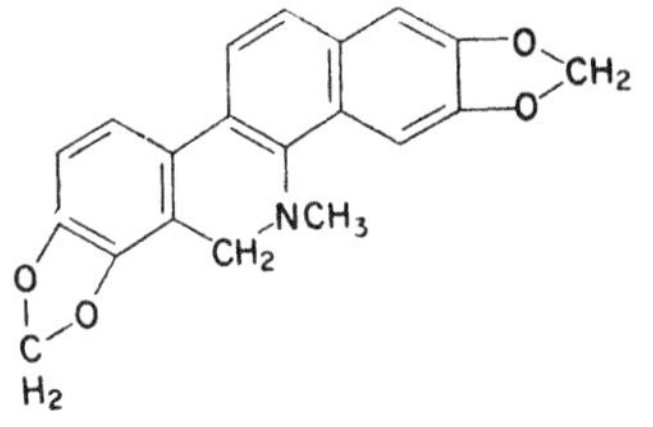

This was confirmed with an authentic specimen kindly sent by Professor Spaeth. It is worth recording that this bald description does little justice to the labour required to identify this compound. It was easily oxidized in the presence of HCl to the orange-red compound sanguinarine hydrochloride.(37) Sanguinarine killed albino rats at a concentration of 2 mg/k injected intraperitoneally (see Tables 5.4 and 5.5). The dihydro compound was much less toxic. In brain tissue from the pigeon, sanguinarine has a marked toxic effect upon the pyruvate oxidase system at 0·041 mM; it also inhibited lactate and succinate oxidation though less severely. These effects are consistent with the effect upon pyruvate amounts in the blood. Thiols such as monothioethylene glycol and BAL could prevent its toxicity, but they were unable to reverse the toxicity.

TABLE 5.5. INHIBITION OF ENZYMES PRODUCED BY SANGUINARINE (CHLORIDE)

Pyruvate oxidase system	% inhibition at	
	0·082 mM	0·246 mM
Brain { homogenate	59	84
Brain { brei	17	73
Succino dehydrogenase	—	65
Glycolysis	14	52
D-amino acid oxidase	—	13
Cytochrome oxidase	—	39

Data of Sarkar.

The chemistry of these compounds has been worked at by Bailey and Robinson(4). The pharmacology was originally investigated in 1892 by Meyer, who first worked upon mustard gas. More recently, the pharmacology has been further investigated by Hakim, in Oxford,(18) to a large extent in collaboration with E. H. Leach for the histology, and the eye specialist Mr. Lloyd;(26) the hope was that this work would shed light upon the development of the glaucoma detected by Kirwan in some 10 per cent of cases of epidemic dropsy, and that an extension of a solution of this might help, even in the general problem of glaucoma. Hakim found that argemone oil itself was considerably more toxic than sanguinarine, suggesting that the oil contained some other toxic compounds. But sanguinarine when injected into rabbits by the subconjunctival route in the eyes did produce a marked rise in tension. Some experiments with monkeys were suggestive though not decisive. From the effects on the pyruvate oxidase system, one would think that sanguinarine combines with some thiol group in the pyruvate oxidase system. The other effects described by Hakim were an antiadrenaline action, histopathological effects on tissues of the eye and the skin, and effects on the tissue fluid balance. It is unknown whether these are also actions on —SH groups.

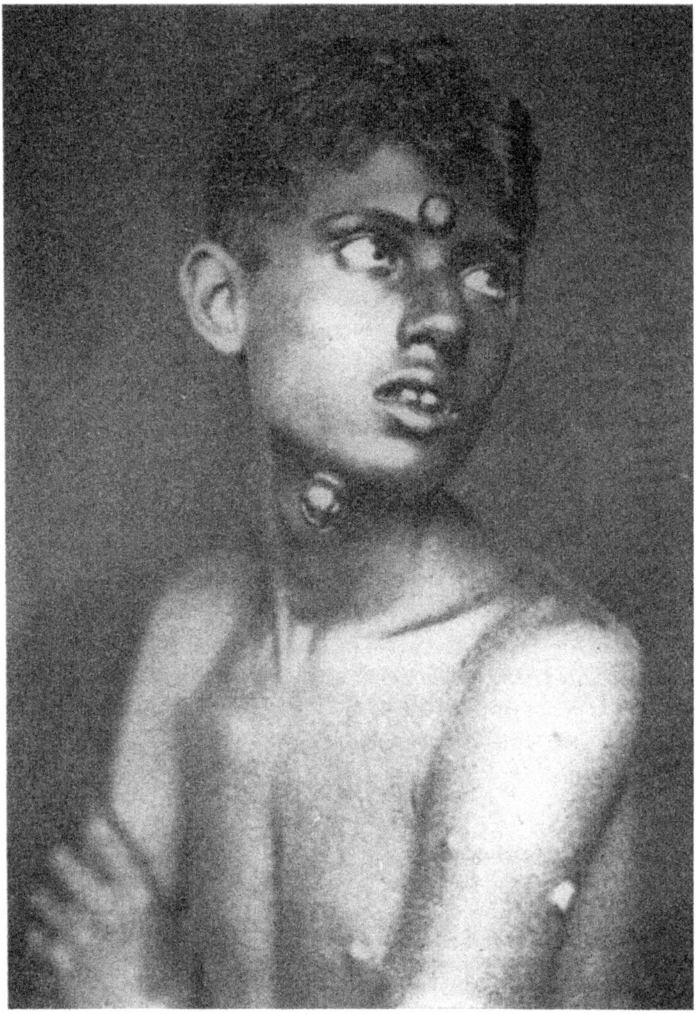

FIG. 5.2. Epidemic dropsy. The case shows the distribution of nodules in the upper part of the body. These growths appeared 2–3 days after the attack of epidemic dropsy. Those on the face were fairly big, some of them attaining the size of a lemon within one month. They were not painful or tender but used to bleed often. The nodule was fleshy in consistency. Histological examination revealed the picture of angioblastic tissues. (De and Chatterji) Calcutta Med. J. **29**, 536, 1935

What is so striking about this story of epidemic dropsy apart from the great progress made in understanding the nutritional nature of the disease, is that the irritation induced by a known alkaloid should give rise to an angiomatous proliferation of blood vessels. We ask ourselves what is the biochemical lesion involved? Does it accelerate proliferation by blocking —SH groups, or does it induce a leakage of some feedback substance normally stopping further growth?

Summary

Concluding the remarks about thiol groups in this chapter, it may be said that these groups are very important in the initiation of biochemical lesions with toxic agents. The activation of the —SH in different enzymes differs much, so that some enzymes react more readily with some classes of inhibitors than others. If one wishes to pick up toxicity with unknown substances, the pyruvate oxidase system in brain is very useful, because it is a complex of a number of —SH enzymes with groups having different activation, so that if the agent tried does not upset one, it may be inhibitory to another. Any interference in the citric acid cycle can cause a biochemical lesion.

Note. Further work in progress by Sohrab Hakim and his colleagues is of considerable interest and importance (Hakim, Mijović and Walker; *Nature*, 189, 198 and 201, 1961). They have found that sanguinarine is widely distributed in 53 species of Poppy-Fumaria weeds and a few Berberidaceae weeds which may be eaten upon occasion by animals. Further the sanguinarine can be converted in the tissuesinto 3, 4-benzacridine, which can pass into the milk. The possible connection of this with malignant disease requires no special comment.

References

1. Bacq, Z. M. (1947), *Actualités biochemiques*, No. 8. Le BAL, Ed. Desoer, Liège.
2. Bacq, Z. M. (1946), *Experientia (Basel)* **2,** 349; 385.
3. Bailey, K. and Perry, S. V. (1947), *Biochem. J.* **41,** xxii.
4. Bailey, A. S. and Robinson, R. (1950), *J. Chem. Soc.* p. 2277.
5. Barron, E. S. G., Miller, Z. B. and Meyer, J. (1947), *Biochem. J.* **41,** 78.
6. Barron, E. S. G., Miller, Z. B., Bartlett, G. R., Meyer, J. and Singer, T. P. (1947), *Biochem. J.* **41,** 69.
7. Beatty, R. A. (1951), *Proc. Roy. Soc. B.* **138,** 575.
7a. Bremer, T. and Greenberg, D. M. (1961), *Biochem. Biophys. Acta.* **46,** 217.
8. Cecil, R. and McPhee, J. R. (1959), *Advances in Protein Chemistry* **14,** 255.
9. Desreux, V., Frédericq, E. and Fischer, P. (1946), *Bull. Soc. Chim. Biol. Paris* **28,** 493.
10. Dickens, F. (1933), *Biochem. J.* **27,** 1141.
11. Dickens, F. (1946), *Biochem. J.* **40,** 145.

12. DIXON, M. (1937), *Nature, Lond.* **140,** 806.
13. DIXON, M. (1948), Biochem. Soc. Symposia No. 2.
14. GILMAN, A. and PHILLIPS, E. S. (1946), *Science* **103,** 409.
15. GILMAN, A., PHILLIPS, E. S., KOELLE, E. S., ALLEN, R. P. and ST. JOHN, E. (1946), *Amer. J. Physiol.* **147,** 115.
16. GOFFART, P. L. and FISCHER, P. (1948), *Arch. Int. Physiol.* **55,** 258.
17. GUNLACH, H. G., STEIN, W. H. and MOORE, S. (1959), *J. Biol. Chem.* **234,** 1754 and 1761.
18. HAKIM, S. A. E. (1954), *Brit. J. Ophthalmology* **38,** 193.
19. HELLERMAN, L. (1939) Cold Spring Harbor Symposia VII, 165.
20. HELLERMAN, L., PERKINS, M. E. and CLARK, W. M. (1933), *Proc. Nat. Acad. Sci. Wash.* **19,** 855.
21. HELLERMAN, L., CHINARD, F. P. and DIETZ, V. R. (1943), *J. Biol. Chem.* **147,** 443.
22. HOPKINS, F. G. and MORGAN, E. J. (1938), *Biochem. J.* **32,** 611.
23. HOPKINS, F. G., MORGAN, E. J. and LUTWAK-MANN, C. (1938), *Biochem. J.* **32,** 1829.
24. HUGHES, A. F. W. and FELL, H. B. (1949), *Quart. J. Microscop. Sci.* **90,** 37.
25. HUGHES, A. F. W. (1949), *J. Roy. Microscop. Soc.* **69,** 215–224.
26. LEACH, E. H. and LLOYD, J. P. F. (1955), *Trans. Ophthalmol. Soc.* **75,** 425.
27. LOTSPEICH, W. D. and PETERS, R. A. (1951), *Biochem. J.* **49,** 704.
28. PETERS, R. A. and WAKELIN, R. W. (1946), *Biochem. J.* **40,** 513.
29. PETERS, R. A. (1936), *Nature, Lond.* **138,** 327.
30. PETERS, R. A. and WAKELIN, R. W. (1947), *Biochem. J.* **41,** 545.
31. PETERS, R. A., WAKELIN, R. W. and CECIL, R. T. (1948), *Biochem. J.* **43,** 45.
32. PETERS, R. A., RYDIN, H., and THOMPSON, R. H. S. (1935), *Biochem. J.* **29,** 53 and 61.
33. PETERS, R. A. (1949), *Symp. Soc. Exp. Biol.* **3,** 36. Ed. J. F. Panielli and R. Brown, Camb. Univ. Press.
34. PETERS, R. A. and WAKELIN, R. W. (1949), *Brit. J. Pharmacol & Chemotherapy* **4,** 51.
35. RAPKINE, L. (1931), *Ann. Physiol. Physicochim. Biol.* **7,** 382.
36. RAPKINE, L. (1933), *C.R. Soc. Biol. Paris.* **112,** 790.
37. SARKAR, S. N. (1948), *Nature* **162,** 265.
38. SCHUBERT, J. (1939), *J. Biol. Chem.* **130,** 631 and (1936) **114,** 341.
39. SHEARER, O. (1922), *Proc. Roy. Soc. B.* **93,** 213.
40. WALKER, E. (1922), *J. Physiol.* **59,** xxxii.
41. WILSON, H. E. C. and GHOSH, B. K. (1937), *Ind. Mep. Gaz.* **72,** 147.

CHAPTER 6

Lethal Synthesis and Carbon-fluorine Compounds

Lethal Synthesis

Lethal synthesis is the term which I proposed for a special phenomenon. It may happen that a substance is itself non-toxic, but that it is converted by enzymic synthesis into a substance which will then cause a "biochemical lesion". The idea that a non-toxic substance could be converted into a toxic substance somehow *in vivo* has been long with us. Voegtlin and Smith(79) (1921), for instance, proved that the change from arsphenamine to the toxic tervalent arsenical took place easily in air or oxygen; no intervention is required by the enzymes of the tissues. The degradation of prontosil to sulphonamide by reduction on the other hand is done by "tissue" action. It seems best to call these cases, toxic conversion.

The term lethal synthesis (Croonian lecture 1952(62)) is the opposite of protective synthesis, which is the term applied to the circumstance where a molecule is conjugated with some biochemical constituent like glycuronic acid, and so excreted in what is held to be a less toxic form. So far as I was concerned, the essential facts were encountered first when I was working with the Belgian biochemist, C. Liébecq* in 1947/48 on the substance fluoroacetate ($FCH_2.COO^-$). This is known now to be the active principle of the plant Gifblaar, which is toxic to cattle. Barron and colleagues(6,32) had thought that the action of fluoroacetate was to block the metabolism of acetate; but this idea was not consistent with some of our experiments using kidney particles. Instead there was an accumulation of citrate which we thought due to a block in citrate metabolism, as the result of synthesis of fluoroacetate to an inhibitor. The proof that a synthetic fluorine compound was formed took us 4 years. Between 1948 and 1952, other suggestions of a similar nature were made. Kidder and Dewey(34) (1949) found a toxic action of the purine antagonist, azaguanine, on the protozoon *Tetrahymena geleei*, and suggested that this might be due to a synthesis to a fraudulent nucleotide. Also in 1949 Umbreit and Waddell(78) proved that the inhibition of the growth of *Streptococcus faecalis* by deoxypyridine was due to its synthesis to deoxypyridine phosphate. In 1952, in the same year that our proof of lethal synthesis appeared, it was found (Heinrich *et al.*(31)) that 8-azaguanine was incorporated into the RNA of the protozoa when their growth was slowed by it.

* Now a Professor of Biochemistry in Liège.

It can be recalled that in 1938, it was noticed (Marshall, Cutting and Emerson[43]) that acetylsulphanilimide was more toxic to the mouse than sulphanilimide itself.

To understand the details of this lethal synthesis in the field of carbon-fluorine compounds, it is best to lay out the researches from the beginning.

Carbon-Fluorine Compounds. Na Fluoroacetate ($FCH_2.COONa$)

Carbon-fluorine compounds must be distinguished very clearly from fluorides. When present in excessive amounts in soil and water, fluorides are toxic; they can induce malformations of bone, fluorotic teeth in man and other disorders in cattle. Their action is, however, radically different from that of the carbon-fluorine compounds like Na fluoroacetate, FCH_2COONa. In these the fluorine atom is not to be distinguished as such. The unusual properties are due to the very stable FC group, which is not even destroyed by treatment with concentrated sulphuric acid, a distillation from which is one way of preparing pure fluoroacetic acid (m.p. 31–32°). (p. 119, Note 1). The chemistry of FC compounds really starts with work by Swarts[72] (1896) in Ghent, whose interest was concentrated upon their synthesis as a piece of pure chemistry. It was later, in the 1930's that the toxicity became apparent, and that their possible use as systematic insecticides in plants was raised (Schrader[71]). In 1940 it was necessary to examine their toxicity; it varied greatly from one species to another, but fortunately it was found that the toxicity of fluoroacetates to the monkey (and presumably man) was much less than that to several species of animal; the dog, for instance, is sensitive to a dose of 0·05 mg/kilo; the rabbit to 0·3 mg/k, the rat to 5 mg/k, the monkey to 20 mg/k, and the frog to 500 mg/k.[13] The toxic effects vary in different animals; in some it affects the heart; in others it causes tonic convulsions like those due to eating strychnine. During the examination of the FC compounds, it transpired that, unlike the arsenicals, they had virtually no effect on enzymes. Many FC compounds were made in Cambridge during the early 1940's (Saunders, B. C.[70]), and this work has been extended in more recent systematic work in London, Ontario (Pattison[52]). In much work by Barron and Bartlett and by others including ourselves, no isolated enzyme has been found which is sensitive to fluoroacetate.*

Up to 1940 no one suspected that fluoroacetate had been in existence before Swarts made it. In 1943, Marais[41] in brilliant work identified it as the main toxic component in the South African plant, *Dichapetalum cymosum*. (Fig 6.1) Found in the Transvaal, the leaves are very toxic to animals. By its isolation and identification, Marais proved that Nature

* There is only one exception known to me (see page 110).

FIG. 6.1. *Dichapetalum cymosum* (Hook). Engl. cuttings showing the rare seeds. Photo kindly sent by Dr. Dyer, Pretoria, Nov. 1959. Also called "Gifblaar".

had made fluoroacetate long before it fell into the hands of scientists!

The plant is interesting in several ways; it has roots going down some 50 ft, so that what appears is the top of an underground tree, which can reproduce by an elaborate branching stem system, (personal communication from Professor Badenhuizen). This and the fact that it comes up among rocks makes it difficult to eradicate. At the pH of the cell sap, some of the compound may be ionized.

As indicated briefly, investigation of the mode of action of fluoroacetate has revealed a new type of toxicity. The actual biochemical lesion is also in this case an attack upon an enzyme, but only after the compound has undergone a synthesis carried out by the enzymes of the tissues. The first step in our understanding of this was to distinguish the action from that of fluoride, which was easily done with a homogenate from the brain of the pigeon.(37) Fluoride induced a fall in respiration in contrast to fluoroacetate (Table 6.1). Brain tissue from the pigeon does not activate fluoroacetate; as a matter of fact we have not been able to show that brain tissue from the pigeon, rat or guinea pig can even activate acetate so as to produce an increased uptake of oxygen *in vitro*. The behaviour of

TABLE 6.1. COMPARISON OF THE EFFECT OF SODIUM FLUOROACETATE AND SODIUM FLUORIDE UPON THE PYRUVATE OXIDASE SYSTEM (PIGEON BRAIN HOMOGENATE). SUBSTRATES: SODIUM PYRUVATE AND SODIUM FUMARATE

	O_2 uptake (μl in 30 min)	
no addition	426	
NaF (24 mM)	262	38 per cent inhibition
NaFl Ac. (75 mM)	439	no inhibition

Data from Liébecq and Peters, 1949.

kidney tissue from the guinea pig proved to be quite different. Exposure of kidney particles to fluoroacetate after washing well and suitably reinforcing, reduced the oxygen uptake with fumarate as a substrate; but there was no accumulation of acetate; hence the Barron hypothesis that fluoroacetate blocked the metabolism of acetate could not be the main effect. We took particular care to exclude other substrates by thorough washing of the tissue. Further search proved that the fall in oxygen uptake with fluoroacetate was accompanied with an accumulation of citrate. That this was not due to a direct effect of fluoroacetate upon the enzymes concerned with synthesis and degradation was proved by setting up systems embodying various enzymes of the citric acid cycle. Since citrate was normally oxidized by our kidney preparation, it was clear that the accumulation of citrate was due to a block in its metabolism rather than to an increased formation of citrate. There was an apparent contradiction; fluoroacetate was non-toxic to isolated enzymes including those of the "citric acid cycle"; at the same time it was toxic to citrate metabolism. We resolved this contradiction by the hypothesis that fluoroacetate

TABLE 6.2. EFFECT OF SODIUM FLUOROACETATE (3·3 mM) UPON THE OXYGEN UPTAKE, ACETIC ACID PRODUCTION AND THE ACCUMULATION OF CITRATE OF A CENTRIFUGED KIDNEY HOMOGENATE (GUINEA-PIG) RESPIRING IN PRESENCE OF SODIUM FUMARATE (REINFORCED WITH Mg^{2+} AND ATP)

	O_2 uptake (μl)	inhibition (%)	acetic (mg)	citric acid (mg)
No addition	436	—	0·11	0·18
Fluoroacetate	246	43	0·13	0·75

Experiment of Liébecq and Peters, 1949.

was converted by synthesis in the citric acid cycle into some other substance which was inhibitory; this conversion did not take place either in an acetone powder from liver or in a particle preparation from brain. The enzymic synthesis to a toxic compound would mean that fluoroacetate was deceiving the condensing enzyme responsible for condensing the C_2 fragment with oxaloacetate.* (Table 6.2)

Martius(44) (1948) reported in a lecture which he gave a fortnight after my lecture on our work, some independent experiments with minced ox heart (unwashed), from which he concluded independently that fluoroacetate blocked citric acid metabolism because it was transformed into some other substance, which he thought might be fluorocitric or fluoroisocitric acid.

We had suggested fluoro-*cis*-aconitic acid, as at that time citric acid was believed to be out of the "tricarboxylic acid cycle" owing to evidence from radioactively marked compound. We were careful not to go beyond our facts in advancing our hypothesis, because it was still possible that in the organized mitochondrial complex, even fluoromalate might have been inhibitory to aconitase by blocking some tightly coupled chain of reactions.

So far these enzymic experiments had not been related to events *in vivo*; it was therefore an important step when Dr. Buffa† and the author(7) proved that injections of fluoroacetate into animals such as the rat and

* About the same time, Kalnitsky(33) working with undialysed kidney found increases in citrate in their medium when fluoroacetate was added, which they attributed to an increased formation of citrate.

† Now Professor of General Pathology in Milan.

guinea pig caused increases of citric acid in several tissues. Rises took place quickly, up to a maximum in 1–2 hr. This connection again of events *in vivo* with those occurring *in vitro* gave confidence in a pursuit of the problem on a wider scale, and incidentally provided the best proof that the "citric acid cycle" is operating *in vivo*.

TABLE 6.3. FLUOROACETATE POISONING; ACCUMULATION OF CITRATE IN THE RAT

Citric acid, μg/g wet tissue

	Control	Poisoned
Kidney	14	1036
Heart	25	677
Spleen	0	413
Stomach	37	386
Small intestine	36	368
Large intestine	21	248
Lung	9	257
Brain	21	166
Blood	3	50
Liver	0.8	31
Diaphragm	0	400
Uterus (virgin)	217	207

Animals killed 1 hr after an intraperitoneal injection of NaFlAc, 5 mg/kg body weight. (Buffa and Peters 1949.) Controls injected with saline.

The fact that citrate accumulates *in vivo* has been confirmed and extended by others (Potter and Busch[66]), and by using sub-maximal doses,[38] it was found that there was a rise in citrate in heart and kidney up to 4 hr after which a gradual decline occurred which lasted for 24 hr. The effects of fluoroacetate in liver may be anomalous.

It was now possible to make concrete the hypothesis that there was direct interference with the metabolism of citrate, because Ogston[51]* had pointed out in the meanwhile the fact that with a three point attachment to the enzyme, asymmetric labelling of citrate might happen, and so it was not excluded any longer as a "citric acid cycle" intermediate. The obvious substance likely to form the block was fluorocitrate, formed by an enzymic synthesis from fluoroacetate; the latter would then deceive the condensing enzyme into the "lethal synthesis" which could prove fatal (Fig. 6.2).

* Now Professor of Biophysical Biochemistry in Australian National University, Canberra; formerly Fellow of Balliol College and Reader in Biochemistry, Oxford.

Proof of this hypothesis required two facts, (1) the isolation of monofluorocitric acid from a poisoned preparation, and (2) the demonstration that this affected the enzyme aconitase *in vitro*, and that it is toxic *in vivo*.

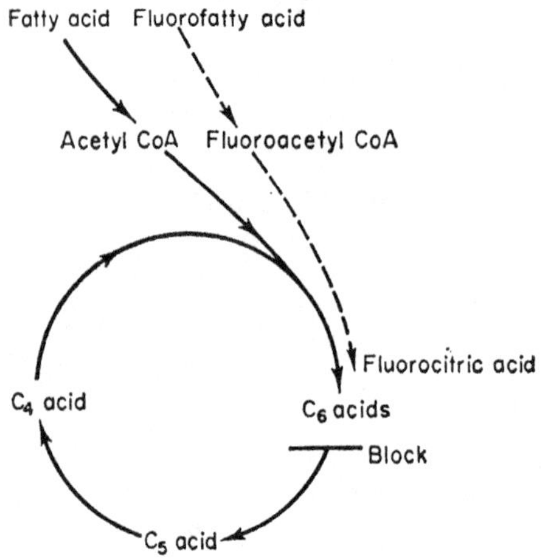

FIG. 6.2. The figure illustrates the hypothesis for the mechanism of lethal synthesis and shows the channel by which the fluoroacetate becomes synthesized to fluorocitrate.

These points may now be considered in turn, as they are most important in establishing this biochemical lesion:

(1) The isolation of monofluorocitric acid was accomplished by us from the kidneys of guinea pigs, rabbits and dogs, using particle preparations suitably reinforced with ATP and Mg, and treated with fluoroacetate[8]. Some of the initial stages were carried through in Rome from the kidneys of dogs. As Professor Vernoni then remarked. The preparations brought back to Oxford for processing were not the kind of trophy usually coming from Rome! The concentration of the postulated fluorocitrate was followed during the isolation by a *rough* test, in which a given amount of kidney particles were exposed to the unknown solution in the presence of 10μmoles of citrate. In 30 min at 38°C normal particles usually metabolize about 9·0μmole; the reduction in the amount metabolized forms an estimate of the amount of the inhibitor added. We were careful to express it in this pedantic way, because we found early in the work that in general the amount of oxygen taken up does not correspond to the citrate metabolism; some respiratory control is possibly present, but increase of citrate disappearance may occur without increase of O_2 uptake. Some facts are known about this empirical test, which are not yet published; for instance the main effect of the inhibitor takes place during the last 20 min of the test.

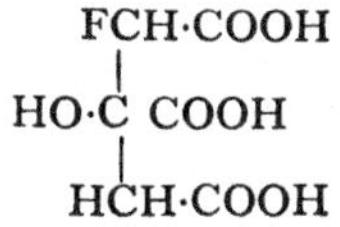

The crystals of fluorocitric acid are shown in Fig. 6.3.

In the course of concentrating the fluoro compound, it was soon found to have pronounced acidic properties; for instance the lead salt was precipitated at a more acidic pH than that of citric acid, though of course there was some overlap in the precipitation. This was confirmed later by finding that after a barium precipitation, N/10 HCl was needed to decompose the barium salt. At one stage we had to go on in the face of contrary evidence, when it seemed that the most active preparations contained virtually no fluorine. This was because we did not realise how little of the fluorocitrate was required for inhibition, so that we had been taking too little for our fluorine analyses!! Fluorocitric acid was finally separated in very small amounts in crystalline form after using a resin column. A titration curve of this showed that it had one acidic group with a strength almost equal to that of HCl.* In the kidney test, as little as 1·25 μg would inhibit the metabolism of 2 μmoles of citrate. It is appropriate to mention that some help can be obtained in the estimation of fluorocitric acid in the presence of fluoroacetate by the use of particles from pigeon brain, because these do not react to fluoroacetate.

The Biochemical Lesion Induced by Fluorocitrate

Fluorocitrate should be more toxic than fluoroacetate upon injection into the animal, if it is the actual toxic agent. This is true when it is injected into the sub-arachnoid space in the anaesthetized pigeon or young rat, or in rabbits through the cisterna or a bore hole in the skull. With a large dose convulsions come on within 15 min. In the pigeon even 11μg has induced convulsions (Fig. 6.4). The convulsions are not due to a reduction of the ionized calcium in the cerebrospinal fluid, because the injection of calcium will not cure them.† When fluoroacetate is injected into the brain in a similar way, in much larger amounts, no convulsions are induced in these animals (Table 6.4 and Fig. 6.5). If fluorocitrate is injected into animals by the intraperitoneal route, the difference is less striking between the dose of the two compounds to produce toxic signs, which we believe to be due to difficulties in penetration to the mitochondria.[30] Even so fluorocitrate is slightly more toxic upon a fluorine basis, and it has a much larger effect upon the accumulation of citrate in the liver than fluoroacetate.[23] The phosphoesters in brain are mainly unchanged with the convulsions, though there is a slight fall in creatinephosphate (Dawson and Peters.[18]

* This is so acidic that when Professor Baird Hastings first looked at our data, he said that our compound must be contaminated with HCl, a suggestion disposed of by the absence of a precipitate with silver nitrate.

† The figure shows the results of an experiment which Professor Baird Hastings and I carried out in Boston on a rabbit in Dr. Sweet's laboratory.

(a)

(b)

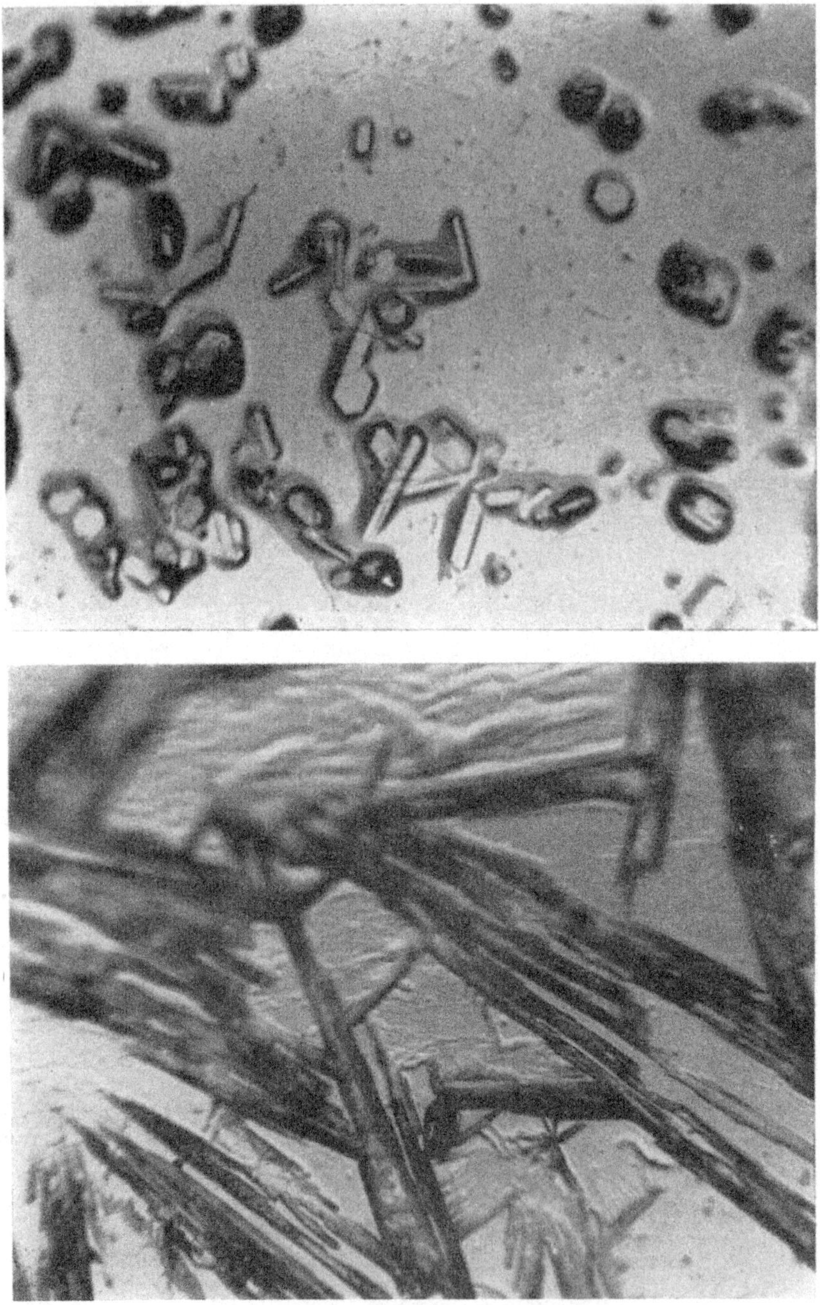

Fig. 6.3. Photographs of crystalline fluorocitric acid (a) Hydrated. (b) Anhydrous and very hygroscopic; photo taken immediately after removal from P_2O_5 desiccactor.

(a)

(b)

FIG. 6.4. Effect of intracranial injection (under ether) of fluorocitrate (a) Convulsion commencing, (b) Head retraction.

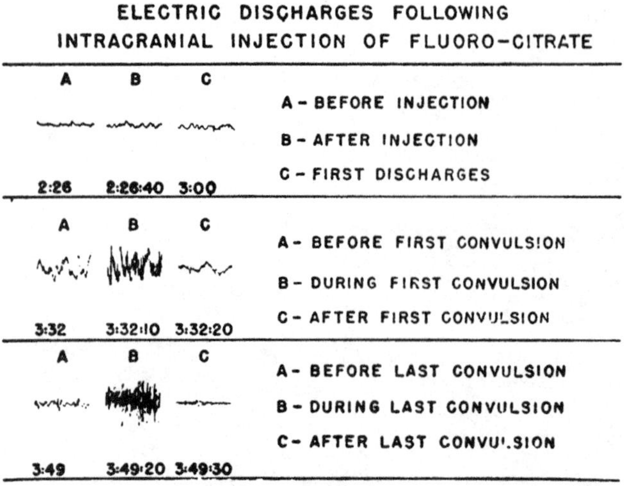

FIG. 6.5. Electric discharges following intracranial injection of fluorocitrate into a rabbit under nembutal anaesthesia. Time below waves. (Hastings and Peters.)

TABLE 6.4. EXAMPLE OF AN EXPERIMENT BY HASTINGS, AND PETERS UPON A RABBIT, NOVEMBER, 1953

Rabbit Anaesthetized with Nembutal

0 min	Fluorocitrate injection (2 μmoles).
10 min	Mild signs, head retraction and rapid respiration, arched back.
30 min	Injected 11 μmoles Ca^{++}. Breathing became regular and legs relaxed quiet. (This effect was considered to be due to the effect of the citrate ion upon the Ca^{++} in the cerebro-spinal fluid.)
40 min	Muscle tremors increasing.
64 min	General convulsions. Two doses of 8 μmoles Ca were given without effect.
77 min	Death after convulsions at frequent intervals.

(1) The citric acid content of the brain was raised as has always been found to be the case in fluorocitrate poisoning. Cerebral hemispheres 121 μg/g and rest 212 μg/g.

(2) In the final stages, there was no effect of Ca^{++}.

(3) The normal for citric acid content of rabbit brain is about 30 μg/g in the cerebrum and rest. The average for the poisoned is Cerebrum 109, and rest 178. In two cases analysed, for the poisoned rabbit, the surface of cerebrum was 243 and the remainder 114; optic lobes, surface 224 and rest 145, so the effect seems to be a surface one.

The Enzyme Concerned

When we found no effect of fluorocitrate upon the isocitrate dehydrogenase (Lotspeich *et al.*[39] 1952), it was clear that it inhibited aconitase. This is an interesting enzyme discovered by Knoop and Martius[36]; it sorts out an equilibrium between the three tricarboxylic acids, such that

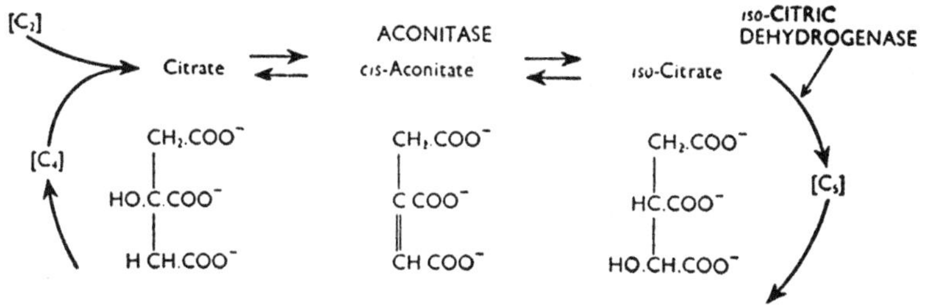

Fig. 6.6. The reactions of the tricarboxylic acid cycle effected by the enzymes aconitase and isocitric dehydrogenase. At equilibrium about 90 per cent is citrate.

about 90 per cent is present as citrate (Fig. 6.5) and the rest as *cis*aconitate and *iso*citrate. Its function is to transfer the —OH group from the centre carbon atom of citric acid to the β position so that decarboxylation can take place. As mentioned, in a mitochondrial kidney preparation, quite small amounts of fluorocitrate will block the aconitase present. At first it was thought that the amounts required to block a soluble preparation of

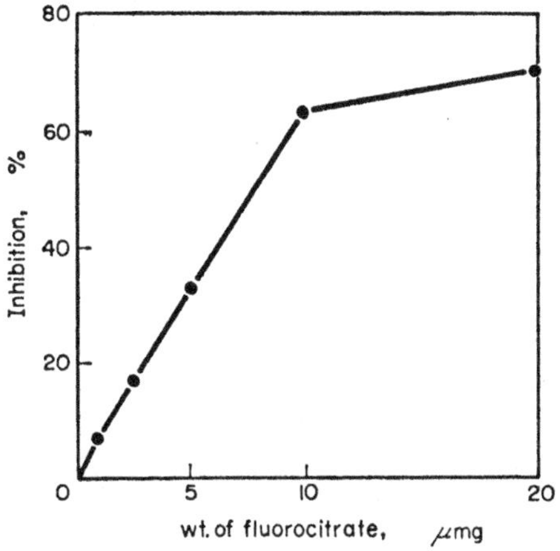

Fig. 6.7. Inhibition of aconitase by enzymic fluorocitrate (activated). Pre-incubation with fluorocitrate 10 min. : substrate DL-isocitrate. Ordinate change in absorption at 240 mpl over 3·5 min as percentage inhibition. (Peters, 1961.)

aconitase were much greater than those needed to block the aconitase in mitochondria.[46] In recent work, however,[57] using a different preparation of aconitase, very large inhibitions have been found; a 50 per cent inhibition was induced by as little as 10^{-8} to 10^{-9} M of enzymic fluorocitrate which is an inhibition of the same order as that reached with the action of insecticides upon cholinesterase (Fig. 6.7). (For the details of some of this work the reader is referred to the Notes at the end of this chapter).

By isolating fluorocitric acid and determining its effects *in vivo* and *in vitro*, we now have firm grounds for proposing the theory that the toxic effect of fluoroacetate is due to a lethal synthesis. If fluoroacetate did not deceive the enzymes concerned with activation and synthesis it would not be toxic. Since the fluorocitrate does not appear to be metabolized with any speed, it becomes responsible for a biochemical lesion at the aconitase stage in the citric acid cycle. Even numbered ω-fluorofatty acids with longer carbon chains also seem to form a FC_2 stage which passes to fluorocitrate and so blocks citrate metabolism. This is quite consistent with the view of Saunders and colleagues[70] in 1947, who had good evidence for the idea that even numbered fatty acids were toxic whereas those with

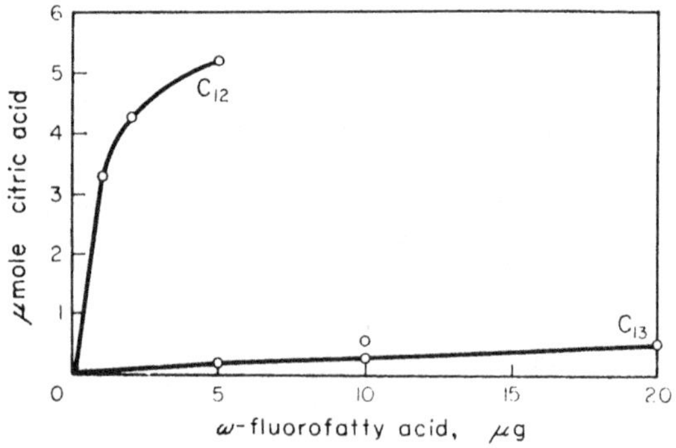

FIG. 6.8. Comparison of inhibitory effect on citrate metabolism by kidney mitochondria of C 12 and C 13 ω-fluorofatty acid ($FCH_2(CH_2)_n$ COOH). C 12 blocked citrate metabolism; C 13 had no effect. (Peters, 1958).

chains containing an odd number of carbon atoms were comparatively harmless. By a degradation of two carbon atoms at a time, these would come down to the three carbon fluoropropionic acid, which is comparatively harmless. This theory of alternate toxicity is supported by subsequent synthetic work (Pattison[52]). Figure 6.8 shows that it holds *in vitro* for two synthetic acids kindly sent to me by Professor Pattison.[63]

Other Enzymes

So far no other enzyme has been found to be inhibited by fluorocitrate. There was a claim that the malic enyzme is poisoned, but on subsequent investigation this appeared to be a non-specific action upon Mn, though fluoromalate appears to be active here (Gal *et al.*[24]).

Synthesis of Fluorocitric Acid

From the above remarks it is clear that kidney particles synthesize fluorocitric acid from fluoroacetate. Some of the details of this synthesis in kidney *in vitro* have been studied by estimating the fluorocitrate formed enzymically with suitable controls; the amounts of fluorocitric acid formed are so small that they must be estimated enzymically; this is simple theoretically, but is complicated in practice. The tests were made either with kidney particles blocked with malonate to avoid re-cycling or with brain particles, which do not synthesize fluorocitrate from fluoroacetate.[61] In experiments of this type, we have shown that particles representing about 150 mg kidney tissue can synthesize in 40 min about 20μg fluorocitric acid from fluoroacetate. For some reason, which is at present unknown, this synthesis is larger when 1-malate is the substrate than when this is oxaloacetate. From experiments with pigeon brains, we have calculated that the citric acid cycle can be blocked *in vivo* with as little as 0.2μg/g of wet tissue, so that the amount synthesized in kidney particles was probably rather larger than is necessary to get the block in citrate metabolism found experimentally in brain.

It was interesting to study in this kidney system the effect of acetate upon this biochemical synthesis. Even 1/20th of the concentration of acetate gave some inhibition of the synthesis of fluorocitric acid, which soon became maximum with an increase of acetate (Fig. 6.9). Hence, the experimental facts are the reverse of the hypothesis of Barron and Bartlett,[6] who considered that fluoroacetate blocked acetate metabolism. At the same time the experiment makes it clear that the inhibition takes place at some centre which can activate acetate, but which is not the normal activating centre for acetate usually studied. The evidence for this statement comes from several sources; acetoned extracts from the liver of pigeons, which readily synthesize citrate from oxaloacetate in the presence of Mg, ATP and cysteine, produce no fluorocitrate, when the acetate is replaced by fluoroacetate. Similarly additions of activating extracts from bacteria do not activate fluoroacetate to produce hydroxamates from hydroxylamine under conditions where acetate gives excellent activation. It is comparatively easy to get an extract of acetoned rabbit kidney, which will activate acetate; but which will do nothing to fluoroacetate. The FAc activating system is unstable to freezing or to hypotonic

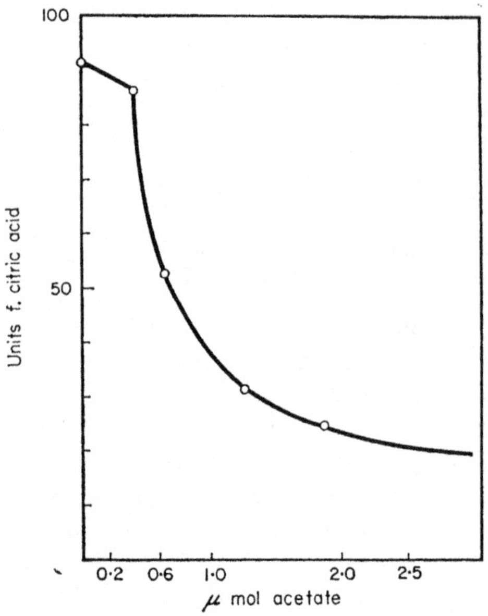

FIG. 6.9. Inhibitory effect of acetate upon the synthesis of fluorocitric acid by kidney particles as determined with kidney particles; substrate, malate and fluoroacetate 20 *u*mol/3 ml.
(Data of Peters and Wakelin, 1954.)

solutions. The only way to preserve it is to adopt the technique of Brady[5] and to dialyse against cysteine. There is no doubt that the substance which is condensed with oxaloacetate is fluoroacetyl S.CoA, as has been so well demonstrated by Brady; hence this is an intermediary in the lethal synthesis. The possible paths for the behaviour of the different fatty acids is shown in Fig. 6.10.

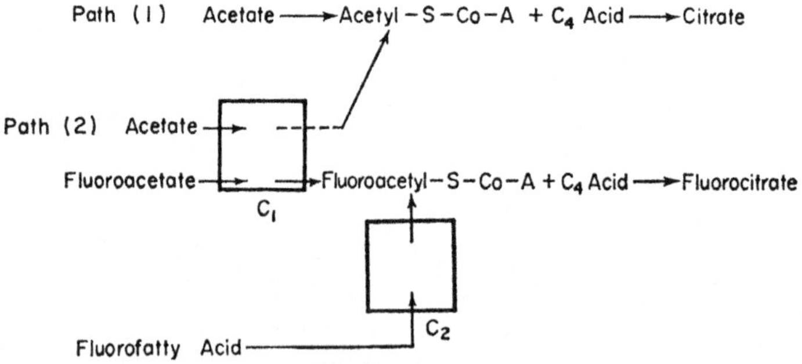

FIG. 6.10. Centre G, appears to be blocked by acetate, monoacetin and acetamide; the latter two act as antidotes. Centre C2 is blocked by butyrin. Path (1), the normal path does not seem to be blocked by the FC compounds.

Other Carbon-fluorine Compounds in Nature

Until recently fluoroacetic acid in the South African plant was the only FC compound known in Nature. There are however in Africa other species of *Dichapetalum*; of these *D.toxicarium* is well known, because it has been used in the Sierra Leone district as a rat killer, "ratsbane", the

Fig. 6.11. *Dichapetalum toxicarium*. Photographs of the dry drupe, the nut-like fruit containing the seeds and of the separated seeds.

ground seeds are toxic, and the description of their effects upon dogs much resembled that of fluoroacetate.* The matter seemed to be worth settling; accordingly I got samples of the seeds of *D. toxicarium* through the help of Sir John Simonsen and the help of government agencies.[67]

* In man, the eating of the seeds can induce the condition known as "broke back". W. Renner (1904) described a case where a native ate some fish poisoned with powdered seeds, which was intended for rats. He got a neuronitis lasting three weeks, that is to say paralysis and sensitivity of the lower limbs. Renner goes on to say that he had seen this before in young people and been puzzled by it, until this clue suggested that it was due to injurious practices of the "local" doctors.

Sir Henry Dale, O.M., F.R.S., was associated with some work upon these seeds in 1906.

When the ground seeds incorporated into rat cake were given to rats, the animals died with oedema, and analysis showed an increase of citrate in the tissues. It was natural to look for fluoroacetate, and an isolation of this was attempted by standard methods. Though very small amounts of impure K fluoroacetate were isolated by fractionating the water soluble fraction, the main toxic component was found to be soluble in lipid solvents[60]; it was isolated finally and was proved to be a C_{18} fatty acid containing one F atom and one double bond.[64] By the use of the modern techniques of nuclear magnetic resonance, of infrared spectroscopy, and gas chromatography after treatment with ozone to split the double bond, the compound was identified as fluorooleic acid (Peters, Hall, Ward* and Sheppard,[58] 1960).

ω-FLUOROOLEIC ACID

$$FCH_2{\cdot}(CH_2)_7{\cdot}CH{:}CH{\cdot}(CH_2)_7{\cdot}COOH$$

In this identification synthetic compounds made by Dr. Pattison and his colleagues played a very important part. Quite recently Pattison and Dear[53] have synthesized F oleic acid and proved its identity with our natural specimen (see Note 2, p. 122).

This was not the only fluorocompound present in these seeds. When the acid became available on a larger scale, made possible by an apparatus for chromatographic separation by R. J. Hall in my laboratory at Babraham, it became clear that there was another acid present in much smaller amounts, with a high melting point of approximately 70°C, with no double bond and a fluorine content higher than that of fluorooleic acid. This saturated acid behaves to a kidney preparation as if it had an even number of carbon atoms. Its structure is now known to be fluoropalmitic acid (Hall and Ward).

During the course of this work since fluoroacetate is present in the leaves of *D. cymosum* the question arose whether the leaves as well as the seeds of *D. toxicarium* contained the F oleic acid. It was found that they did not (Table 6.5). A test on a few seeds of *D. cymosum* also gave no indication of the presence of the long chain compound. Its presence in the seeds of *D. toxicarium* is not therefore due to storage of fluoroacetate in the seeds of the genus as an oil, but it appears to be peculiar to these seeds. Several other species of *Dichapetalum* tested from East Africa had little toxicity.[54]

Fluorooleic acid was the second carbon-fluorine compound to be isolated in Nature. A closer examination of its clinical effect and a comparison of this with the action of several long chain compounds kindly given us by Professor Pattison including ω-fluorodecanoate showed that the

* Dr. P. F. V. Ward is now in the Biochemistry Department of the Agricultural Research Council Institute of Animal Physiology, Babraham.

TABLE 6.5. FLUORINE COMPOUNDS IN LEAVES AND SEEDS OF *D. cymosum* AND *D. toxicarium*

	Fluorine μg/g dry wt				
	Total	F_t†	F_{alk}‡	F acetate	F oleic*
Leaves					
D.cymosum					
Sample (1)	270	36	217	+ +	0
Sample (2)	70	14	9	+	+
D.toxicarium	0	12	8	0	0
Seeds					
D.cymosum	0			0	0
D.toxicarium	2035			trace	+ +

Data of Peters and Hall, 1960.

0 = not measurable.

* Or long-chain fluorofatty acid.

† Extracted at pH 1·0.

‡ After hydrolysis with conc. sodium hydroxide; represent fluoroacetate and any other alkali labile FC compounds.

effects on the citrate metabolism in some tissues were different from those of fluoroacetate; there was a definite tendency for the largest citrate accumulations to occur in the heart rather than in the kidney with the longer chain compounds; there was also an increase in citrate accumulation in the liver. *In vitro* the long chain fluoroacids also behave differently; they are more easily activated by particles from the kidney than fluoroacetate; even as little as 1–2μg produces as large an effect on the citrate metabolism as 25μg of fluoroacetate. Nevertheless, the final channel of action is the same, the biochemical effect is due to a lethal synthesis to fluorocitrate, which will occur after a prior conversion to a fluoroacetyl fragment.

The approximate toxicity for this compound[55] for various animals has been obtained and is included in Table 6.6. One point of interest is that the toxicity to the sheep is relatively less than that to the rabbit, the sheep/rabbit ratio for the dose is much more favourable for the fluoro-fatty acid than it is for fluoroacetate. This makes it worth considering whether the natural compound should not be used instead of fluoroacetate in ridding Australia and New Zealand of rabbits now that myxomatosis is no longer effective. One point in favour of the fluorofatty acid is that the animal dies from a sudden heart attack or so it appears. This may occur in the middle of eating greenstuff. The death, therefore, seems to be more humane.

TABLE 6.6
Estimates of LD50
(mg/kg for fluorooleic acid)

Animal	LD_{50}
Rat	6
Guinea pig	1·5
Pigeon	2·5
Rabbit	0·5
Sheep	2·0

Dose ratio rabbit/sheep 1/4. For fluoroacetate this is about 1/1.
Data of Peters and Hall, 1960.

OTHER CARBON-FLUORINE COMPOUNDS

Some interesting points in relation to our main themes of biochemical lesions and lethal synthesis have arisen during a study of certain synthetic compounds, for which we are indebted to Dr. P. W. Kent* and his colleagues.[73]

Deoxy-fluoro-glycerols. The compound with the F in the (1) position behaves differently from that with the F in the (2) position.

DEOXY-FLUORO-GLYCEROLS

1-Deoxy- -1-fluoroglycerol   2-Deoxy- -2-fluoroglycerol

1-*deoxy*-1-*fluoroglycerol.* When injected into an animal, after a rather long interval the animal dies, and there is an accumulation of citrate in various organs; this shows that the compound is converted to fluoro-acetyl–S–CoA. Dr. R. D. O'Brien[49]† showed in my laboratory that the conversion does not occur in the heart, because in a simultaneous perfusion of the heart and the liver, there is no poisoning of the heart unless

* Dr. P. W. Kent is a Lecturer in the Department of Biochemistry, Oxford, and a student of Christchurch.

† Now Professor of Entomology at Cornell University, Ithaca, after residence in University of Western Ontario, London, Ontario.

the liver is in the circulation. The steps for this conversion are not known yet, partly because the conversion in the liver seems to be too slow to show up any effects with even tissue slices in various media (Fig. 6.12).

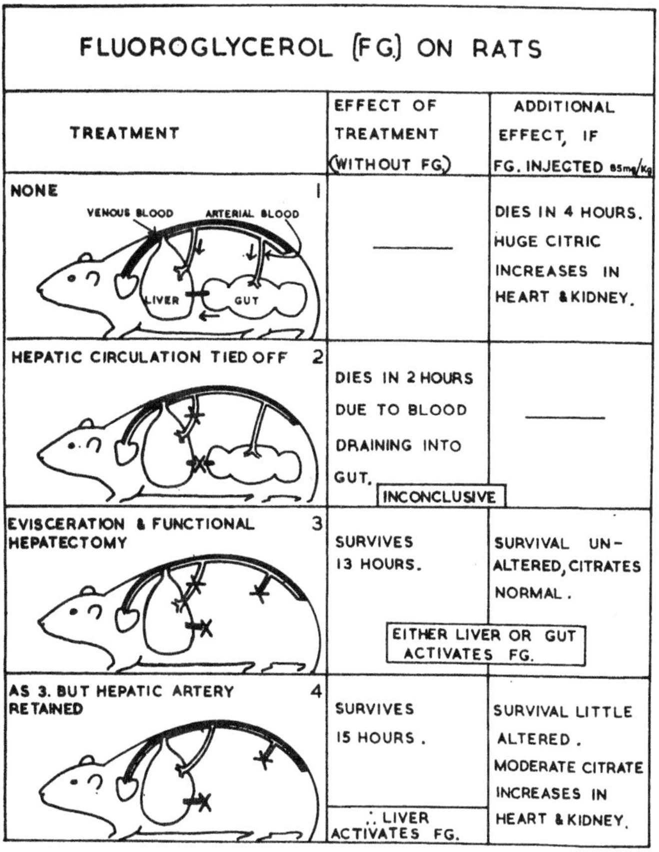

FIG. 6.12. Diagram to illustrate the metabolism of fluoroglycerol (1-deoxy-1-fluoroglycerol). O'Brien and Peters.)

It is, however, an interesting phenomenon, because there is a double lethal synthesis necessary for the final death. If the liver did nothing to the compound nothing would happen, and further if the tissues did not convert the FC_2 fragment to fluorocitrate, the compound would be harmless.

2-deoxy-2-fluoroglycerol. This compound also causes citrate accumulation. A dose of 100 mg per kilo is needed for the mouse. It is probable that the first step in its metabolism is a conversion to 2-deoxy-2-fluoroglyceraldehyde, the metabolism of which has been studied (O'Brien and

Peters[48], Treble and Peters[75]). Leaving aside a reaction of the aldehyde with —SH groups,* liver aldehyde dehydrogenase oxidized the aldehyde strongly to the 2-deoxy-2-fluorogylcerate; this will cause accumulation of citrate in kidney particle preparations, which is very toxic to the kidney mitochondria, when it is the (-)-2-deoxy-2-fluoroglycerate; it appears to be converted to fluorocitrate via the action of serine hydroxymethylase. We were indebted to Dr. P. W. Kent for a specimen of the fluoroglycerates (see formulae).

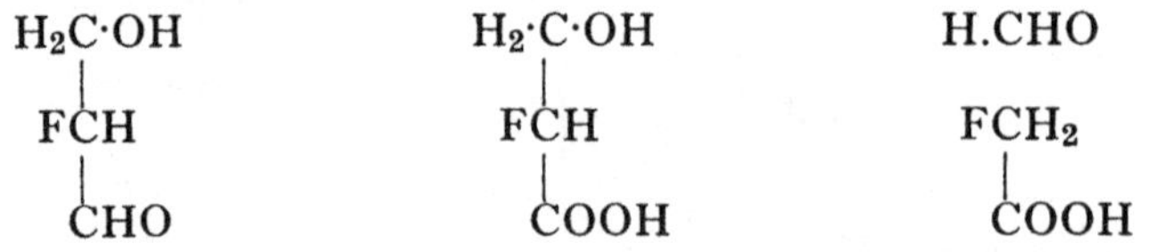

2-deoxy-2-fluoro-glyceraldehyde | 2-deoxy-2-fluoro-glyceric acid | fluoro-acetic acid

2-Fluoroethanol ($F{\cdot}CH_2{\cdot}CH_2{\cdot}OH$)

It has been accepted for a long time that this compound is as toxic as fluoroacetate, and there has been the tacit assumption that the metabolic path was the same. But there were some contradictory reports in the literature. On the one hand, it was thought to be oxidized by the liver, and on the other hand, it was found to be an inhibitor of the usually described alcohol dehydrogenase (alcohol: NAD oxidoreductase.) This matter has been clarified in recent work by Dr. Treble (in Cambridge); the alcohol dehydrogenase is inhibited by fluoroethanol,[76] but there is also present in liver a dehydrogenase, which precipitates during the process of purification at a lower concentration of ammonium sulphate than the usual alcohol dehydrogenase, and which is capable of oxidizing fluoroethanol. Like the observations upon 1-deoxy-1-fluoroglycerol, conversion of fluoroethanol to fluoroacetyl SCoA seems to require the liver specifically. This is probably the reason why the effect of the compound on rats differs clinically from that of fluoroacetate. The experience in our laboratory (Treble) is that the effect of fluoroethanol is slow, perhaps taking some 6 hrs to kill a rat. This must be mentioned because of a newspaper report suggesting that the compound causes a rapid death.

Fluoromalonate has been made in Israel (Chari–Bitron[12]), and its toxicity has been found to be less than that of fluoroacetate, about one-tenth by subcutaneous injection, which suggests that it does not involve a direct decarboxylation to fluoroacetate, unless this is less rapid than might be supposed.

* This was interpreted wrongly as a phosphorylation in our first paper.[48]

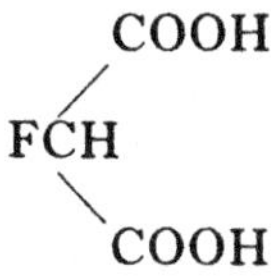

FLUOROMALONIC ACID

Summary

As a result of what has been said up to the present time, we can put forward the following scheme for the behaviour of these carbon–fluorine compounds, in which the true biochemical lesion appears to be the blocking of aconitase by the synthesized fluoro-citrate.

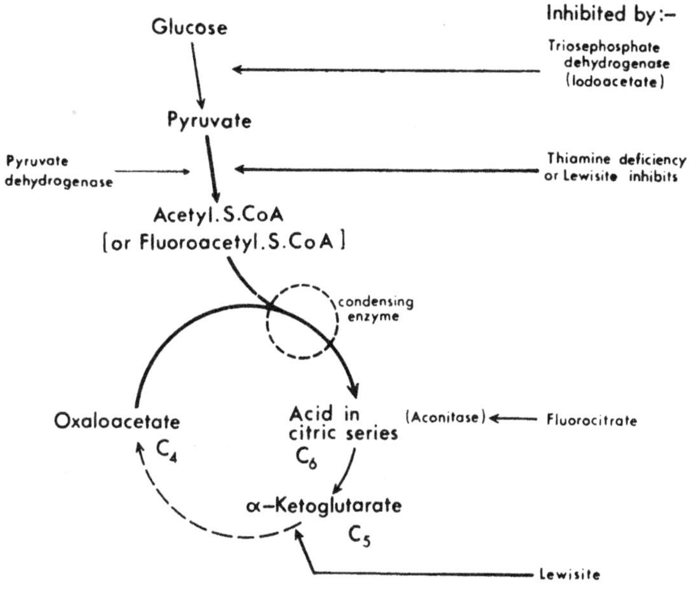

Fig. 6.13. The diagram indicates the site at which thiamine deficiency, lewisite toxicity and fluorocitrate interfere with the citric acid cycle. Fluoroacetyl CoA is also formed from the fatty acids by C_2 degradation.

The diagram (Fig. 6.13) summarizes the effect on the citric acid cycle of the FC compounds and of thiamine deficiency and lewisite toxicity. It is to be noted that all these conditions also cause convulsions.

Other Possibilities

As stated above, there is no evidence that fluoroacetate or fluorocitrate is toxic to any other enzyme in the higher animal than those mentioned. The origin of the convulsions induced by fluorocitrate will start as a blocking of aconitase, but how this induces convulsions is still rather obscure. It seems probable that it is the depression in the metabolism of the cells of the brain concerned which gives rise to the necessary lowering of potential required to produce the stimulation. It has been often suggested that there is a formation of fluoroacetylcholine, which would then be the toxic agent. The matter has been investigated by ourselves and others; but no support has been found for the view that it is the "choline" compound which is the active agent. In particular, fluoroacetylcholine made synthetically (Salle and others[69,63]) has not special properties over and above those given by fluoroacetate, and by choline. Furthermore, there is no evidence of an accumulation of acetylcholine. Again it might be expected that like an activated C_2 fragment, fluoroacetate might be built into some fat or into cholesterol, to form fluorocholesterol. There is evidence that it is built into some lipid substances in the liver.[65] More recently Gal, Drewis and Taylor[25] have found that radioactively marked fluoroacetate ($F^{14}CH_2 \cdot COO^-$) gives rise to a small CO_2 evolution in the animal after injection, and that it enters the lipid fraction of the liver and the cholesterol. I am not aware, however, that these compounds have been shown to be toxic; wherever there has been toxicity there always is an accumulation of citrate. The only exception to this is one experiment recorded by Chenoweth, and it is difficult to know what weight to attach to this against all the others which have been done. In microorganisms there has been an instance recorded recently (Gotto and Kornberg[27]) in which a glyoxalase preparation has been found to be inhibited by fluoroacetate. So far this is a complete exception, and suggests that from the point of view of the enzyme concerned there is not much to choose between an —F or an —OH group.

Effects Upon the Metabolism of the Whole Animal

Much interest has centred lately upon the question of oxaloacetate and its relation to the metabolism of the ketone bodies; if for any reason the supply of oxaloacetate via the citric acid cycle fails, then the acetyl–S–CoA which accumulated might be diverted towards the metabolism of fat. It might be thought therefore that with poisoning of the aconitase by fluorocitrate, this would happen. Colour certainly is given to this by some work by Engel and colleagues[15,21] who have been injecting fluoroacetate into rats to get a condition which they call sodium fluoroacetate diabetes, and in which there is a rise in the concentration of the sugar of the blood and also of lactic acid, as well as of ketone bodies (Fig. 6.14.)

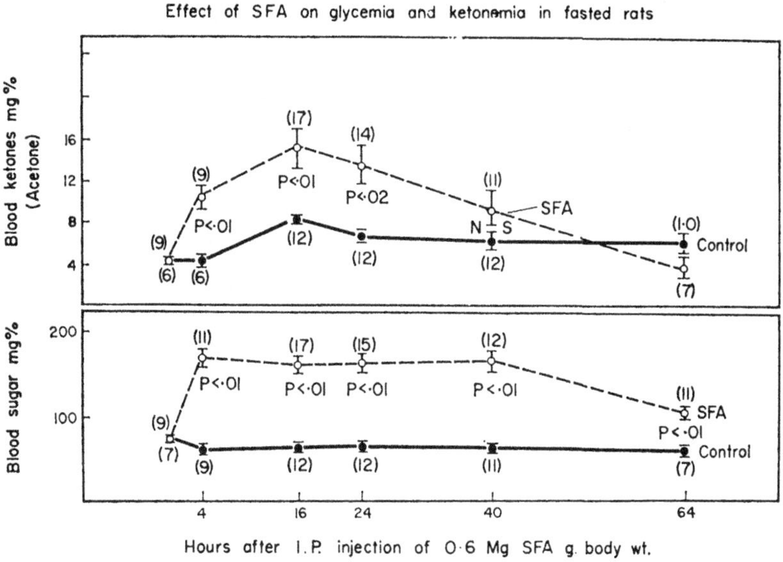

FIG. 6.14. Glycaemia and ketonaemia following intraperitoneal injection of 0·6 mg sodium fluoroacetate per 100 g in fasted rats. (Data of Cole, Engel and Fredericks, 1955.)

This latter rise occurs very readily in the urine upon injection of fluoro-*citrate* (Gal *et al.*[23]). If it be true, that it occurs as a result of the block of aconitase, it suggests that the enzymes of the cycle must resemble an assembly line more than some situations in which there is a pool of intermediates at each stage. This would mean that each stage in the cell would be to some extent self-limiting, and this would be in addition to the feed back effects which have been recently revealed. In connection with these ideas some experiments are relevant which were made on sheep by Annison,* Hill, Lindsay* and Peters.[3] They found that the toxic dose for sheep at Babraham is 0·4 mg/kilo. Less than this however, 0·3 mg/kilo, when given in divided doses at intervals may cause death. Recent work in Australia by Whittem (personal communication) suggests that this may be due to toxic-effects upon the muscle of the heart, the damage being sufficient with a first dose to make the second dose lethal. There is often a terminal or temporary rise in sugar and lactate in the

* Dr. Annison and Dr. Lindsay were at Babraham, A. R. C. Institute of Animal Physiology, and are now at the University of New England, Armidale, Australia.

blood (Fig. 6.15), and an increase in ketone bodies has been observed. But as the latter is not seen always, it may be due not to a block in the synthesis of acetyl S–CoA to citrate, as Dr. Lindsay suggests, but to some secondary action of a hormone. The matter could be explored further with advantage.

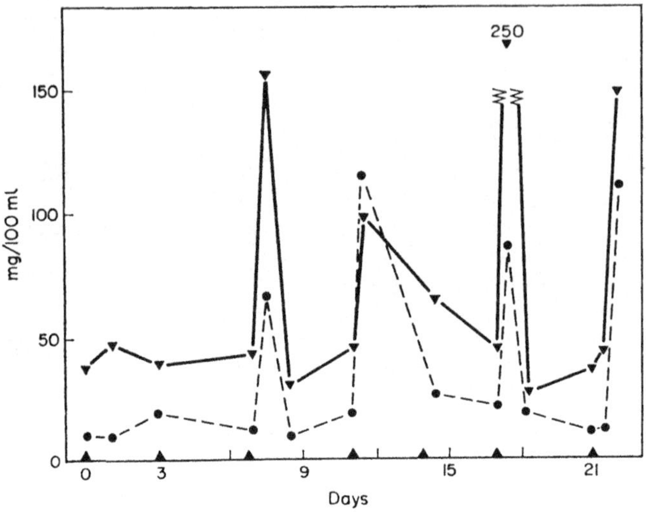

FIG. 6.15. Changes in blood glucose and blood lactate following successive administration of a non-lethal dose of sodium fluoroacetate (0·1 mg/kilo). Injection of sodium fluoroacetate ▲; Blood glucose ▼; Blood lactate ●. (Annison *et al*, 1960.)

SOME PRACTICAL CONSIDERATIONS

The carbon–fluorine compounds are potentially so toxic that it would be very desirable to have an antidote, and it is clear that an antidote must be a substance which will displace fluorocitrate from its combination with aconitase. It is true that monoacetin[14] and perhaps acetamide[26] if given within a few minutes of poisoning with fluoroacetate, will avert the toxic effect in rats; but the most acceptable interpretation of this is that these C_2 compounds stop the synthesis of the fluorocitrate.

Many compounds have been tried as antidotes by others without success, and I have also tried with colleagues* a very large number of substances which for various reasons we thought might be of value. However, I came to the conclusion that the only method of approach which could give a more certain result would be to study the effect of possible

* Particularly with my assistants Mr. R. W. Wakelin, Miss R. Shawdon and Miss P. Ivall.

compounds at various levels of organization of aconitase. Accordingly a series of compounds was tested both upon the soluble aconitase previously mentioned, and upon the aconitase in kidney particles. The plan has been to test these substances upon the reaction *iso*citrate→*cis*-aconitate, with or without fluorocitrate. As a first approximation it was thought that some base of unusual constitution might combine more readily with the fluorocitrate attached to the aconitase than its combination with the active centre. A series of bases tried all gave negative results. Other compounds were then tested, and to our surprise, it was acids which were effective, and especially among these certain substituted malonates, and malonate itself. Figure 6.16 gives an example of one experiment. These compounds were also tested with positive results upon guinea pig kidney particles.

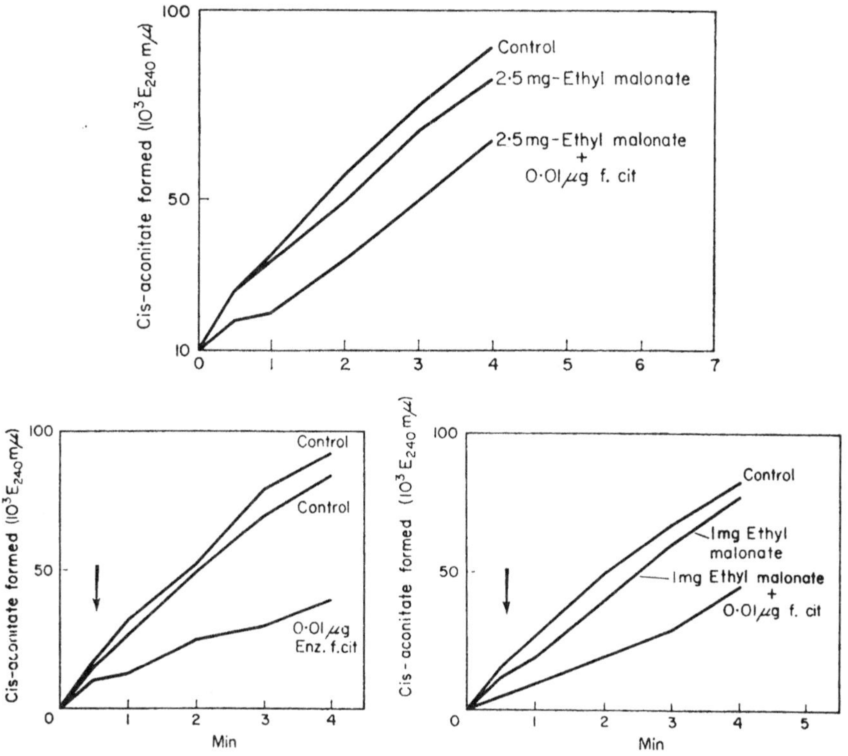

FIG. 6.16. Aconitase. Reversal of inhibitory effect of fluorocitrate by malonate (ethyl). (Not previously published.)

After pre-incubation with and without fluorocitrate *iso*-citrate is added and the reaction is studied over 4 min, *iso*-citrate → *cis*-aconitate; The calculations are made over the time 0·5–4·0 min, owing to the non-specific disturbance during the first 0·5 min. (Peters, with assistance of P. Legge.)

With these facts behind us, it was reasonable to say that the problem of finding an antidote was solved *in vitro*. Tests of these compounds and of several other combinations have been done *in vivo*, and they can be shown to prolong life as compared with untreated, poisoned animals. Unfortunately, there is something still to be learnt before we can have an effective antidote *in vivo*, because it is an invariable rule that animals treated with the malonates, get weaker and weaker and eventually die.

At the moment the only practical treatment for fluoroacetate poisoning is to inject monoacetin to stop any further synthesis of fluorocitrate, to give small doses of barbiturates to reduce convulsions, and to keep the animal alive by supportive therapy. The rationale for this is the fact that in rats a sublethal dose causes a rise in citrate which disappears in 24 to 48 hr.[38]

Other Plants

Until recently, there was no other genus known which could form this type of compound. However, one has now turned up in Australia in the Northern Territory and in part of Queensland. I was most interested to visit Mr. Whittem and his colleagues in 1961 in Alice Springs. When cattle eat the *Acacia georgina* (Gidgea tree), they may die, and there has been a loss in some districts of up to £50,000 of cattle. The tree is not always poisonous; but when toxic the leaves contain a fluoro compound behaving just like fluoroacetate[47]; it has not yet been isolated as a crystalline salt, but it was found to induce citrate accumulations in guinea-pigs.* The toxic compound in the leaves of *Acacia georgina* from Queensland has been identified as fluoroacetate by gas chromatography, (Oelrichs and McEwan[50]). The formation of the fluorine compound is still a puzzle, because the tree may or may not be toxic, and there is nothing to distinguish the toxic plant from the non-toxic one, so far as the botanist is concerned. One thing is clear, which on general grounds might be suspected, there exist other genera of plants which are capable of carrying, out the same synthesis of the FC bond.

Specific Effects of Fluorocitrate *in vivo*

In view of the need for liver tissue to activate the deoxyfluoroglycerol we thought it would be interesting to see whether a hepatoma could be inhibited in its growth by 1 or 2-deoxy 1 or 2-fluoroglycerol. Accordingly

* This was a striking example of the effect of modern air transport in improving contacts. The trichloracetic extracts of tissue made by Mr. Murray and myself in Alice Springs were sent by air via Darwin to Cambridge, where they were analysed for citrate by my assistant Mrs. Alison Davies. Since the above, L. R. Murray has isolated the K salt of fluoroacetate.

Dr. Cater and I[11] investigated the effect of this on a transplantable hepatoma in the August strain of rats. We observed no difference; but at the same time we thought it would be interesting to find out what fluorocitrate itself could do, and during routine autopsy it was noticed that the kidneys were pale and large in an animal which had died 2 hr after poisoning with fluorocitrate. Examination showed that there was very extensive damage of the proximal tubules, which somewhat resembled the effects of cadmium. The whole epithelial structure of the tubule was shed and it appeared as casts in the urine. The interest of this observation lies in the fact that, as far as we know, fluorocitrate only blocks one enzyme, aconitase. This being so, we have an instance of the fatal damage which can ensue from the blocking of one enzyme. In this case the biochemical lesion has not been reversed. Figures L1 and L2 in the Linacre lecture (pages 175 and 176 and Fig. 16c) show the degree of kidney damage. If a non-lethal dose was first given followed after some days by a lethal dose, extensive fatty degeneration may be found. The histological picture resembles that found in the kidneys of patients dying with lipoid nephrosis. The continued block causes accumulation of lipid. (Figs. 6.16 b & c.)

Summary

The biochemical lesion produced by these compounds of carbon and fluorine is of a special type. The compounds are not toxic themselves; they become toxic because the tissues convert them by enzymic synthesis, lethal synthesis, to another compound which has a blocking effect upon one enzyme; this enzyme is that known as aconitase which blocks the metabolism of the two carbon stage arising in the course of fat and carbohydrate metabolism at the citrate stage of the citric acid cycle. In some cases two stages are needed to produce the toxic fluorocitrate, a stage of degradation by the liver to a FC_2 compound followed by the conversion of this by lethal synthesis into the fluorocitrate which blocks the enzyme. The exploration of this biochemical lesion has thrown light upon various natural poisons, and has provided biochemists with a new tool for exploring intermediary metabolism, because so far as is known fluorocitrate is specific in its blocking action on aconitase.

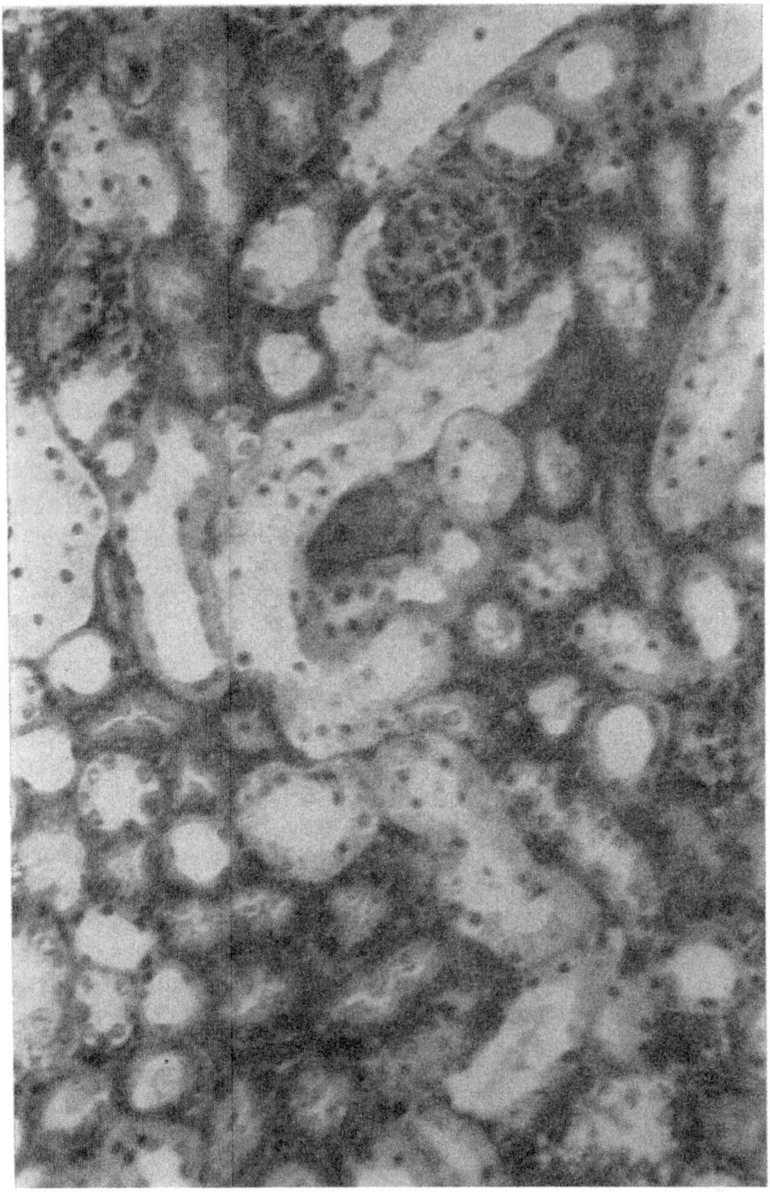

FIG. 6.16(a). Kidney of non-tumour-bearing rat, acute death 2 hr after fluorocitrate (synthetic) 80 mg/kg ip. Distension of capsular space and dilatation with necrosis of first part of the proximal convoluted tubule. The middle portions of the proximal convoluted tubules show swollen cells and blocked lumens. Picro-Mallory stain.

FIG. 6.16(a). (Cater & Peters 1961).

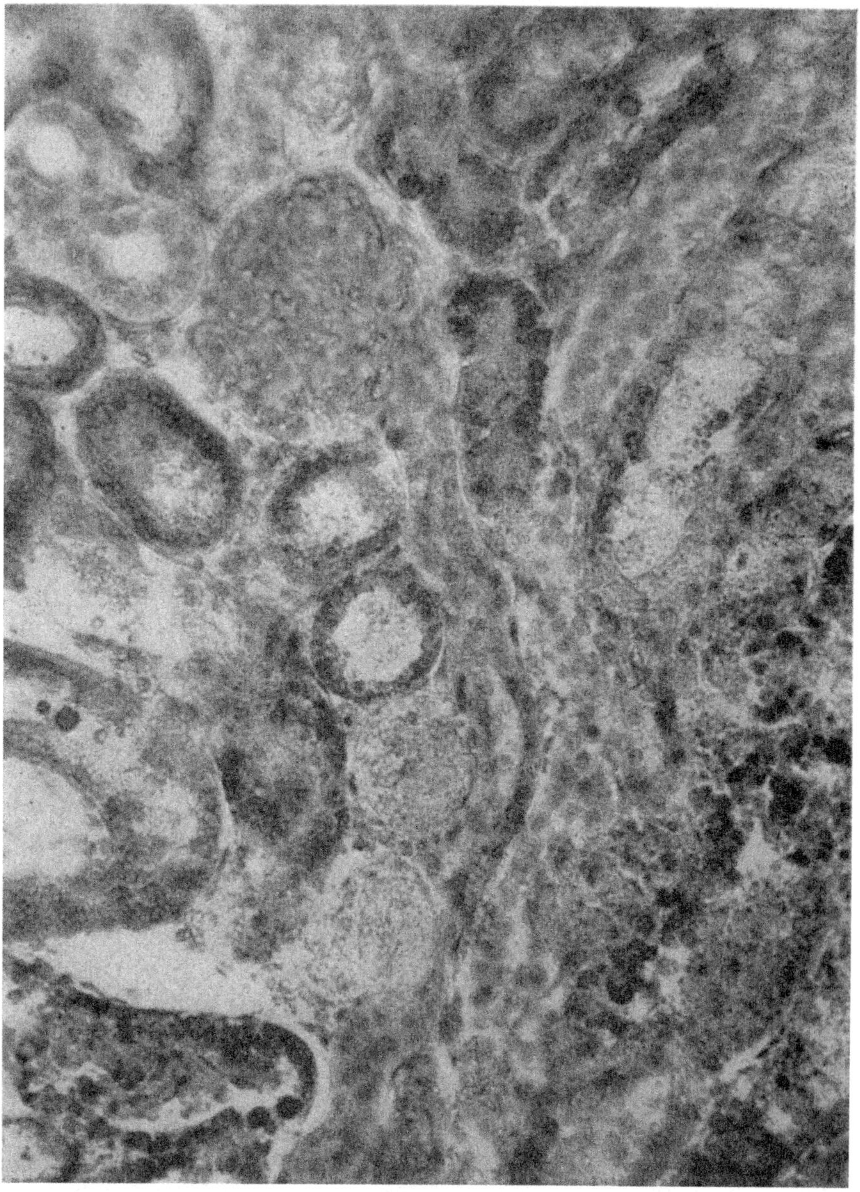

FIG. 6.16(b)

FIG. 6.16(b). Kidney of non-tumour-bearing rat, which was killed 48 hr after second dose of fluorocitrate; 40 mg/kg given day O; 15 mg/kg day 4; killed day 6. Note many lipid droplets in basal part of tubular cells. Some of the tubules are dilated and the cells are degenerate. Frozen section, stained with Scharlach R.

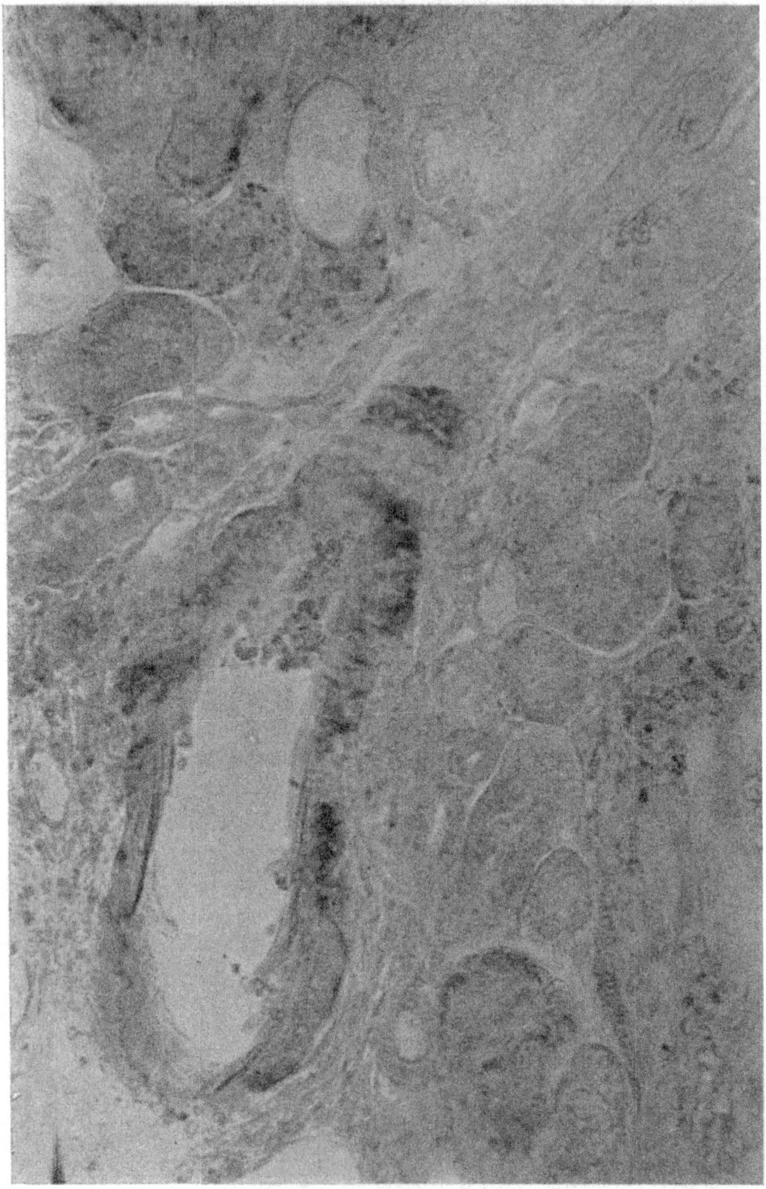

FIG. 6.16(c). Arcuate artery and interlobular artery of non-tumour-bearing rat which died 3½ hr after a second dose of fluorocitrate (total 60 mg/kg). Note heavy lipid deposits in muscular wall of artery and its branch. Some lipid is present in walls of tubules but some tubules are full of a granular deposit. Frozen section, stained with Scharlach R.

NOTES TO CHAPTER 6

Some Further Considerations

The facts given so far form an outline of work in this field. There are several other points, which, though not of much significance for the general reader seem worth including for those who may be specially interested. These facts are summarized under the following headings:

(1) The stability of mono-fluoro carbon compounds to alkali.
(2) The preparation of fluorooleic acid.
(3) The behaviour of aconitase with fluorocitrate.
(4) Reversibility.
(5) Influence of various factors on aconitase.
(6) Purification of fluorocitric acid.
(7) The reactions involved in building up fluoroacetyl–S–CoA.
(8) The effect of fluoroacetic acid upon the metabolism of yeast.
(9) Is fluorocitrate toxic to leaves?
(10) Radioprotection and miscellaneous.

1. *The Stability of Mono-fluoro Compounds (FC) to Alkali*

When the FC linkage is in the ω position in a long chain fatty acid, it can only be broken by heating with sodium in a bomb to 500°C. It was thought that it was stable also to alkali in fluoroacetate, but now it is known that an exposure to 30 per cent alkali at 100°C will split off the F as fluoride; our experience in this respect coincides with that of Saunders.[70] Fluorocitrate is also unstable in aqueous alkali.

When fluoropyruvic acid was first obtained by Mager and Blank, it was found to block the pyruvate oxidase system, without however giving an accumulation of citrate.[40] It was then found to combine with the —SH group in cysteine with the elimination of fluoride ion.[64]* A ring compound may also be formed. The pyruvate oxidase system is blocked; but in this case there must also be other blocks before pyruvate is reached because there is no accumulation of pyruvate upon poisoning animals with fluoropyruvate. It is interesting that the vicinal keto group makes such a profound change in the reactivity.[23] [1]

Fluoropyruvic acid ($FCH_2 \cdot CO \cdot COOH$)

2. *Fluorooleic Acid*

The preparation finally worked out followed the lines for seperation of an unsaturated fatty acid. It was extracted from the finely ground seeds with ether, and the ether extract treated for 1 min with methanolic potash. After this we took the fraction soluble in acetone at −19°C. It was then subjected to reversed phase chromatography in our first experiments. Latterly it was separated in a large-scale rectangular chromatography tank, designed by Mr. R. J. Hall, using powdered paper and acetic acid—formic acid—for development, and a marker of oil red 4B which ran at about the same speed as the F compound. A paper chromatographic method modified from the Schlenk method for fatty acids was

* The estimation of F^- was made by Hall's method.[28] He now has an improved micromethod for estimating F^- (in press).

also useful for determining the purity of small fractions. After separation of the paper from the tank, the oily fraction of crude F oleic acid was concentrated and then purified with a calcium precipitation.

Analysis of the fluorine content and titration indicated that the acid had 18 C atoms and one F atom. The position of the F atom was fixed by Nuclear Magnetic Resonance to be in the ω-position, by Dr. N. Sheppard. Infrared spectroscopy showed that the compound resembled oleic acid rather then elaidic acid, suggesting that it has a cis configuration. By splitting it with ozone, two fragments were obtained, which were identified by gas chromatography as fluorononanoic acid and azelaic acid. In this identification extensive use was made of synthetic long

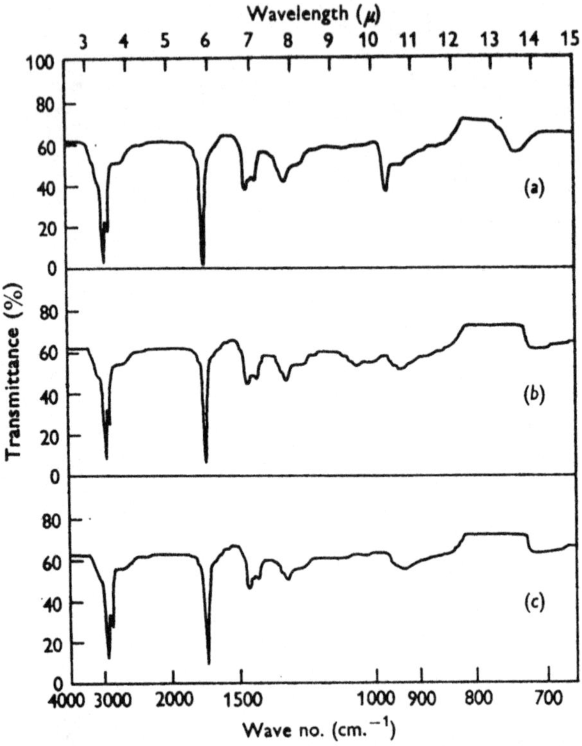

FIG. 6.18 (a). Infrared spectra of (a) elaidic acid, (b) ω-fluorooleic acid, and (c) oleic acid (5 per cent, w/v in carbon tetrachloride).

chain compounds containing FCH_2 group which were kindly supplied by Dr. F. L. M. Pattison from his extensive collection. With the above information in our hands the compound was identified as fluorooleic acid.(58) (Fig. 6.18a & b.)

The structure of the compound isolated has now been further proved by a synthesis of fluorooleic acid by Dr. Pattison and R. E. A. Dear. The method which they have used takes the following course.(53)

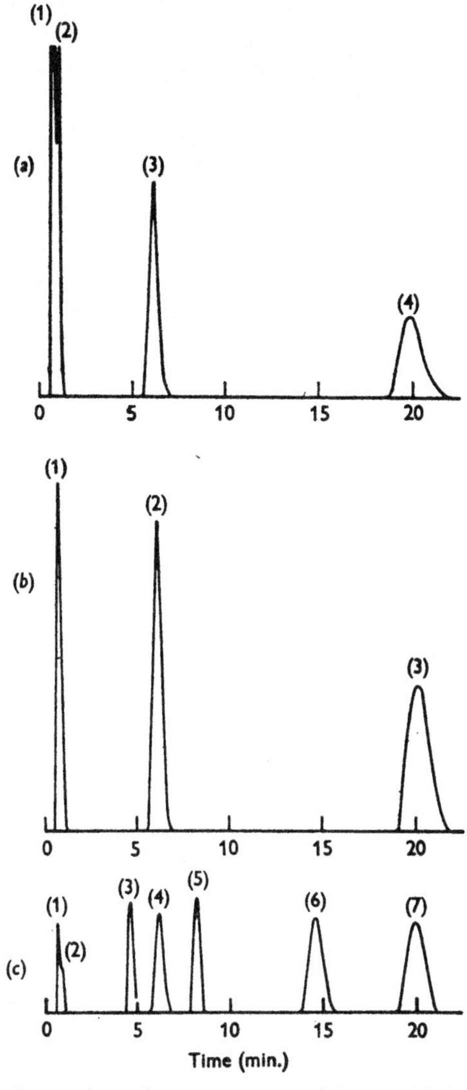

FIG. 6.18 (b). Separation of methyl esters of fatty acids by gas chromatography at 197° on a 4 ft 3 in. column with Reoplex 400 (40 per cent w/w) as stationary phase and 60–80 mesh Embacel as support. N_2 pressure: 23 cm, Hg; N_2 flow: 120 ml/min. (a) Methyl esters of products of ozonolysis of fluoro-octadecenoic acid from *D. toxicarium*. Peaks in order of appearance: (1) ether, (2) methanol, (3) and (4) methyl esters of acids produced by ozonolysis. (b) Methyl esters of products of ozonolysis of fluoro-octadecenoic acid from *D. toxicarium* mixed with methyl ω-fluorononanoate and methyl azelaate: (1) ether, (2) methyl ester of acid produced by ozonolysis + methyl ω-fluorononanoate, (3) methyl ester of acid produced by ozonolysis + methylazelate. (c) Mixture of samples of methyl esters: (1) ether, (2) methanol, (3) methyl ω-fluoro-octanoate, (4) methyl ω-fluorononanoate, (5) methyl ω-fluorodecanoate, (6) methyl suberate, (7) methyl azelaate.

8-fluoro-octyl bromide → 10-fluoro-1-decyne → 1-chloro-17-fluoro-8-heptadecyne
→ $C_{18}H_{30}FN.$ → $C_{18}H_{31}FO_2$

$$F(CH_2)_8Br + NaC \equiv CH \rightarrow F(CH_2)_8\cdot C \equiv CH + I\cdot C(H_2)_7\cdot Cl \rightarrow$$
$$F(CH_2)_8.C \equiv C\cdot(CH_2)_7Cl + NaCN \rightarrow F(CH_2)_8C \equiv C\cdot(CH_2)_7\cdot CN \rightarrow$$
$$F(CH_2)_8.C \equiv C\cdot(CH_2)_7CO\cdot NH_2 \rightarrow F(CH_2)_8\cdot C \equiv C.(CH_2)_7\cdot COOH \rightarrow$$
$$F(CH_2)_8\cdot CH = CH\cdot(CH_2)_7\cdot COOH(cis)$$

3. *The Behaviour of Aconitase (Aconitate Hydrate) and with Fluorocitrate*

At present the evidence is that the enzyme aconitase in the animal is one enzyme, which can catalyse each of the two steps involved. As shown originally by Dickman and Cloutier,[19] when the iron has been removed the enzyme must be activated with ferrous iron and a reducing compound, the most effective being cysteine. No other metal can replace the iron. When highly purified, the iron has been removed from the enzyme, and must be added for reactivation as Morrison[45] found in his highly purified preparation. When used in this state we found that aconitase in this soluble form was sensitive to (enzymically prepared) fluorocitrate, but that it required a relatively large concentration to inhibit, much larger than that required for the mitochondrial preparations from kidney. This fact was troublesome; and it weakened the last stage in our argument that aconitase was attacked specifically. More recently, owing to an accidental observation, we have been using a partly purified preparation of the enzyme, in which some 50 per cent is still present in a form containing the iron, and therefore which does not need treatment with Fe^{2+} and a reducing agent. With this preparation as mentioned,[57] much more powerful inhibitions have been obtained with enzymic fluorocitrate, so that 50 per cent inhibition has been reached with 10^{-8} to 10^{-9}M. So now the logic is that the fluorocitrate is at least as active in solution as it has been shown to be in the particle preparations.

4. *Reversibility*

It is very important whether the effect of fluorocitrate upon aconitase is reversible. Upon this must turn to some extent the question whether any antidote can be obtained. In our previous tests with the purer aconitase, we got the impression that it was reversible. This was derived from experiments with soluble

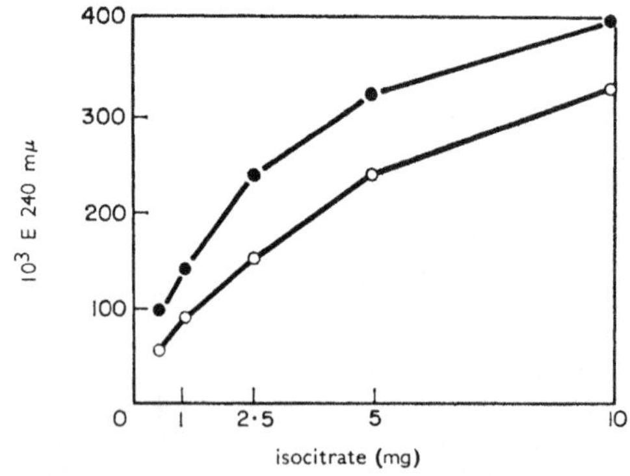

FIG. 6.19. Reversibility of fluorocitrate inhibition, initial rates of action of aconitase with increasing substrate (isocitrate) concentration. 10 min incubation plus or without fluorocitrate before addition of the substrate. The results are given for the period 0·5 to 4·0 min of action. ● No fluorocitrate. ○ Plus fluorocitrate. (Peters, 1961.)

aconitase in which we ran the reaction first for 10 min, at the end of which time we added more of the *iso*-citrate. The effect of this addition was to increase the slope of the "inhibited" enzyme so that it became parallel with that of the control. At that time we did not realize how inhibitory small amounts of *cis*-aconitate could be to the reaction *iso*-citrate→*cis*-aconitate. In more recent work we have found that even as little as 10μg of *cis*-aconitate is enough to produce an appreciable inhibition, an amount which could well be present within 4 min as a difference between the control and the inhibited preparation. If the experiments are started with different amounts of isocitrate, after the usual pre-incubation with fluorocitrate, it is found that even as much as 40 μmoles of isocitrate is not enough to give complete reversal. Hence we must now believe that it is difficult to reverse the inhibition.(57)

5. *Influence of Various Factors on Aconitase*

The influence of several factors has been studied(57) in order to see whether change of conditions would affect the use of the partly purified enzyme for determination of fluorocitrate, In this way it has been found that there is relatively little effect of a change in pH of +0·5. Likewise a change of ± 5°C does not make much difference in the inhibition. The effect of some salts is interesting. Using the "tris" buffer, the addition of both Na salts and K salts induces an increase in the rate of reaction of the aconitase present of approximately 100 per cent, but there was not an appreciable effect upon the activity of fluorocitrate. This statement in regard to K reverses an opinion expressed in a recent paper, in which there was an increase in rate of reaction with the first additions of K salts, which thereafter decreased rapidly. As this does not occur with a thrice recrystalized specimen of Kahlbaum KCl, the writer is obliged to conclude that it was due to an unknown impurity in the Analar specimen used (Davies and Peters, unpub.). The effect of some other monovalent cations is to produce a similar increase in activity.

6. *Fluorocitric Acid* (*Synthetic*)

Fluorocitric acid, which can exist in 4 isomers owing to the two optically active centres, has been synthesized by D. E. A. Rivett(68) from the esters of fluoro-oxaloacetate and bromoacetate; for the reaction see Fig. 6.20. The yields are small but the triethyl fluorocitrate is obtained in reasonably pure condition; this has, been confirmed by E. M. Gal, who used the method for the synthesis of radio-

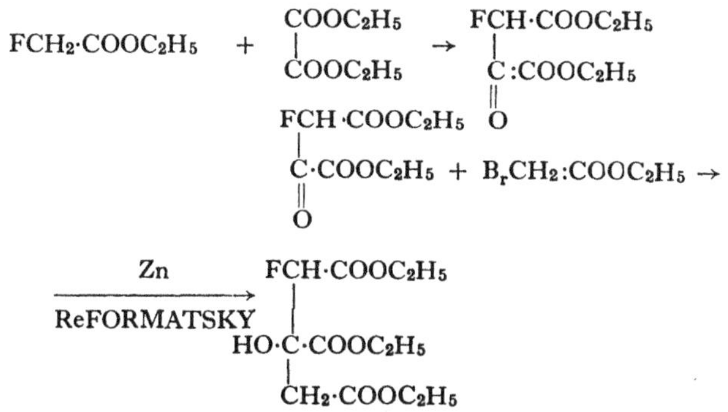

FIG. 6.20. Rivett's synthesis of fluorocitric acid (triester).

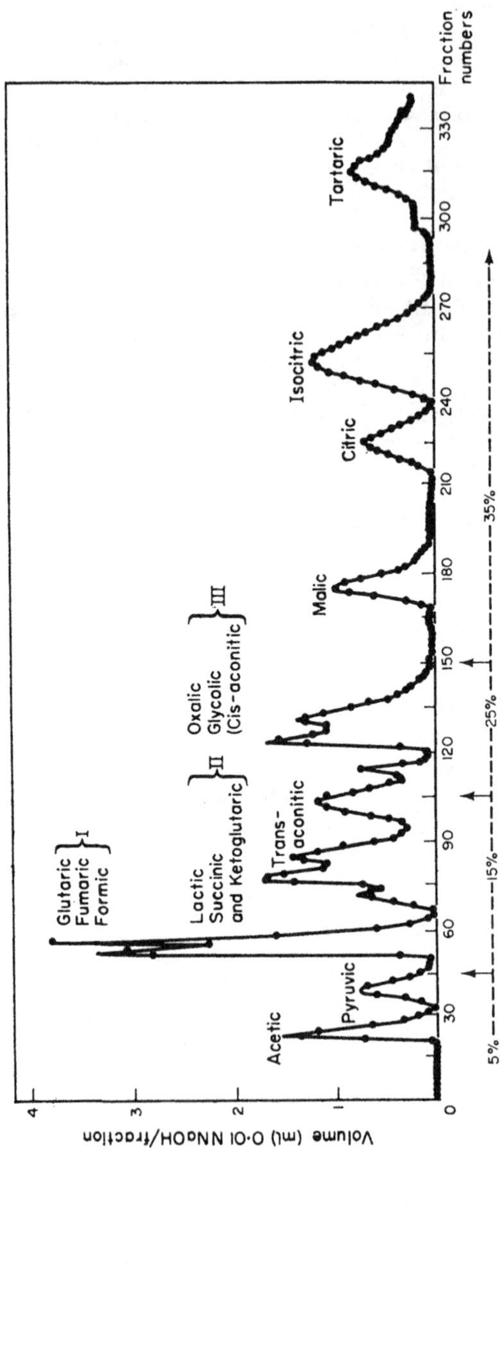

FIG. 6.21. Silicic acid-column chromatography by method of Bulen, Varner and Burrell (1952) for separation of acids. Solvent 5 per cent butan-1-ol chloroform, etc.

actively labelled fluorocitric acid. Though the ester is pure attempts to convert it to the sodium salt result in considerable change. After much labour Dr. P. F. V. Ward has found that after treatment with alkali, the silicic acid column of Bulen, Varner and Burrell[9] (1952) can be used to separate the 5–12 substances present. Rivett purified by making the Ba salt, a method which in fact eliminates some of the impurities. Some of the peaks obtained on the silicic acid column have not yet been identified; but the two final peaks appear to be those of two pairs of

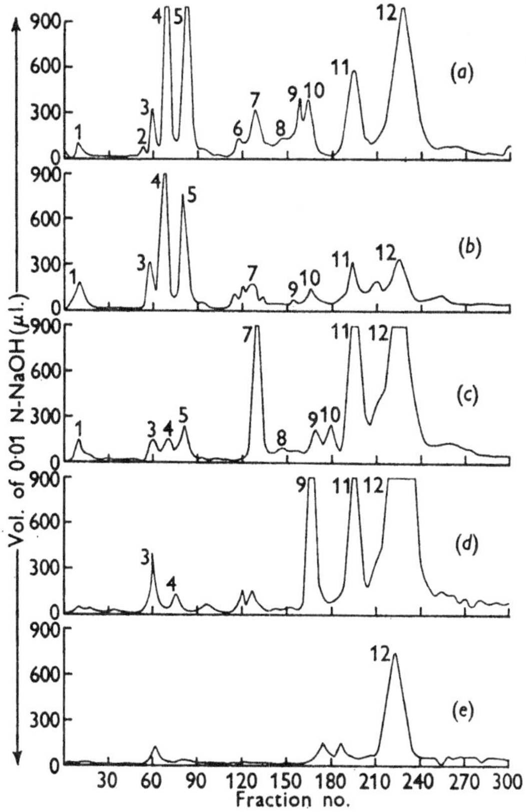

FIG. 6.22. Silicic acid-column chromatography of samples of fluorocitric acid. Sample (a): synthetic triethyl fluorocitrate hydrolysed by 0·1N NaOH at 20° for 24 hr. Sample (b): a second synthetic triethyl fluorocitrate hydrolysed by 0·1 N NaOH at 20° for 48 hr. Sample (C): a third synthetic triethyl fluorocitrate hydrolysed by N NaOH at 20° for 48 hr. Sample (d): a fourth synthetic triethyl fluorocitrate hydrolysed by a 3N HCl at 100° for 6 hr and precipitated as the barium salt. Sample (e): biosynthetic fluorocitric acid. Fraction volumes were 2·2 ml and eluting solvents were used as follows: fractions 1–48, butan-1-ol-chloroform (5 : 100, v/v); fractions 49–108, butan-1-ol-chloroform (15 : 100, v/v); fractions 109–157, butan-ol-chloroform (25 : 100, v/v); fractions 158–300, butan-1-ol-chloroform (35 : 100, v/v).

isomers, the last peak being the one which coincides with that of the enzymically synthesized acid. Further treatment of the sample with alkali results in the decrease of peaks 3 and 4 with the simultaneous increase of peaks 11 and 12; this is interpreted as due to the conversion of a monoester into the fully ionized compounds. When the alkaline treatment is prolonged, there is a further change to peaks 7 and 10, which appear to be degradation products of fluorocitrate. It seems that the only method at present available for getting pure and active fluorocitrate is to separate the peak concerned; so far no crystalline derivative has been found. Careful tests with kidney particles and with soluble aconitase have shown that peak 12 from synthetic fluorocitrate has about half of the activity of the compound prepared with enzymes.[(80)] (See Fig. 6.21 and 6.22)

The presumed monoester in peak 4 can be reactivated by standing in an appropriate volume of N/2 NaOH for 1 hr at 50°C or at room temperature overnight. This is necessary for any test upon soluble aconitase; when injected into an animal *in vivo* (rat), the alkaline treatment is not necessary, apparently the esterases of the body can remove the monoester group, whereas they are unable to deal with the triethylester.

There is one matter which has not been solved. In the original experiments with Morrison, it was found that the synthetic fluorocitrate was more toxic to the soluble aconitase. This was observed once with one of the peaks from the column which was certainly not fluorocitrate. The peak has, however, never appeared again.

7. *The Reactions involved in Building up Fluoroacetyl* CoA

The work of Brady has shown very clearly that the compound which condenses is fluoroacetyl CoA; he has also studied some reactions of this. The question then arises how the fluoroacetate becomes fluoroacetyl CoA. This has been explored by Marcus and Elliott.[(42)] By the use of glutathione in the "acetoned" pigeon liver preparation, they were able to prove that fluoroacetate would also undergo COOH activation, so that it could react with, for example *p*-nitro-aniline. They confirmed Brady's observation that fluoroacetyl–S–CoA was less stable than S–CoA. Following up this work, they made fluoroacetyl phosphate and studied some of its reactions.

Using a bacterial system, the reaction should go in the following two stages:

$$\text{Fluoroacetate} + \text{ATP} \rightarrow \text{fluoroacetyl·P} + \text{ADP} \qquad (1)$$

$$\text{Fluoroacetyl·P} + \text{CoA} \rightarrow \text{fluoroacetyl-CoA} + \text{ortho-P} \qquad (2)$$

Using a purified *E.coli* extract, there was an arsenate activation dependent on addition of CoA, which suggested that fluoroacetylphosphate was active in the transacetylase reaction. It also proved to be active in the acylation of *p*-nitroaniline. When transacetylase and CoA were in excess, F-acetyl phosphate reacted 3·2 times as rapidly as acetyl-P, whereas when amine acetylase was in excess the rates were identical. In the first case the amine acetylase is limiting, in the latter the transacetylase. It was coupled with the condensing enzyme from pig heart, adding l-malate, DPN, and malic dehydrogenase. The reaction with acetyl phosphate was in this case more rapid than that of F acetyl-P, but the product contained fluorocitrate, as shown by enzymic tests. They also studied the behaviour in the acetokinase reaction: viz.

$$\text{F-acetyl-P} + \text{ADP} \rightarrow \text{fluoroacetate} + \text{ATP} \qquad (3)$$

We found in our experiments that the fluoroacetate will not replace acetate in this reaction, and this was confirmed; they felt that the reaction was likely to be far towards ATP at equilibrium, and therefore they studied it from the other end adding ADP. By adding an ATP utilizing system from glucose and hexokinase plus *E.coli* extract and ADP, they got substantial increase in hydrolysis of F-acetyl phosphate. The results led them to try the forward reaction. With high concentrations of fluoroacetate, hydroxamic acid was formed, but when the concentration was raised beyond 0·25 M, there was a definite inhibition, not due to the presence of fluoride as an impurity. From their results they concluded that the mammalian systems must be essentially the same, but the experiments were not reported for the accepted reactions: viz.

$$\text{Fluoroacetate} + \text{ATP} + \text{CoA} \rightarrow \text{F-acetyl-CoA} + \text{AMP pyrop.} \quad (4)$$

Fluorophosphate. The above observations must be distinguished very clearly from fluorophosphate, which was shown by Ochoa *et al.*[74] to be formed from fluoride and phosphate by pyruvokinase. So far neither we nor others appear to have any evidence that this reacts with these systems.

8. *The Effect of Fluoroacetic Acid upon the Metabolism of Yeast*

A curious phenomenon has been found by Aldous and colleagues.[2] Yeast takes up fluoroacetic acid according to the pH of the medium in which it is placed, which has been proved to mean that it takes it up in proportion to the concentration of undissociated acid in the medium. The fluoroacetic acid is retained by the cells and therefore appears to be bound in some form. After the exposure the cells show a different metabolism at acid and neutral pH. Their oxygen uptake in presence of glucose is blocked at pH, 2·5. but proceeds normally at neutral pH. In a personal communication, Professor Aldous tells me that he does not think the effect on the metabolism is due to a formation of fluorocitrate. It seems that the metabolism of yeast may vary with the pH to which the yeast is exposed.

9. *Is Fluorocitrate Toxic to the Metabolism of Leaves in the Plant?*

I have been asked frequently whether we have any explanation of the curious lack of toxicity of fluoroacetate to the plant. It is known, for instance, that peas do not show injury until dosed with 50–100 times the amount of fluoroacetate required to kill *Aphis fabae* feeding upon the plants.[17] The matter has been tested on the aconitase extracted from cell suspensions of sycamore (*Acer pseudo-platinus* L.) by Treble, Lamport and Peters[77] who found that 2000-fold concentrations of fluorocitrate were needed to inhibit the aconitase from the plant as compared to that in the animal aconitase used. This means that if the results can be transferred to conditions *in vivo*, any fluorocitrate formed will produce much less inhibition in the plant. The reason for this curious difference is not known.

10. *Radioprotection and Miscellaneous*

There has been a most interesting application of fluoroacetate poisoning as a protective agent against radiation. It has been claimed[22] that the citrate formed from fluoroacetate given in sublethal doses protected against lead poisoning. Now Bacq *et al.*[4] have shown that small doses of fluoroacetate can protect mice against radiation when given 5 hr before irradiation; it is the citrate formed which does this. They suggest that the citrate might complex the Mg ions wanted for ribonuclease activity.

Another important observation has been investigated by Lynen *et al.*[35] Squalene synthesis is inhibited by fluoromevalonic acid.

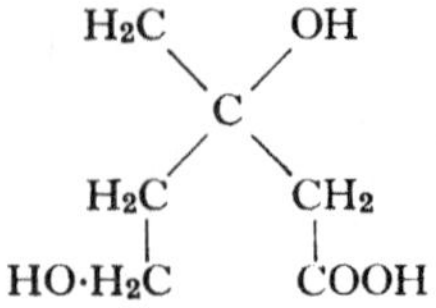

MEVALONIC ACID

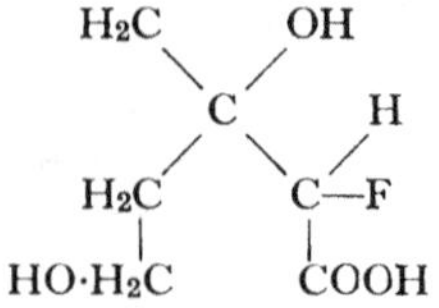

FLUOROMEVALONIC ACID

REFERENCES

1. AVI-DOR, Y., and MAGER, G. (1956), *Biochem. J.* **63,** 613.
2. ALDOUS, J. G., and ROZEE, K. R. (1956), *Biochem. J.* **62,** 605; (1958), *J. Cell. Comp. Physiol.* **52,** 45.
3. ANNISON, E. F., HILL, K. J., LINDSAY, D. B., and PETERS, R. A. (1960), *J. Comp. Path. & Ther.* **70,** 145.
4. BACQ, Z. M., LIÉBECQ-HUTTER, S. and LIÉBECQ, C. (1960), *Radiation Research* **13,** 286.
5. BRADY, R. O. (1955), *J. Biol. Chem.* **217,** 213.
6. BARTLETT, G. R. and BARRON, E. S. G. (1947), *J. Biol. Chem.* **170,** 67.
7. BUFFA, P. and PETERS, R. A. (1949), *J. Physiol.* **110,** 488.
8. BUFFA, P., PETERS, R. A. and WAKELIN, R. W. (1951), *Biochem. J.* **48,** 467.
9. BULEN, W. A., VARNER, J. E. and BURRELL, R. C. (1952), *Analyt. Chem.* **24,** 187.
10. BULLWINKLE, R., FALLON, R. H., FLEMING, T. C. and HASTINGS, A. B. (1955), *Quart. J. Exp. Physiol.* **40,** 269.
11. CATER, D. B. and PETERS, R. A. (1961), *Brit. J. Exp. Path.* **42,** 278.
12. CHARI-BITRON, A. (1961), *Biochem-Pharmacol* **6,** 169.
13. CHENOWETH, M. B. (1949), *Pharmacol. Rev.* **1,** 383 (Review).
14. CHENOWETH, M. B., KANDEL, A., JOHNSON, L. B. and BENNETT, D. R. (1951), *J. Pharmacol & Exper. Ther.* **102,** 31.
15. COLE, B. T., ENGEL, F. L. and FREDERICKS, J. (1955), *Endocrinology* **56,** 675.
16. DALY, I. DE BURGH (1957), *Proc. Roy. Soc. B.* **146,** 202 (Lecture).
17. DAVID, W. A. L. and GARDINER, B. O. C. (1951), *Ann. Appl. Biol.* **38,** 91.
18. DAWSON, R. M. C. and PETERS, R. A. (1955), *Biochem. Biophys. Acta* **16,** 254.
19. DICKMAN, S. R. and CLOUTLER, A. A. (1951), *J. Biol. Chem.* **188,** 379.
20. ELLIOTT, W. B. and KALNITSKY, G. (1950), *J. Biol. Chem.* **186,** 487.
21. ENGEL, F. L., HEWSON, K. and COLE, B. T. (1954), *Amer. J. Physiol.* **179,** 325.
22. FRIED, J. E., ROSENTHAL, M. W. and SCHUBERT, J. (1956), *Proc. Soc. Biol. Med.* **92,** 33.
23. GAL, E. M., PETERS, R. A. and WAKELIN, R. W. (1956), *Biochem. J.* **64,** 161.
24. GAL, E. M., FAIRHURST, A. S., SMITH, R. E. and PETERS, R. A. (1956), *Biochem. Biophys. Acta* **20,** 583.
25. GAL, E. M., DREWIS, P. A. and TAYLOR, N. F. (1961), *Arch. Biochem. Biophys.* **93,** 1.
26. GITTER, S. (1956), *Biochem. J.* **63,** 182.
27. GOTTO, A. M. and KORNBERG, H. L. (1961), *Biochem. J.* **79,** 9P.
28. HALL, R. J. (1957), *Analyst* **82,** 979, 663.
29. HALL, R. J. and WARD, P. F. V. (Personal communication).

30. HASTINGS, A. B., PETERS, R. A. and WAKELIN, R. W. (1953), *J. Physiol.* **120,** 51P.
31. HEINRICH, M. R., DEWEY, V. C., PARKS, R. E. and KIDDER, G. W. (1952), *J. Biol. Chem.* **197,** 199.
32. KALNITSKY, G. and BARRON, E. S. G. (1947), *J. Biol. Chem.* **170,** 83.
33. KALNITSKY, G. (1948), *Arch. Biochem.* **17,** 403.
34. KIDDER, G. W. and DEWEY, V. C. (1949), *J. Biol. Chem.* **179,** 181.
35. KIRSCHNER, K., HENNING, U. and LYNEN, F. (1961), *Liebigs. Ann. Chem.* **644,** 48.
36. KNOOP, F. and MARTIUS, C. (1937), *Z. Physiol. C.* **246,** 1.
37. LIÉBECQ, C. and PETERS, R. A. (1949), *Biochem. Biophys. Acta* **3,** 215.
38. LINDENBAUM, W., WHITE, M. R. and SCHUBERT, J. (1951), *J. Biol. Chem.* **190,** 585.
39. LOTSPEICH, W. D., PETERS, R. A. and WILSON, T. H. (1952), *Biochem. J.* **51,** 20.
40. MAGER, G. and BLANK, I. (1954), *Nature, Lond.* **173,** 126.
41. MARAIS, J. S. C. (1944), *Onderstepoort, J. Vet. Sci. Animal Ind.* **20,** 67.
42. MARCUS, A. and ELLIOTT, W. B. (1958), *J. Amer. Chem. Soc.* **80,** 4287 and (1959), *J. Biol. Chem.* **234,** 1011.
43. MARSHALL, E. K. J., CUTTING, W. C. and EMERSON, K. J. (1938), *J. Amer. Med. Assoc.* **110,** 252.
44. MARTIUS, C. (1949), *Liebigs Ann.* **561,** 227.
45. MORRISON, J. F. (1954), *Biochem. J.* **65,** 99.
46. MORRISON, J. F. and PETERS, R. A. (1954), *Biochem. J.* **56,** xxxvi and **58,** 473.
47. MURRAY, L. R., MCCONNELL, J. D. and WHITTEM, J. H. (1961), *Austr. J. Science* **24,** 41.
48. O'BRIEN, R. D. and PETERS, R. A. (1958), *Biochem. Pharmacol.* **1,** 1.
49. O'BRIEN, R. D. and PETERS, R. A. (1958), *Biochem. J.* **79,** 188.
50. OELRICHS, P. B. and MCEWAN, T. (1961, *Nature, Lond.* **190,** 808.
51. OGSTON, A. G. (1948), *Nature, Lond.* **162,** 963.
52. PATTISON, F. L. M. (1959), *Toxic Aliphatic Fluorine Compounds Elsevier.*
53. PATTISON, F. L. M. and DEAR, R. E. A. (1961), *Nature, Lond.* **192,** 1284.
54. PETERS, R. A. and HALL, R. J. (1960), *Nature, Lond.* **187,** 573.
55. PETERS, R. A. and HALL, R. J. (1960), *J. Sci. Food & Agriculture* pp. 608–612.
56. PETERS, R. A. (1948),*Proc. Roy. Soc. Med.* **41,** 781; (1955), *Disc. Faraday Soc.* **20,** 189.
57. PETERS, R. A. (1961), *Biochem. J.* **79,** 261.
58. PETERS, R. A., HALL, R. J., WARD, P. F. V. and SHEPPARD, N. (1960), *Biochem. J.* **77,** 17.
59. PETERS, R. A., WAKELIN, R. W. and HASTINGS, A. B. (1953), *Proc. xix Inter. Congress of Physiology*, p. 674. Also *Quart. J. Exp. Physiol.* **40,** 258.
60. PETERS, R. A. ,WAKELIN, R. W., BIRKS, F. T., MARTIN, A. J. P. and WEBB, J. (1954), *Biochem. J.* **58,** xl and (1959), *Biochem. J.* **71,** 245.
61. PETERS, R. A. and WAKELIN, R. W. (1957), *Biochem. J.* **67,** 280.
62. PETERS, R. A. (1952), *Proc. Roy. Soc. B.* **139,** 143. Croonian lecture.
63. PETERS, R. A. (1957), *Advances in Enzymology* **18,** 113–159; (1958), *Biochem. Pharmacol.* **1,** 354.
64. PETERS, R. A. and HALL, R. J. (1957), *Biochem. Biophys. Atca* **26,** 433.
65. PHILLIPS, A. H. and LANGDON, R. G. (1955), *Arch. Biochem. Biophys.* **58,** 247.
66. POTTER, V. R. and BUSCH, H. (1950), *Cancer Res.* **10,** 353.
67. RENNER, W. (1904), *Brit. Med. J.* **1,** 1314.

68. RIVETT, D. E. A. (1953), *J. Chem. Soc.* **4,** 3710.
69. SALLE, J. (1952), *Arch. Intern. Pharmacodyn.* **91,** 339.
70. SAUNDERS, B. C. (1947), *Nature, Lond.* **160,** 179. Also (1957), *Phosphorus & Fluorine*, p. 212, Camb. Univ. Press.
71. SCHRADER, G. (1946), Brit. Intelligence Objectives Sub-committee Report No. 1095, H.M. Stationery Office, London.
72. SWARTS, F. (1896), *Bull. Acad. Roy. Belg.* III^{e}s XXXI, 675.
73. TAYLOR, N. F. and KENT, P. W. (1954), *Nature, Lond.* **174,** 401.
74. TIETZ, A. and OCHOA, S. (1958), *Arch. Biochem. Biophys.* **78,** 477.
75. TREBLE, D. H. and PETERS, R. A. (1961), *Biochem. J.* **80,** 15P.
76. TREBLE, D. H. (1962), *Biochem. J.* **82,** 129.
77. TREBLE, D. H., LAMPORT, D. T. A. and PETERS, R. A. (1962), *Biochem. J.* (in press).
78. UMBREIT, W. W. and WADDELL, J. G. (1949), *Proc. Soc. Exp. Biol. N.Y.* **70,** 293.
79. VOEGTLIN, C. and SMITH, H. W. (1920), *J. Pharmacol.* **15,** 475.
80. WARD, P. F. V. and PETERS, R. A. (1961), *Biochem. J.* **78,** 661.

CHAPTER 7

Other Work upon Lethal Synthesis, and Lethal Incorporation

ABOUT the same time that we were proposing the possibility of a synthetic action to explain the toxicity of fluoroacetate, others were thinking upon the same lines. For instance, Work and Work(33) (1948) guessed such a possibility in regard to the synthesis of protein, a possibility proved by Levine and Tarver(16) (1951). Ethionine, an analogue of methionine studied originally by du Vigneaud, was found to inhibit the growth of rats in a competitive manner, the inhibition being relieved by methionine. In detail they used ethionine labelled with C^{14} in the ethyl group. A large part of the dose was excreted in the urine during the first day, the ethionine reducing the rate of growth by 50 per cent. Nevertheless, there was an uptake of ethionine into the proteins after making due allowance for any ethionine adsorbed; the liver and kidneys were the chief tissues studied.

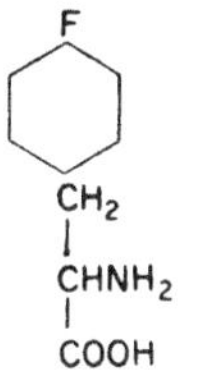

p-fluorophenyl-alanine

$CH_3{\cdot}S{\cdot}CH_2{\cdot}CH_2{\cdot}CHNH_2{\cdot}COOH$

methionine

$C_2H_5{\cdot}S{\cdot}CH_2{\cdot}CH_2{\cdot}CHNH_2{\cdot}COOH$

ethionine

In the protein field also it was proved(22) that *p*-fluorophenylalanine was incorporated into the proteins of *E.coli*, and that this interfered with growth and was in fact lethal to the organisms. Synthesis into the proteins is not, however, always deleterious(30); for instance, Richmond(24) (1962), working with *E.coli W* organisms found that addition of *p*-fluorophenylalanine had only a slight inhibitory effect on the rate of synthesis of alkaline phosphatase, though the fluoro compound largely replaced the phenylalanine in the enzyme, the specific activity of which was the same as that of the normal enzyme. Kerridge(14) found that bacterial flagellae containing the compound were not motile.

It is evident that wide variation takes place in what may happen with these analogues, which raises a question of definition, and was discussed in a recent review (Matthews[18]). In this the author prefers the wider term, "Biosynthetic incorporation of metabolite analogues". Arguments about terms are apt to be sterile, but there are certain events to be distinguished here. There may be synthesis of a foreign molecule into a non-toxic substance, which has usually been called "protective synthesis". Or there may be synthesis of a substance to form a lethal product, as in the case of fluoroacetate; this is lethal synthesis. Synthesis may also take place with the incorporation of a foreign molecule into proteins or (say) nucleic acids, which may produce a product which is toxic or non-toxic; in the latter case there is simple incorporation. Inactive proteins may be made which dilute the enzymes of the growing cell. When toxic we have a lethal synthesis accompanied by a lethal incorporation. In some cases it has been proved that some of the analogues may act also as enzyme inhibitors.

Very extensive and interesting work has been done upon the nitrogen and fluorine analogues of purines and pyrimidines, which is of theoretical interest in the nucleic acid field and also of practical importance. Though this is not the place for an extensive review, the reader should have some account in order to appreciate the potentialities of this approach (for review, see refs., 7, 17).

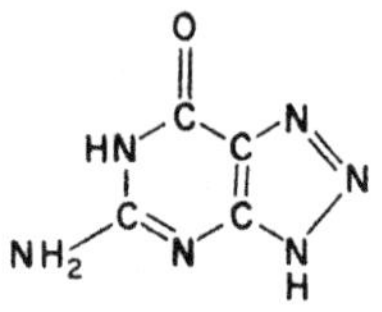

8 - azaguanine

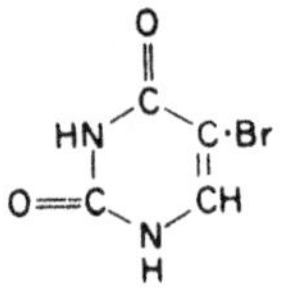

5-bromouracil

At the outset it should be said that these compounds are built into nucleic acids. 8-azaguanine[19] enters the RNA of tobacco virus and also that of *B.cereus*, and bromouracil[31] the RNA of *S.faecalis*.[20]

Again 5-halogenated pyrimidines were built into the other nucleic acid, DNA, both that of *E.coli* and of its bacteriophages. 5-bromouracil was shown by Weygand *et al.* (1952) to be incorporated into the nucleic acid of a *Streptococcus faecalis* stopping its growth. In their systematic study, Dunn and Smith[4] showed that there was a replacement of thymine in strains of *E.coli* by 5-bromouracil, 5-iodouracil and 5-chlorouracil, again with an inhibition of growth. All three compounds were incorporated into the DNA of the inhibited bacteria. Thymine annulled the inhibition due to bromouracil and iodouracil; but both thymine and uracil were required to annul the inhibition of *E.coli*.15T, an organism requiring addition of thymine. Curiously they did not find the iodo- and bromo-compounds in the RNA, only the chlorouracil (see also Bessman *et al.*[(2)]).

The main facts for 5-bromouracil have been confirmed recently for a thymine requiring mutant of *E.coli*, and new facts added.[(21)] The analogue is incorporated immediately into DNA with a slowed growth of the bacteria, blocking the uptake of thymine. There was a 20–30 per cent increase in the incorporation of some amino acids.

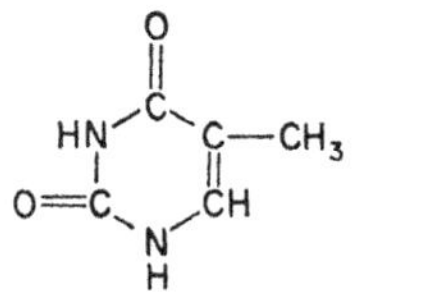

thymine 3 (5)-methyluracil

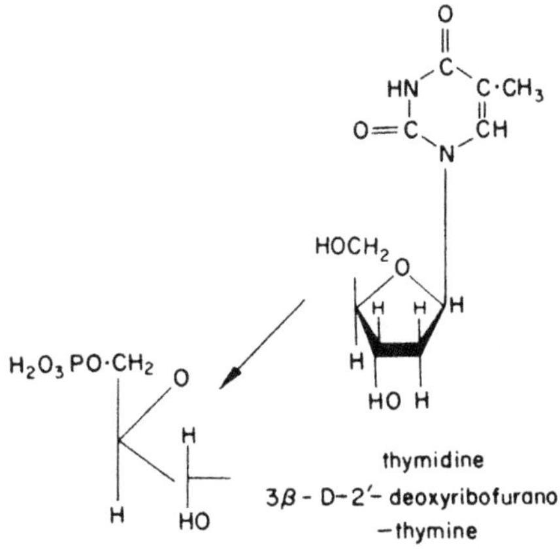

thymidine
3β - D-2'- deoxyribofuranosido
—thymine

thymidylic acid

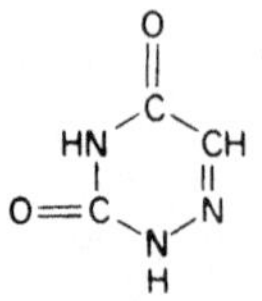

4 (6)-azauracil

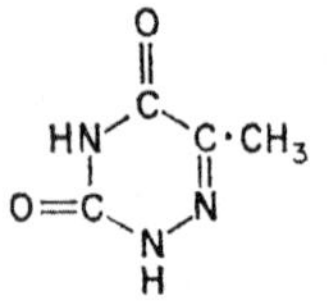

4 (6)-azathymine

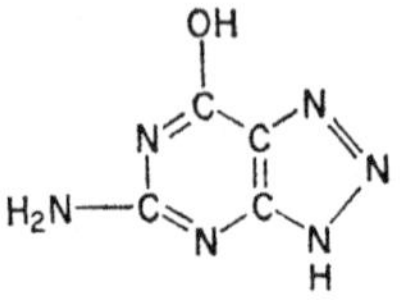

guanine

The main effect of these pyrimidines seems to be due to the lethal synthesis of "fraudulent" nucleotides; but it is not the only effect, as it has been proved that in some cases, there is an inhibition of the enzymes concerned with synthesis at some points in the synthetic pathway. One may give in illustration the researches of A. D. Welch and colleagues in Yale, who obtained good effects against tumours with 6-azauridine and 5-iododeoxyuridine. Their bacterial studies on 6-azathymine are again of much interest[5]; it is converted to the deoxyribonucleoside, azathymidine, which is then incorporated into the DNA of the cells; but it is not this incorporation which inhibits the growth of the microbe. Azathymine is metabolized by the whole mouse to several metabolic derivatives, of which the ribonucleoside has been identified.[5] Even liver slices or homogenates can convert the compound to the ribonucleoside and to the analogous 2′-deoxyribonucleoside, azathymidine. The latter, however, does not get incorporated into the DNA, probably because it has an unfavourable inhibition index, and so requires 1000 times the concentration to prevent the utilization of thymidine itself for the biosynthesis (see also refs. 13, 26).

In a study of the effects of 8-azaguanine on the specificity of protein synthesis, Chantrenne[3] found that this compound suppresses protein synthesis in *B.cereus*, but that it leaves untouched the formation of the material of the cell wall and some peptides. The inhibition of protein synthesis is reached within 10 min, and the growth of the bacteria stops within 20 min. Protection and even reversal of the toxic action of the aza compound can be induced by addition of guanosine. Even so, if the protein synthesis in general starts again, it does not mean that all proteins are synthesized; for instance, catalase is restored, but the synthesis of penicillinase is still stopped. This action only happens with those cells which incorporate the azaguanine into the RNA and nucleotides. The incorporation of the abnormal compound seems to interfere with the passage of amino acids from the transfer RNA into both soluble and ribosomal RNA. "It would seem that the presence of azaguanine in the ribonucleic acids prevents some normal interaction between the soluble and the ribosomal RNA which is important for protein synthesis." Though this point may not have yet been rigidly defined, there seems no doubt that there is an effect here of much subtlety.

Following up their earlier observations on aza compounds and Tetrahymena geleei, Kidder *et al.*[11,15,32] have studied the effect of azaguanine on various cancers, finding that no tumour of the sarcoma series was inhibited by it. They got early evidence that the mitotic index for some tumour cells was markedly reduced by azaguanine, whereas that for intestinal epithelium and for testis was not affected. This inhibitor was incorporated into the nucleoproteins of the protozoon in the place of guanine.

The study, again in the school of Dr. Welch by Pasternak and Handschumacher[23] (1959) has opened up a further interesting facet of the action of these substances. The compound in question is a much more potent inhibitor of the growth of the transplantable mouse tumour than 6-azauracil. The general effect of this is to depress the incorporation of orotic acid into the nucleic acids of the liver, intestine and tumours. But the incorporation of uridine is increased, the tumours accumulating orotidine. The metabolism of 6-azauracil *in vivo* and *in vitro* by liver and a lymphoma in mice shows only a conversion to 6-azauridine-5′-phosphate. So the conclusion is reached that the biochemical lesion in this case is due to an inhibition of orotidylic acid decarboxylase caused by the 6-azauridine-5′-phosphate synthesized from the 6-azauridine given. It is to be noted, therefore, that this is a case of a one step lethal synthesis, not involving incorporation. Additional actions have been described by others. Skoda and Sôrm[27] found that azauridine competes with uridine for the uridine kinase of *E.coli*; and Skoda[28] concluded that 6-azauridine-5′-diphosphate is a specific inhibitor of the biosynthesis of ribonucleic acids catalyzed by polynucleotide phosphorylase in *E.coli* and *B.cereus*.

Fluoropyrimidines

The pyrimidines substituted with a fluorine atom have proved to be most interesting. The problem of synthesizing these compounds was solved by Dr. C. Heidelberger, with a group from Mercks.(10) In a summary paper(9) they have reviewed the results of several years work, much of which has been confirmed by others.(8) Both fluoroorotic acid and 5-fluorouracil are inhibitory to some tumours; fluorouracil is the most

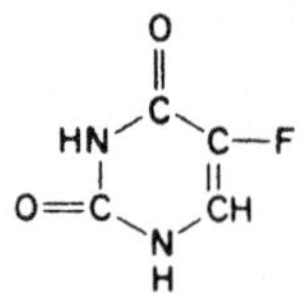

5 - fluorouracil

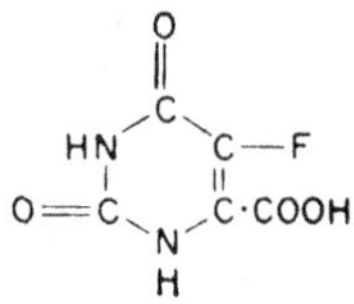

5-fluoroorotic acid

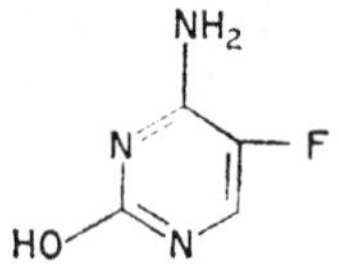

5- fluorocytosine

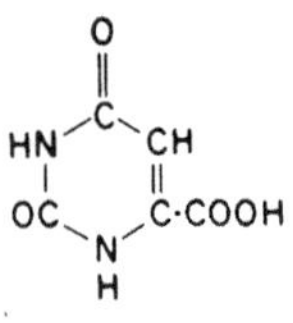

orotic acid

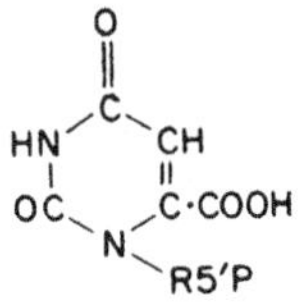

orotidine - 5'- phosphate

useful and is being tried out on patients with carcinoma. The complex metabolism includes a lethal synthesis to fluorouridine and to further compounds on the way to the nucleic acids, so that blocks may be produced at several stages of metabolism. The main effects have been summarized under four headings.

(1) The compound is converted into ribonucleotides of mono, di- and tri-phosphates and to deoxynucleoside monophosphate; and this also occurs with the nucleosides.

(2) F-uracil and F-orotic acid are incorporated into the RNA of liver, spleen and mouse tumours. Further the F-uridine and F-D uridine are also incorporated in the same way going into the RNA of *E.coli* and of tobacco mosaic with inhibition of virus formation.

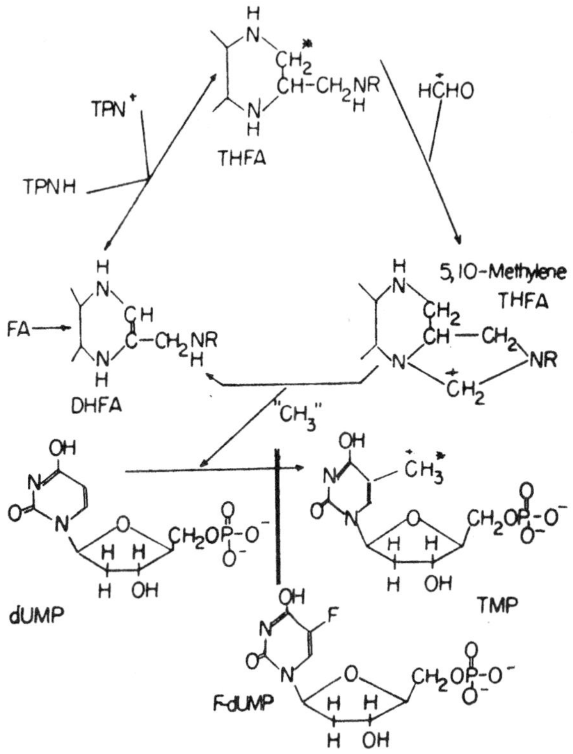

Scheme from Hartmann and Heidelberger, (*J. Biol. Chem.* **236,** 3006, 1961) showing the site of interference with synthesis of thymidylate (TMP) in Ehrlich ascites tumour cells by 5-fluoro-2'-deoxyuridine-5'- monophosphate (F-dUMP) coming from 5-fluorouracil. dUMP = deoxyuridylate FA = folic acid.

(3) The fluorinated pyrimidines, in particular F-uridine, inhibit the incorporation of labelled uracil and orotic acid into RNA, which they think is due in the main to inhibition by fluoropyrimidine of the biosynthesis of RNA at the level of the polymerization reaction.

(4) Finally Hartman and Heidelberger[(9)] place much stress upon the inhibition of the methylation of deoxyuridylic acid to thymidylic acid, which is due to the competitive inhibition of the thymidylate synthetase by 5-fluoro-2′-deoxyuridine-5′-monophosphate, a powerful inhibition occurring at concentrations as low as 10^{-8} M.

Much of this can be included in the general statement that the F compounds make fraudulent nucleotides; but the work has also made it clear that there are large variations as between one cell and another (and also between different kinds of tumours) in the emphasis upon one or the other of these actions.

Another complication has been raised lately[29]; the fluorinated pyrimidines affect feedback mechanisms also. *In vitro* using an *E.coli* sonicate, it was found that in the reaction where the cyclization of carbamyl aspartic acid to dihydroorotic acid takes place, fluoroorotic acid was 4 times more inhibitory than dihydroorotic acid.

In the case of phage (Aronson[1]), both uracil and thymine are required to reverse the effect of 5-fluorouracil, the effects being different if either is left out. When thymine is added alone, there is an accumulation of phage precursor DNA, whereas if uracil is not added there is an accumulation of phage precursor protein. F-uracil enters the smaller ribosomes.

With the same analogue and *E.coli*, Gros and Naono[6] had conditions in which there was a continued synthesis of RNA and protein, with no synthesis of DNA. Up to 50 per cent of the uracil residues in the RNA were replaced by the F-uracil. Though the proteins behaved in the same way when examined by chromatography, they were different in their incorporation of labelled proline, tyrosine and arginine; the latter was increased in the protein molecules, whereas the content of the two former amino acids was reduced, so that the protein molecules formed were different. This difference showed up in a concrete way when the properties of the enzymes were tested. Both the synthesis of β-galactosidase and of β-glucuronidase were inhibited, whereas alkaline phosphatase, serine deaminase and glucose-6-phosphate dehydrogenase continued to be made, though at a slower rate. The proteins of the alkaline phosphatase and of β-galactosidase were altered. The β-galactosidase so made had no enzymic activity, though it was antigenically related to the active enzyme by immunological methods. The alkaline phosphatase had the same catalytic activities as the normal enzyme, but it was much less stable to heat. The metabolic effects were almost immediate. One minute after the addition of the F-uracil to the *E.coli*, it was found to be exclusively in the fraction of the RNA which was distinct from the ribosomal and the s-RNA. From this it might be concluded that the incorporation of the foreign molecule into the RNA acted as a template which was metabolically unstable.

Other interesting effects of the same compound have been reported. In studying the synthesis of the mucopeptide of the cell wall of Staph.-aureus, Rogers and Perkins[25] (1960) tried several compounds such as 5-bromouracil. The effect of F-uracil was unique, as it stopped the formation of the mucopeptide from 30 to 60 per cent at a concentration as low as 0·1 μ mole/ml; at the same time there was 10 times more N-acetylhexosamine formed than in the untreated cells. These effects were stopped by the presence of uracil or uridine, so that they were connected with this phase of the action of the fluoro compound. By careful fractionation of the compounds which accumulated, they found that the fluorouracil was built into three compounds.

(1) A compound containing F-uracil as well as phosphorus, muramic acid, alanine, glutamic acid and lysine.

(2) and (3) contained fluorouridine and (2) N acetyl glucosamine (3) muramic acid as well as other residues.

Acute Leukaemia and Folic Acid Antagonists

There has been a most interesting application by Jacobson[(12)] to the problem of acute leukaemia in children. He has found that the folic acid antagonist, aminopterin, arrests cells from the marrow which are dividing.

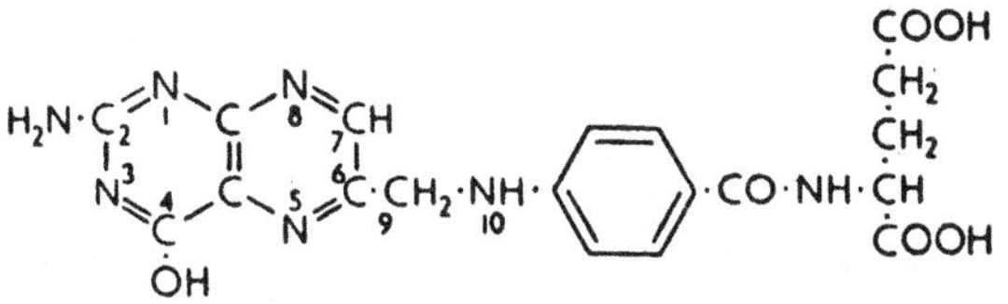

Folic acid

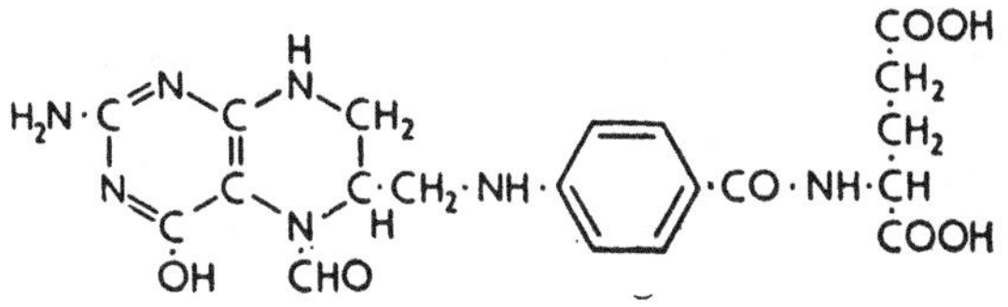

Folinic acid

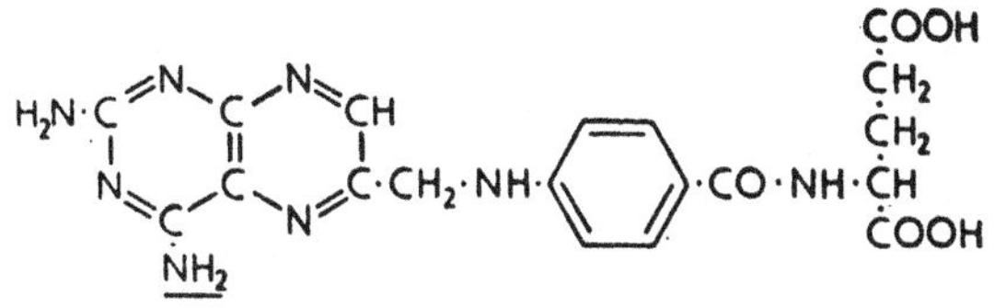

Aminopterin

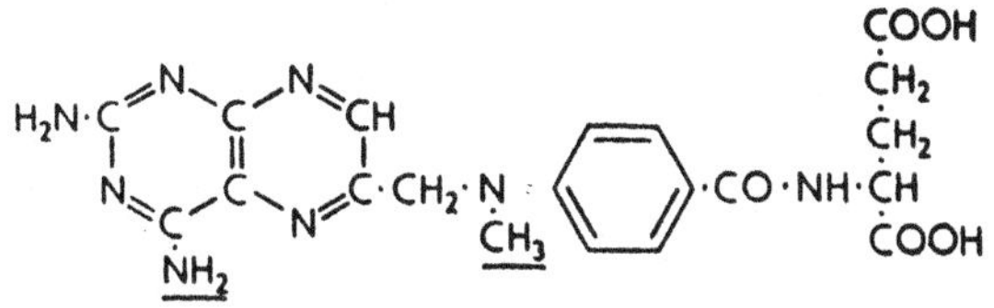

A-methopterin

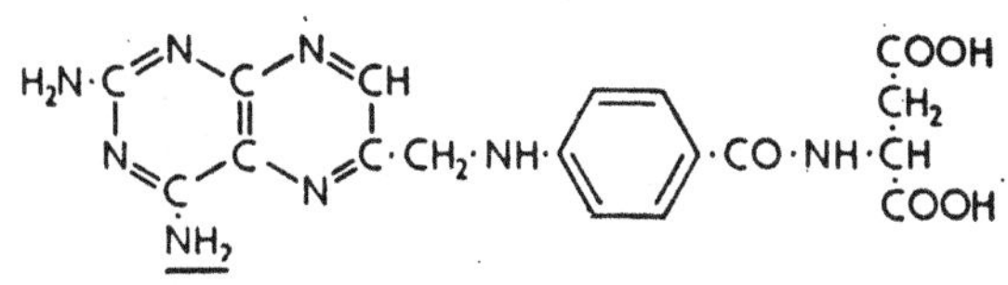

Amino-an-fol

This arrest takes place in metaphase, and does not occur when folinic acid is added, from which he concludes that folinic acid is essential for the passage of the metaphase stage into that of anaphase. Among other reactions folinic acid is necessary for the incorporation of formate into carbon atom 8 of the purine ring and of carbon atom 2; it is also wanted for the methylation of uracil in position 5 to form thymine. Folic acid cannot be built up by normal marrow cells from its precursors, but this can evidently be done by leukaemic cells, which can use 2-amino-4, 6,-dihydroxypteridine and *p*-aminobenzoylglutamate for this purpose. Dr. Jacobson therefore thought that the leukaemic cells might be induced to perform a lethal synthesis and so to build up an antagonist from suitable precursors. This happened when he used 2,4-diamino-5-hydroxypteridine and *p*-aminobenzoylglutamate, which were synthesized to an antagonist. These facts are most interesting from the scientific standpoint and very important in clinical medicine. The practical outcome of this idea is awaited with much interest.

It is interesting to note that when the malignant cells from leukaemic children are treated with aminopterin, they are able to inactivate this; as also A-methopterin. Jacobson thinks that this inactivation is due to ring closure of the terminal glutamic acid.

Normal marrow cells must be rather an exception, because he also found that some normal tissues can carry out this conversion, as for instance chicken embryo, osteoblasts and fibroblasts and the red blood cells of the fowl.

In conclusion it should be said that there are other examples of lethal synthesis and lethal incorporation to be found in the literature; some analogues are incorporated and some form an intermediate compound which may act as an antimetabolite. There seems to be no doubt that the method of approach is proving of value. It is also revealing that incorporation can take place without a lethal effect, and in proteins without entire elimination of enzymic activity.

REFERENCES

1. ARONSON, A. I. (1961), *Biochem. Biophys. Acta* **49,** 89.
2. BESSMAN, M. J., LEHMAN, I. R., ADLER, J., ZIMMERMAN, S. B., SIMON, E. S. and KORNBERG, A. (1958), *Proc. Nat. Acad. Sci.* **44,** 633.
3. CHANTRENNE, N. (1961), *Biological Structure and Function*, Ed .Goodwin and Lindberg, vol. **1,** p. 281.
4. DUNN, D. B. and SMITH, J. D. (1957), *Biochem. J.* **67,** 494.
5. GATTO, R. A. and PRUSOFF, W. H. (1962), *Biochem. Pharmacol.* **11,** 323.
6. GROS, F. and NAONO, S. (1961), *Protein Synthesis*, Ed. R. J. C. Harris, p. 195.
7. HANDSHUMACHER, R. E. and WELCH, A. D. (1960), in *The Nucleic Acids* (Eds. E. Chargaff, and J. N. Davidson,), **3,** 512. Academic Press (Review).
8. HARBERS, E., CHAUDHURI, N. K. and HEIDELBERGER, C. (1959), *J. Biol. Chem.* **234,** 1255.
9. HARTMAN, K. U. and HEIDELBERGER, C. (1961), *J. Biol. Chem.* **236,** 3006.
10. HEIDELBERGER, C., CHAUDHURI, N. K., DANNEBERG, P., MOOREN, D., GRIESBACK, L., DUSCHINSKY, R., SCHRITZER, R. J., PLEVEN, E. and SCHEINER, J. (1957), *Nature* **179,** 663.
11. HEINRICH, M. R., DEWEY, V. C., PARKS, R. E. and KIDDER, G. W. (1952), *J. Biol. Chem.* **197,** 199.
12. JACOBSON, W. (1960), *Biochem. Pharmacol.* **5,** 130 and 143; (1961), *Biological Approaches to Cancer Chemotherapy*, p. 149, Ed. R. J. C. Harris, Acad. Press; (1958), Symp. *The Evaluation of Drug Toxicity*, p. 76, Eds. A. L. Walpole and A. Spikes, J. & A. Churchill Ltd.
13. JAFFE, J. J., HANDSCHUMACHER, R. E. and WELCH, A. D. (1957), *Yale J. Biol. Med.* **30,** 168.
14. KERRIDGE, D. (1959), *Biochem. Biophys. Acta* **31,** 579.
15. KIDDER, G. W., DEWEY, V. C., PARKS, R. E. J. and WOODSIDE, A. L. (1951), *Cancer Research* **11,** 204.
16. LEVINE, M. and TARVER, H. (1951), *J. Biol. Chem.* **192,** 835; GROSS, D. and TARVER, H. (1955), *J. Biol. Chem.* **217,** 169.
17. MARKHAM, R. (1958), 8th Symp. *Chemical Microbiology*, pp. 163–177, Camb. Univ. Press, Eds. S. T. Cowan and E. Rowatt, (Review).
18. MATTHEWS, R. E. F. (1958), *Pharmacol. Rev.* **10,** 359–406. (Review).
19. MATTHEWS, R. E. F. (1953), *J. Gen. Microbiol.* **8,** 277; (1954), **10,** 521.
20. MATTHEWS, R. E. F. and SMITH, J. D. (1955), *Advance Virus Res.* **3,** 49. (Review).
21. MICHAELSON, I. A. and MANDEL, G. (1962), *Biochem. Pharmacol.* **11,** 243
22. MUNIER, R. and COHEN, G. N. (1956), *Biochem. Biophys. Acta* **21,** 592 and (1959), **31,** 347 and 379.
23. PASTERNAK, C. A. and HANDSCHUMACHER, R. E. (1959), *J. Biol. Chem.* **234,** 2992.
24. RICHMOND, M. H. (1962), *Biochem. J.* **84,** 110P. Also (1959), **73,** 261.
25. ROGERS, H. J. and PERKINS, H. R. (1961), *Biological Structure and Function*, Eds. Goodwin and Lindberg, Vol.1 p. 289. Also *Biochem. J.* **77,** 448 (1960).
26. SCHINDLER, R. and WELCH, A. D. (1957), *Science* **125,** 528.
27. SKODA, J. and ŠORM, F. (1959), *Collection Czechoslov. Chem. Communs.* **24,** 1331.
28. SKODA, J. (1961), *Protein Biosynthesis*, Ed. R. J. C. Harris, p. 185.
29. SMITH, L. H. Jr. and SULLIVAN, M. (1960), *Biochem. Biophys. Acta* **39,** 554.
30. WESTHEAD, E. W. and BOYER, P. D. (1961), *Biochem. Biophys. Acta* **54,** 145.

31. WEYGAND, F., WACKER, A. and DELLWEG, H. (1952), *Z. Naturf.* **7b,** 19.
32. WOODSIDE, G. L., KIDDER, G. W., DEWEY, V. C. and PARKS, R. E. (1953), *Cancer Research* **13,** 289–91.
33. WORK, T. S. and WORK, E. (1948), *The Basis of Chemotherapy*, Oliver and Boyd.

CHAPTER 8

The Biochemical Lesion in Burns and the Effect of Temperature

DURING the war I was asked by the Medical Research Council through Sir Alan Drury, C.B.E., F.R.S. to investigate the problem of burns. This problem is one of great complexity. Most research progress then was directed towards the alleviation of secondary complications, as I think; but these were so serious that it was absolutely necessary to find out without delay how to stop the external infection (a research undertaken by Colebrook and colleagues[7]) how to make good the plasma loss, and how to improve skin grafting (a research which had interesting developments in the hands of Medawar).

It seemed clear that our function should be to explore the biochemistry; but as is usually the case, the biochemical approach had a longer range with less immediate practical applications. One should first consider the burn and its site, then the spread of any effect from this to the general circulation, the problem of shock, and then the whole series of events concerned with the healing.

In this way I decided to work upon the primary biochemical lesion, as distinct from the secondary pathological, physiological and other changes. When we concentrated upon the initial local change, it brought us up against the problem of the burn toxin, and much of our efforts in Oxford went in this direction. Was Underhill right in thinking that the haemo-concentration was the main pathological change,[31,32] or were those who believed in the toxin theory (Robertson and Boyd[24]) to be supported? How could biochemistry help? If it was only haemo-concentration and local oedema, then we merely had to remove the haemo-concentration and to stop the local oedema. Though the latter comes on with great speed after the burn, Rossiter[25] in our group found that it could be stopped by applying pressure of 10 to 20 mm Hg, applied for about 3 hr. The question of a burn toxin has always been bedevilled by the septicaemia which appears so soon, so that progress could not be made until the local infection could be cured. The means of doing this only became available during the war with antibiotics and the detergent CTAB, cetavlon.

In addition to the absorption of toxins from bacterial infection of the burned site, there is the problem whether there is also a systemic burn toxin, a long standing problem upon which valuable work was done in

Edinburgh shortly before the Second World War[33,34] (1937). In these experiments made on anaesthetized rabbits, with full aseptic technique, the burn fluid was not found to be toxic when injected intravenously 4 hr after burning. The toxicity, however, increased gradually up to 48 hr after burning when injected. No evidence was obtained that the effects were due to the growth of microorganisms; if the oedema fluid was passed through a Seitz filter, which should remove bacteria, the fluid was toxic though less so. Toxicity was associated especially with the globulin fraction. so that the toxin appeared to be formed during autolysis of the tissue. The condition which developed resembled somewhat closely that of acute toxaemia, especially in so far as there was necrosis and fatty degeneration of the liver cells in the centres of the lobules.

Study of the Biochemistry of the Burn Site

The problem may be stated as follows (for Review see ref. 23).

The Problem

If we go back to first principles, the problem of the burn presents itself as follows. In any burned area of the skin, there must be two zones at least (Fig. 8.1). In one of these, Zone I, the skin may have reached such

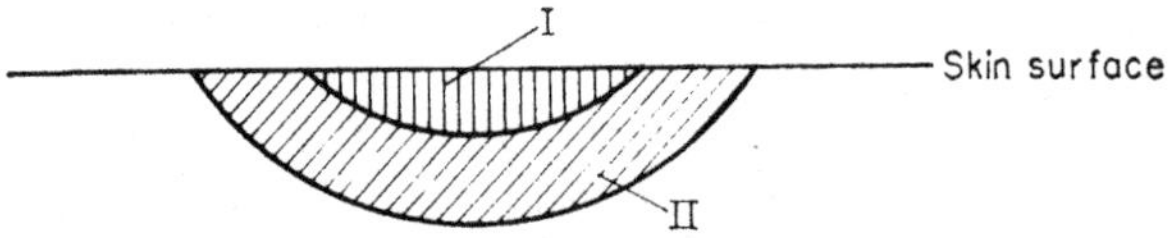

Fig. 8.1

a temperature that all biochemical reactions are stopped and most proteins coagulated; the skin is dead. On the other hand there will usually be a zone surrounding this, Zone II, where the temperature will be less; where the heat has been enough to upset some of the biochemical mechanism; but not enough to destroy most of the enzymes. Reflection soon shows that Zone II is potentially dangerous, because undamaged constituents from skin cells could enter the circulation. The effects in this zone will naturally be a combination of time and temperature; but it might well be that changes would occur in some lipo-proteins on the surface of the cell, which would permit the passage of some proteinase or other enzyme. The liberation of such factors if they existed could account for the so-called toxins.

The first point to decide was the exact temperature required to produce thermal injury. As this information was not available we set out to get it.

The apparatus shown in Fig. 8.2 was designed with J. T. Cox to determine the limiting temperatures for this damage. (The apparatus turned out to be rather similar to a glass one used by McMaster and Hudack(11) (1932) of which I was unaware at the time.) Sevitt(26) has more recently used a similar design. The Table 8.1 shows our findings.(15)

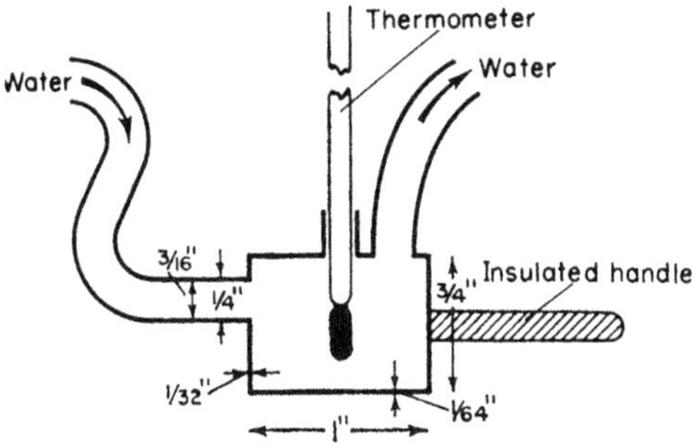

FIG. 8.2

TABLE 8.1

Shows the rate of development and subsidence of signs after application of heat to skin of guinea-pigs. E = erythema, Fl. = flare, Bl. = blanching. Œd. (Œdema) expressed as per cent water content of tissue (average related to normal 100 per cent) found 2 hr after application. Observations at each temperature made on 7 or more animals.

Heat applied for 1 *min. Temperature* 45°–80° *C.*

Temp. °C	Time after application (min)					Œdema, per cent normal, 120′	Standard deviation	Histological[3] damage, 120′
	0′	1′	5′, 10′, 15′	30′	60′			
45	Nil	Nil	Nil	Nil	Nil	103	± 27	A
50	Nil	Nil	E. +	Nil	Nil	97	± 9	A
55	Bl.	E. +	E. + + Fl. +	E. + Fl. +	E. +	119	± 28	D, E
60[1]	Bl.	Bl.	E. + + Fl. + +	E. + + Fl. + +	E. + Fl. +	180	± 62	F
65[1]	Bl.	Bl. Fl. +	Bl. Fl. + + +	Bl. Fl. + +	Bl. Fl. + +	..	..	G or H
70 and 80	Fl. + (Heat fixation)[2]	Fl. + +	Fl. + + +	Fl. + +	Fl. + +	(70) 199 (80) 202	± 18 ± 45	H

[1] At 60°–65°C for 1 min there was a separation down to the prickle cell layer, so that the epidermis could be wiped off.

[2] Heat coagulation to form a stiff yellow plaque.

[3] Three animals observed.

Leach, Peters and Rossiter, 1943.

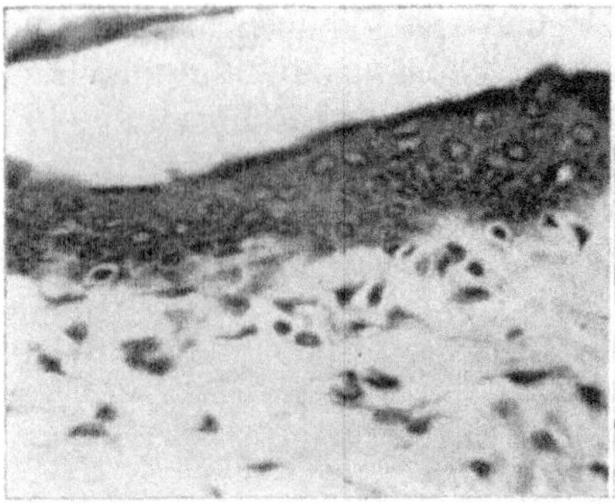

1. Normal skin of guinea pig stained with Weigerts iron haematoxylin and eosin. Magnification ×470. Same stain magnification in figs. 4–9.

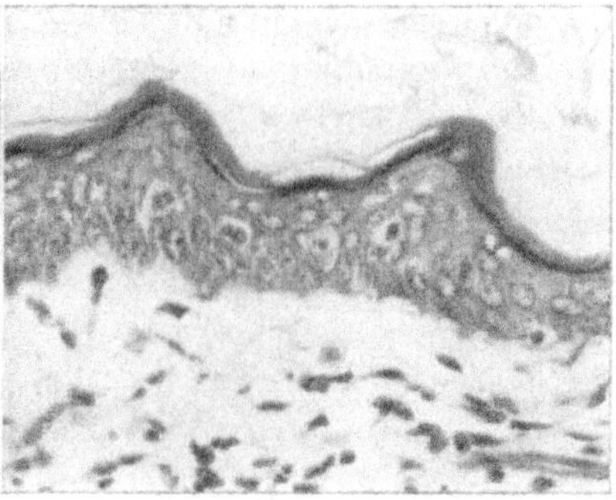

2. Stage A. 50°C for 1 min. Removed 2 hr later.

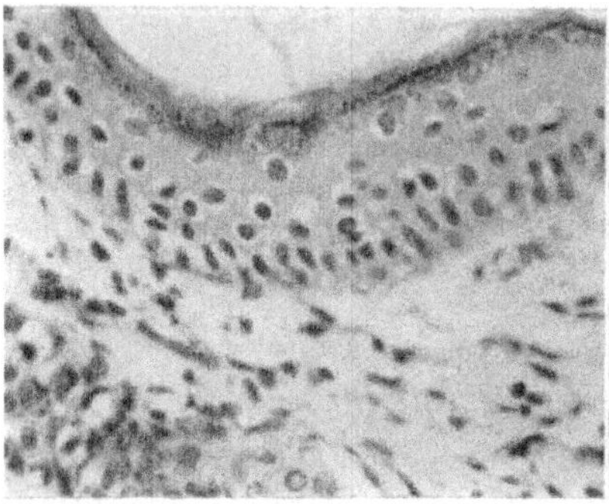

3. Stage B. 55°C for 1 min. Removed 5 min later.

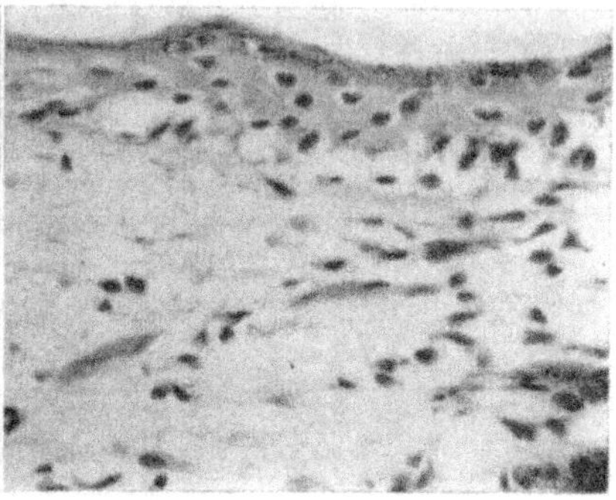

4. Stage C. 55°C for 1 min. Removed 1 min later.

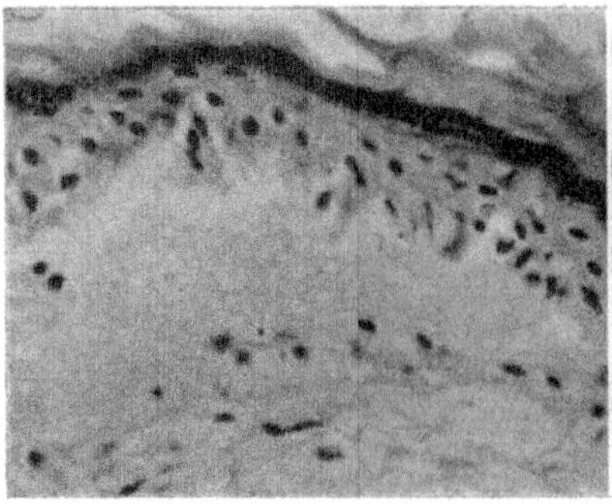

5. Stage D. 55°C for 1 min. Removed 2 hr later.

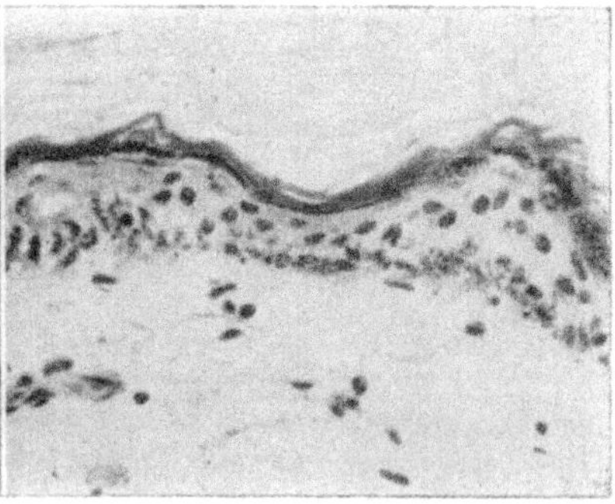

6. Stage E. 55°C for 1 min. Removed 2 hr later.

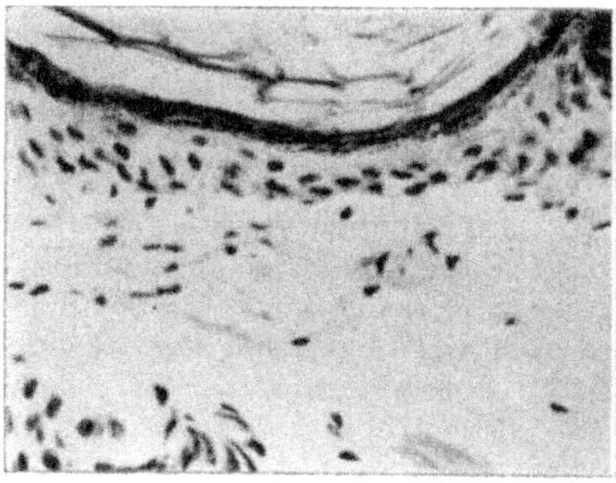

7. Stage F. 60°C for 1 min. Removed immediately.

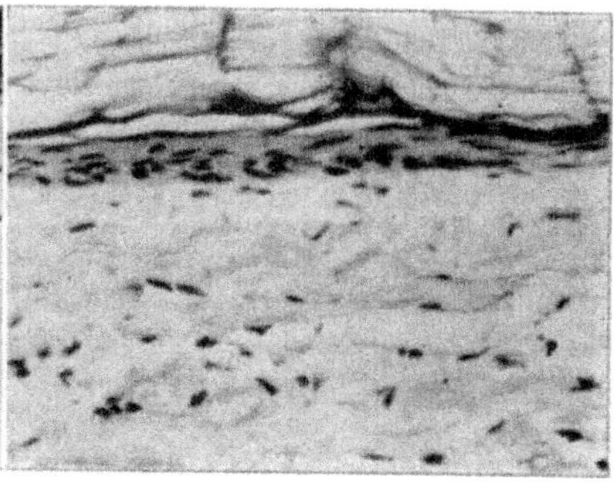

8. Stage G. 70°C for 1 min. Removed 2 hr later.

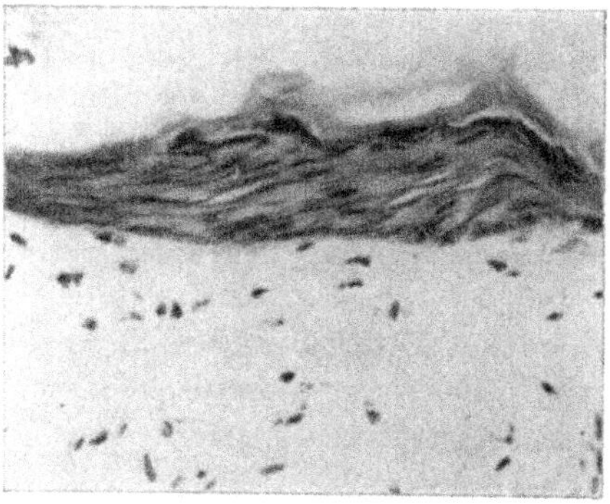

9. Stage H. 80°C for 1 min. Removed 2 hr later.

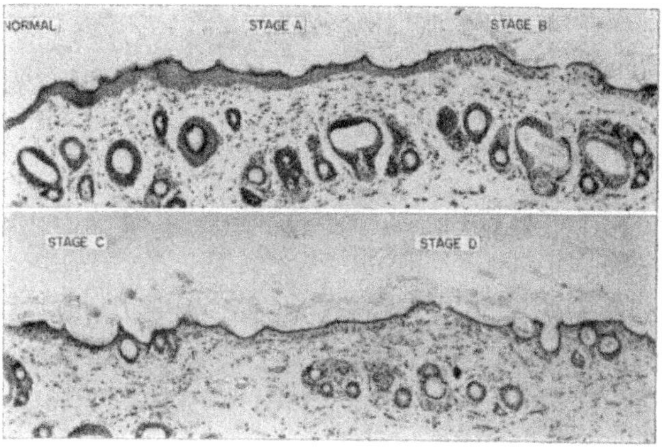

10. 55°C for 1 min. Removed 2 hr later. Showing junction of normal and burnt areas. Stained Weigert's iron haematoxylin and eosin. Magnification $\frac{1}{6}$ of other figures.

HISTOLOGICALLY OBSERVED STAGES IN BURNS—ALL 1 MIN EXPOSURE TO THE "IRON"

(As seen by E. H. Leach)

Stage A. 50°C. Reversible swelling of nuclei of the epithelium.

Stage B. 55°C. Removed after 5 min Nuclei collapsed leaving a perinuclear vacuole, but they remain round. Loss of basophil staining in the cytoplasm. In stages B and C–E the granules staining red with pyronin-methyl green can be seen between some of the prickle cells, and between them and the basement membrane.

Stage C. 55°C. Nuclei collapse further; they become crescented or flattened and small spaces appear between the prickle cells.

Stage D. 55°C. Nuclei stain densely and uniformly—pyknosis. Some large spaces now formed between the prickle cells and the basement membrane contain a coagulum and basophil granules—an irreversible stage.

In stages C and D, nucleoprotein is escaping, and DNA (Feulgen staining) is to be seen in the spaces between the prickle cells and in the incipient blister space.

Stage E. Nuclei crenated and fragmentary—karyorrhexis.

Stage F. 60°C. Partial heat coagulation of epithelial cells which are preserved in a condition not very dissimilar from that of normal cells.

Stage G. 70°C. Heat coagulation of the epithelium with gross distortion of the epithelial cells and their nuclei. Damage to collagen fibres of the dermis.

For exposures of 1 min, to the heating apparatus, the crucial temperature is 50 to 52°C where there is some interference with the skin cells, showing up in the first instance as a reversible swelling of the nucleus. Oedema comes after an exposure for 1 min at 55°C; the peak of damage is reached in about 30 min and there is an escape of DNA and RNA (Feulgen and pyronin staining substances). At 60°C there is a surface heat coagulation of the outer layer. Histologically, therefore, there are in fact two types of burn, and the way is open for the idea that the Zone II would be one which could allow the escape of substances from the cells.

The temperatures mentioned already are those of the heated surface of the apparatus applied to the outside layer of the skin. We had to know what was the actual intradermal temperature reached, and this was investigated by Mendelsohn and Rossiter,[19] who found that for an external temperature of 55°C for 1 min, a temperature sufficient to form a scab, the intradermal temperature was not more than 47·5°C (Fig. 8.3). This appears to be a critical temperature for the mammal, though Kellaway and Rawlinson[14] (1944) found that for prolonged exposures of several hours, the threshold for injury may be as low as 41–43°C. See Table 8.2.

We now come to two questions. (1) Can short exposures of up to 50°C inactivate any known enzymes? (2) Do such exposures induce a leakage of enzymes, and especially of a proteinase?

Taking the first question, some observations were made in Oxford upon the heat inactivation of some enzymes by Dr. Olafsson from Finland[22]

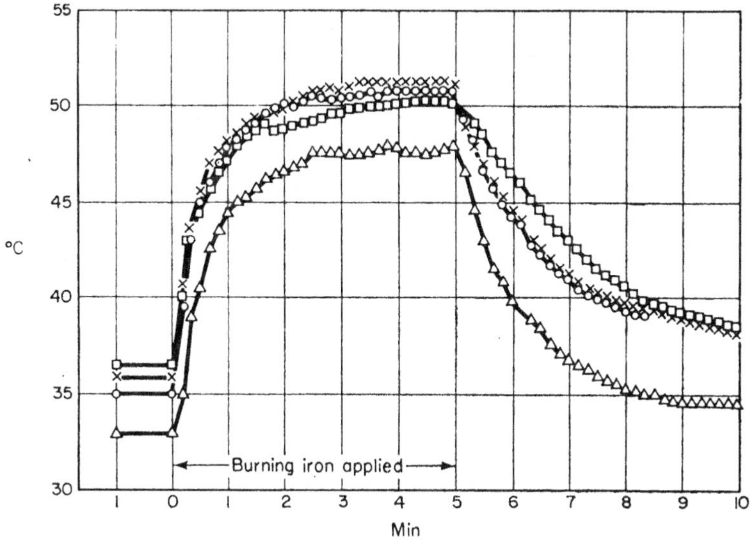

FIG. 8.3. Time distribution of subcutaneous temperature under the burning iron. Temperature of iron, 55°C. Guinea-pig under nembutal anaethesia. (Mendelssohn and Rossiter, 1944.)

TABLE 8.2. TEMPERATURE REACHED AFTER A 1-MIN APPLICATION OF THE BURNING IRON TO THE SKIN OF ANAESTHETIZED GUINEA-PIGS

Temperature of burning iron	No. of observations	Mean temperature reached	Standard error of mean	P.
45	9	38·4	0·2	
47	8	40·1	0·4	< 0·01
50	11	41·1	0·6	< 0·214
55	16	44·8	0·6	< 0·01
60	14	48·4	1·0	< 0·01
70	7	54·4	1·1	< 0·01

Data of Mendelssohn and Rossiter, 1944. Erythema produced by skin cell temperature 40–50°C. Irreversible destruction, etc., between 45° and 55°C.

(1942) and others (Table 8.3). The most sensitive system was that dealing with glucose in rat brain but the pyruvate oxidase system was also affected. There were some slight effects upon other enzymes, but if we take into account that 50°C was a maximum temperature, it is improbable that any of the enzymes tested were sensitive enough to heat to be inactivated at

TABLE 8.3. THE HEAT-INACTIVATION OF SOME ENZYMES BY SHORT EXPOSURES TO TEMPERATURES OF 48–55°C (pH 7·3)

(*A selection from observations from the Department of Biochemistry, Oxford*)

Enzyme	5 min and more			Less than 5 min		
	Temp. °C	Time of heating	Change %	Temp. °C	Time of heating	Change %
1. *Enzymes*						
Cathepsin	55	5′	−6·4			
Skin proteinase	55	5′	−13			
Amino-acid oxidase	55	5′	−27			
Succinodehydro-genase	54·8	10′	−57			
Hexokinase (brain)	50	5′	Nil			
Cytochrome oxidase	49·6	5′	−48	51·4	2′	−78
Malate dehydrogenase	49·8	5′	−71·4	50·7	2′	−77·3
Lactate dehydrogenase	50	5′	−90·3	50	2′	−45·5
2. *Enzyne systems*						
"Pyruvate" (pigeon brain D)	{48 53	10′	−22 −79	53	1′	− 6
"Glucose" (rat brain B)	{48 53	10′	−62 −91			
"Succinate" (rat brain B)	{48 53	10′	−21 −40			
"Glucose" guinea-pig skin				{50 55	1′ (*in vivo*)*	− 16 −43
"Glucose" guinea-pig skin				48·5	3′ (*in vivo*)	−40

Notes. D = dispersion. B = brei (equal volumes cooled liver and ice-cold water). *Burning iron.

the temperature mentioned. 1 min at 53°C induced a decrease in activity of 6 per cent in a dispersion from pigeon brain. Since we could get no evidence that there was inactivation of enzymes by the heat, it seems more reasonable to interpret the first effects as meaning that there is some change induced in a lipoprotein complex of the surface of the cells, altering the permeability.

In 1945–47, careful and accurate work was also being conducted in Boston, by Moritz and colleagues(20); their experiments were done mainly on pigs and some on man. The main objective was to determine the time and temperature necessary for critical exposures, using an apparatus in which water of varying temperature was circulated in immediate contact with the skin. We know from a few experiments that the critical temperature for human skin was much the same as that for the guinea pig; and the Boston work established that there was little difference in different skin cells. Taken with observations in the literature, living tissue seems to behave much the same to heat. It is one of the few conditions against which there seems to be no protection. The Boston work covered a wide range of exposure, including the important field of flash burns.

It may be thought that the delineation of zones in our work was only another expression of the usual terminology. first degree, etc. Though connected with this, I think it is different, because the partially damaged cells need not necessarily come from the same layer of the skin.

Action on an enzyme in the surface membrane of skin cells is unlikely though it is not excluded because temperature might alter some unknown enzymic equilibrium in the surface. More work is needed; but since many enzymes are not destroyed by a low temperature burn, and are therefore capable of liberation in an active state from the tissue, it was worth looking for active enzymes which might be liberated.

In the belief that some proteinase might be one of these, Miss Beloff (now Mrs. Chain) and I searched for one, starting in 1943.[(3,4)] A condition for its activity would be that it had an optimum pH about pH 7·0, which was not the optimum pH of tissue cathepsins or of the leucocyte proteinases then known. We succeeded in finding such a proteinase, which can be called dermoproteinase in the skin of man and of the rat, rabbit and guinea pig.

Our dermoproteinase had some of the properties of trypsin, but it was not trypsin or chymotrypsin, as it will not digest the two specific substrates of Bergmann, viz. benzoyl-arginine-amide or carbobenzoxy-1-tyrosol-glycine amide (Neville-Jones and Peters[(21)]); nor is it the enzyme hydrolysing leucylglycylglycine also present in the skin extracts, as it has a different solubility. It is not the same as leucine aminopeptidase. Dermoproteinase is best prepared from acetone-dried skin, and must then be extracted with 5 per cent KCl. The enzyme fulfils the condition that it could be liberated because it is comparatively stable at temperatures below 70°C.

If the proteinase leaks out of the cells after a burn, then one would expect to find less proteinase in the skin. Table 8.4 shows that this is so, if the tissue is left to stand for 1 or 2 hr. When skin is taken immediately after burning there is no change. At the time that these experiments were reported or shortly after, it was known from our work that there was an exit from the skin of the "pyronin staining" substance which was shown by Brachet[(5)] to be nucleoprotein, as well as of the proteinase; and it was known that there is present the enzyme hydrolyzing the tripeptidase

TABLE 8.4. TOTAL CHANGES IN PROTEINASE-CONTENT OF RAT SKIN DUE TO BURNING FOR ONE MINUTE AT 54–66°C, AS COMPARED WITH CONTROL; ETHER-ANAESTHESIA

(From data of Beloff and Peters, 1944, 1945)

Temperature	Time killed after burn (min)	Average change %
54–55°C	20	− 3·2
	60	−23·5
	120	−32·0
60–61°C	60	−43·1
	120 and 240	−54·5
65–67°C	10–20	−32·1
65–67°C	Before burning (control)	− 4·0

leucylglycylglycine; this appears in the lymph from a burned dog's limb and in blister fluid from a patient. There is also a protein migrating with half the speed of the globulin in the plasma.(23) We tried to get toxic activity by injecting large doses of skin digest, but were not successful. One of the complicating factors, which we realized then, was the presence of an anti-proteinase, which would enter from the plasma; this complicates the interpretation of experiments very much.(2) As little as 0·05 ml of plasma per ml of our enzyme extract was enough to inhibit by 70 to 80 per cent. Further work on this has appeared recently.(17)

That this proteinase enzyme is an independent entity may be judged by a subsequent finding of Martin and Axelrod(16) (1957). They confirmed our results, but found that the activity of the skin proteinase could be much increased by extracting with much more concentrated KCl (10 per cent, i.e. twice our concentration). If the extract stood, it lost activity, which could be restored by a further treatment with concentrated KCl. Their interpretation was that it was associated with antiproteinase, which was dissociated in salt solutions of high ionic content.

Interesting observations have been made by Ungar and Damgaard(30) (1954). They have studied the breakdown of protein in skin slices from guinea pigs, using an optical density measurement at 280 mμ. They find that an exposure to 50°C for 5 min induces an increased breakdown of protein and this occurs more at 55°C. Evidently a rapid discharge of "protein" takes place immediately, and then the incubation shows up the proteolysis over the next minutes. The initial change is not in their opinion an enzymic one. One wonders whether it is protein which comes out, or whether this is not an appearance of the nucleoprotein which was seen in our burns by histological methods. These authors found that trypsin inhibitor stopped the enzymic phase of their observations. They are inclined to interpret our experiments as the conversion of some precursor into an active protease; I do not see how this is consistent with our measurements of the loss of activity from the skin.

Other substances are probably liberated, e.g. both leucotaxine and histamine, but there is no effect upon the burn of injecting antihistamines (Sevitt[27]).

In regard to the mode of formation of the blister, our hypothesis was that the dermoproteinase digested the prickle cell processes which anchor or hold the epithelium to the dermis, and that it is this which allows the blister to form. Medawar[18] had shown already that trypsin separated dermis from epidermis in slices from human skin.

We did not take the matter further, but in the last years a most interesting series of papers has appeared from Louvain,[9,10] and also a thesis from Holland (Burbach, 1957[6]). To deal with the latter first. Burbach has tested our hypothesis of blister formation by finding that hyaluronidase when injected intradermally in the operating room does not cause a blister, but that commercial trypsin will do so. The blister fluid has a higher protein content than the capillary filtrate. By injecting what he considered to be epidermal protease, he got no detachment as with trypsin, so he was unable to establish our hypothesis though he supported our idea; it seems that it will be necessary in view of the work of Martin and Axelrod to make sure that there is no anti-factor present. This will involve more work on the enzyme.

The papers from Louvain by Godfraind and Simonart (1957) make points of great interest in relation to the problem of the systematic toxin in burns. Experiments in frogs and then in rabbits have shown that oedema fluid from burns produces symptoms closely similar to those of a severe burn. Injection of heated proteins equivalent in quantity to that which has been denatured in the case of severe burns gives identical symptoms; but such proteins have to be left some time for the effects to show. In other words their work demonstrates that the toxin is produced as peptones, etc. by the action of some enzyme in the oedema fluid. It would have been nice for the theory of Beloff and Peters if it could have been shown that this was an effect of dermoproteinase; but the Belgian work has advanced the subject of "burns" toxins very considerably. I have been much interested in talking to some surgeons who have had experience of burns, who have told me that they think there is something worse about the burn induced by lower temperatures. Such an idea is quite consistent with the hypothseis of the effect of the second zone.

ADDENDUM (added in January, 1962)

The past paragraphs were written before some of the most recent work in this field. Matters are again in the melting pot so far as bacterial infection is concerned. Upon a recent report by Allgower *et al.*[1] (1960). that oedema fluid was always infected with bacteria, the Louvain workers[13] (1961) have re-examined their experiments. They now find that in rabbits burned at 80°C, the oedema fluid is often infected after 24 hr and always infected after 48hr, the infections being similar to bacteria in the intestinal tract, *E. coli*, cereus, etc. This happens when the skin is unbroken, and is not reduced by sterilizing the intestine with neomycin. By plunging the bodies into chloramine-T solution, they can reduce the infection, which they regard as cutaneous in origin. It is to be noted that these results are contrary to those reported by Wilson *et al.*; even after 6 and 12 hr they had toxicity in their oedema fluid. The Louvain workers found no odour attached to their oedema fluid, and they think that their observations upon the toxicity of peptone could be explained by the idea that their peptone was infected after 48 hr. In a personal communication, Professor Simonart has told me that they are working actively at the problems, and that he thinks indeed that it is probable that the major effects are due to a bacterial endotoxin; which he points out can be active in very small doses. Even as little as 0·55 mg/k of *E. coli* endotoxin can be toxic, though a heat labile factor from plasma also seems needed. What is clear is that the whole problem of burns is even more complex than it appeared to be before. The initial biochemical lesion is certainly the liberation of nucleic acids from the cells and of some enzymes; the protease is probably the factor which by digestion of the prickle cells prepares the way for the lifting of the skin in blister formation. At the same time it is not clear that it is these enzymes which are the toxic factors. Even if the toxic effects are due to bacterial endotoxins, it is still true that in burns there are effects over and above those present in ordinary wounds and surgical trauma. There are of course many secondary changes after burns, such as loss of nitrogen, lipid accumulation, haemoconcentration, etc., which are not considered here as they are secondary to the biochemical lesion. In burns due to heat we have an example of a biochemical lesion which does not appear to be enzymic in origin, but to be due to changes in permeability and to disturbance of nucleic acids.

Any reader who is interested in this subject would be well advised to read the reviews in a recent conference[8] (1961) and especially the report by Fine. Modern work is evidently showing a trend towards the idea that resistance in a damaged tissue may be lowered sufficiently to allow an internal growth of bacteria, which produce endotoxins, an event not occurring in a normal animal. At the same time it is difficult to believe that some of the changes in carbohydrate metabolism which come on relatively quickly in ischaemic shock, Stoner *et al.*[29] are due to the action of bacteria. If they are not then some problems still remain which are not bacterial in origin.

If the temperatures described are detrimental to life generally, it becomes interesting to know whether there is any similar information in the plant world, as to critical temperatures.

Thermal Injury in the Botanical World

For some time now (since 1946). the Cereals Institute at St. Albans, together with the John Innes Horticultural Institution have been following the effect of heat damage in the seeds of cereals, principally in relation to germination.(12) They have been able to work out an equation for heat damage.

$\theta = 122 - 5\log_{10}t - 44\log_{10}m$, where $\left\{ \begin{array}{l} \theta = \text{temperature } t = \text{time of} \\ \text{heat (mins) } m = \%\text{ moisture.} \end{array} \right.$

If grain with 37 per cent moisture content is immersed in hot water for 1 min at varying temperatures, the grain shows what they call a damage zone of 53 to 62°C. At the death temperature of the wheat, there is severe protein coagulation and inactivation of enzymes. But at the damage zone, the behaviour is peculiar; if placed in rather dry conditions the seed germinates and behaves normally, though the onset of germination is delayed. But if the seeds are put into excessively wet conditions, they form nucleolar blisters and may not germinate at all, a fact which they interpret as due to an initial damage to the permeability of the cell. The analogy with human burns is rather remarkable, and the hint about dry conditions. In the mammal the only way in which water can be kept out is to use pressure, the argument for which is strong where it can be used. Again it seems that nature has not provided well for protection against heat, the evolutionary reason for which is not clear.

Summary

The nature of the biochemical lesion due to a thermal burn in the skin has been investigated in animals, mainly in the guinea pig. 50–55°C during a time of 30 to 60 sec is a critical temperature for damage, when the heat is applied externally by a burning iron. There are changes in permeability, with liberation of substances from the skin cells, which can be detected histologically; and the degree of damage can be correlated with increasing temperature. Exposure to higher temperatures produces a different condition, in which there is fixation and coagulation of the protein. Estimates of the actual temperature in the skin cells causing damage in these short exposures gives a temperature of 48°C or less. This temperature applied for a short period cannot certainly damage any enzymes. It is possible that the first change is in some lipoprotein in the surface membrane of the cells; this is accompanied by the exit of some RNA. It is certain that many active enzymes could leave the skin with a change in the permeability which appears. A new proteinase was found to be present and to leave the cells after burning, and this it has been proposed is responsible for the blister formation. Tissue infiltration of fluid from the skin can occur with great speed, and be stopped by a relatively light pressure. It is still not certain what part the escape of some enzymes from the damaged skin cells can play in the toxic condition following burns, because there is some evidence now that bacteria may grow in a wound even when the skin has not been broken.

I have set forth our work on this biochemical lesion in burns, because I feel it to be a good example of the way in which a concentration upon the initial biochemical change can lead into an area of scientific significance. Such an attack is of course long range; but if the views of the toxin are correct, and if it proves to be a proteinase, then the obvious treatment would be with the antiproteinase, and a fruitful line of work is suggested. It will be noted that in this case there is no evidence that the biochemical lesion starts with damage to an enzyme.

References

1. Allgower, M., Aschieri, F. and Hulligor, L.(1960), Report 1st International Congress, Washington.
2. Beloff, A. (1946), *Biochem. J.* **40,** 108.
3. Beloff, A. and Peters, R. A. (1945), *J. Physiol.* **103,** 461.
4. Beloff, A. and Peters, R. A. (1946), *J. Physiol.* **105,** 54.
5. Brachet, J. (1950), *Ann. Soc. Roy. Zool. de Belg.* **81,** 185.
6. Burbach, J. P. E (1957), Doctoral Thesis, University of Amsterdam, Drukkerij, Holland, N.V.
7. Colebrook, L. (1944), Med. Res. Council Special Report, No. 249.
8. Fine, J. (1961), *Federation Proc.* **20,** July, No. 2, Pt. 3, Suppl. 9.
9. Godfraind, T. (1958), Thesis. Univ. Louvrain. L'Auto-intoxication, Après Brûlure.
10. Godfraind, T. and Simonart, A. (1957), (A. Simonart) *Acad. Roy. Med. Belgique* (*Mai*), vi° Sèrce, XXII, 231.
11. Hudack, S. S. and McMaster, P. D. (1932), *J. Exp. Med.* **55,** 431.
12. Hutchinson, J. B., Greer, E. N. and Thomas, P. T. (1946), *Nature* **158,** 120.
13. Jauman, C., Jean Jean, M. and Wauters, G. (1961), *Rev. Belge de Path. et de Médicine Exper.* **28,** 145.
    Also Jean-Jean, M. and Simonart, A. (1962), *Rev. Belge de Path.* **28,** 397.
14. Kellaway, C. H. and Rawlinson, W. A. (1944), *Aust. J. Exp. Biol. Med. Sci.* **22,** 63.
15. Leach, E. H., Peters, R. A. and Rossiter, R. J. (1943), *Quart. J. Exp. Physiol.* **32,** 67.
16. Martin, C. J. and Axelrod, A. E. (1957), *J. Biol. Chem.* **224,** 309.
17. Martin, C. J. (1962), *J. Biol. Chem.* **237,** 2099.
18. Medawar, P. B. (1941), *Nature, Lond.* **148,** 783.
19. Mendelssohn, K., and Rossiter, R. (1944), *Quart. J. Exp. Physiol.* **32,** 301.

20. Moritz, A. R. and Henriques, F. C., Jr. (1947), *Amer. J. Path.* **23,** 695; Moritz, A. R. (1947), *Amer. J. Path.* **23,** 915.
21. Neville-Jones, D. and Peters, R. A. (1948), *Biochem. J.* **43,** 303.
22. Olafsson, P. (1945) (Unpublished), see Peters, *Brit. Med. Bull.* **3,** 81.
23. Peters, R. A. (1945), *Brit. Med. Bull.* **3,** 81 (Review).
24. Robertson, B. and Boyd, G. (1923), *J. Lab. Clin. Med.* **9,** 1.
25. Rossiter, R. J. (1944), *Lancet* **1,** 9.
26. Sevitt, S. (1949), *J. Path. Bact.* **61,** 427.
27. Sevitt, S. (1949), *Brit. J. Exper. Path.* **30,** 540.
28. Spink, W. W. and Vick, J. (1961), *J. Exp. Med.* **114,** 501.
29. Stoner, H. B., Heath, F. and Collins, O. M. (1960), *B.J.* **76,** 135.
30. Ungar, H. and Damgaard, E. (1954), *Proc. Soc.* **87,** 378. *Exp. Biol. Med.*
31. Underhill, F. P., Kapsinow, R. and Fisk, M. E. (1930), *Amer. J. Physiol.* **95,** 302, 315, 325, 330, 334, 339, 348, 364.
32. Underhill, F. P. and Kapsinow, R. (1931), *J. Lab. Clin. Med.* **16,** 823.
33. Wilson, W. C., Jeffery, J. S., Roxburgh, A. N. and Stewart, C. P. (1937), *Brit. J. Surg.* **24,** 601.
34. Wilson, W. C., Macgregor, A. R. and Stewart, C. P. (1938), *Brit. J. Surg.* **25,** 826.

CHAPTER 9

Some Historical Points and other Biochemical Lesions

A Concluding Miscellany

Some remarks are needed in this book about the background of the researches described, and about their dependence upon earlier work by others. Nevertheless, I do not think it is easy to put any scientific work into its proper perspective. All researches must necessarily depend upon what has gone before. A few general observations upon the philosophy of an historical approach may make clearer what I feel, even if some of them are rather obvious.* I suppose that everyone would like both to put forward some original new hypothesis, make it a theory and then prove it themselves. This, however, we can seldom do single-handed with the complicated techniques of modern science; and so it falls to most of us to take part sometimes in proposing new ideas, and sometimes in producing the scientific proof of ideas which have been well adumbrated before. Naturally, however balanced we try to be, we tend to lay stress in our own minds upon the particular piece in which we may have taken a hand; if it be the initial hypothesis, then we like to feel that this has been the most important part; and vice versa, if we have been fortunate enough to prove a theory, it is only human that we should like to have the comfortable feeling that this has been what really mattered. The world itself which should judge these things with academic calm and detachment, seldom does so. It is apt to be fickle as to whether it gives greater credit to the initiator of the idea, or to its final scientific proof, and there are all kinds of variations in what is said. I knew one professor who said that "any fool could put forward an idea, but that it took a really clever man to prove one". On the other hand I have known of cases where someone has gone to meticulous trouble to get a final proof, which is after all "science", and his proof is hardly noticed in the blinding light of the credit given to the originator of the theory. And it is admittedly difficult in many cases to decide which is the most important because both are so necessary. Then we have the oft-repeated remark that "technique is everything". Again there are times when progress seems to depend entirely upon a new technique.

* Varetti[33] left his history to the end.

One may instance the extra power given us by the electron microscope. In ascribing credit to ideas, it is important not to get an early idea out of focus with its time. It would be true historically to say that Jules Verne suggested the possibility of encircling the globe in 80 days; but it would be quite wrong to give the impression that this prophesied the speed of modern transport, and to suggest that his idea became responsible for the development of the modern jet aircraft! It has required not only the discoveries of Faraday, but also a continual engineering invention, the result of independent imagination, intellectual effort and improved technique. In biochemistry, I believe that it is still possible to start something new with rather simple technique; but it takes the panoply of modern machines to bring this to a successful conclusion. From our present standpoint it is difficult to realize that only a short time ago, biochemistry could be defined as a science which the chemists called physiology and the physiologists chemistry.

In our work on biochemical lesions, we have been no exception to the general rules. Sometimes we have been lucky enough to put forward an idea and prove it; but in general we have often taken a hand in parts of the whole. For instance, the researches described have established, I think, the concept of the biochemical lesion, as a useful mode of attack upon some biochemical problems. It was proved for thiamine. The hypothesis of lethal synthesis was put forward and proved for some carbon–fluorine compounds, in the course of which we proved the reality *in vivo* of the citric acid cycle; the idea of lethal synthesis has, however, been applied in a wider field and much developed by others. The fluid cytoskeletal hypothesis (page 216 *et seq.*) has had to wait for proof for the electron microscope.

I hope that I have made it clear in Chapter 2 how the idea of working at the biochemical lesion arose, as an offshoot of work upon the action of a vitamin. The key to the action of thiamine lay in the decision to explore the intermediary metabolism of the pigeon's brain *in vitro*.

It must be recognized, however, that there had been others before that period whose ideas were all important to biochemists. In this relation one could go back at least to Claude Bernard in 1878, who stressed that by a study of poisons one could learn about the "physiological mechanisms of life", as pointed out recently by G. E. Paget.(24) As Florkin(10) has said, Bernard certainly looked forward to the day when it would be possible to reunite in one discipline, Physiology, Pathology and Therapeutics, now happening at the molecular level. The quotation (Note 1, p. 168) indicates that P. Ehrlich(8) (1909) had clear ideas about the action of arsenic inside cells. In 1916, A. R. Cushny, the distinguished pharmacologist, devoted an address to the British Association to "The analysis of living matter through its reaction to poisons".

It must be remembered that one is always influenced by one's teachers.

H. K. Anderson (the physiologist, afterwards Master of Caius College) early introduced me to C. S. Sherrington's "Integrative action of the Nervous System",* probably thereby giving my views a permanent slant towards the wish to see integration in cellular studies. Especially important too was Hopkin's vision of a dynamic equilibrium in the cell, and his faith in the applications of chemistry and physics *inside* the cell. (For portrait of Hopkins see ref. 25.) This had a decisive effect in turning the minds of those whom he taught away from the vague concept of protoplasm into the chemical reality which it is thought to be today. No one is now inhibited by the thought that chemical reactions inside the cell are essentially different from those outside it.

Inborn Errors

Parallel with this, though somewhat earlier (1909), there was an interest shown in the genetical side of biochemistry. To Archibald E. Garrod[(12)] must go the crystallization of thought and experimental observation into the phrase "Inborn errors of metabolism". Garrod often quoted Harvey's faith in the study of new diseases as a key to new knowledge, and indicated further his appreciation of a Rectorial address given by Huppert in 1896 on chemical individuality. "Huppert argued that chemical differences in the constituents of germ cells must determine the attributes of the organisms developed from them" (Garrod, 1930). In a sense Garrod's thoughts given first in 1908, were a precursor of the studies of O. Warburg on cancer tissue[(34)] when the latter set out to show a difference between normal and cancerous tissues in their glycolytic metabolism. In so far as the cancer becomes a separate genetic individual, this would be a kind of "inborn error" of the cell. Garrod† had evidently pondered upon the newer views of metabolism and upon the probable developments of enzyme biochemistry, because at a time when few tissue enzymes were known, he envisaged the inborn error as due to a missing enzyme; the excretion of the substance giving the black urine in alcaptonuria, for instance, would be due to the absence of one stage in the metabolism of tyrosine. Garrod's ideas in this field have been steadily established by modern work, and in especial by the brilliant research upon the mutants of *Neurospora* in California.[(5)]‡

* One of my great privileges was to work for 12 years in a department next to that of Sir Charles.

† Regius Professor of Medicine when I went to Oxford in 1923. I have the happiest memories of this kindly and wise man, whose ideas are now coming into their own.

‡ This work by Beadle, Tatum and others has revealed mutants with one enzymic stage often missing. These mutants have been used to work out stages in metabolism. In reality it may be said that the "inborn error" produces a missing enzyme in some metabolic pathway, and that this in its turn causes a permanent biochemical lesion.

The conception of the inborn error is also being applied to some of the modern metabolic diseases, for instance the glycogen disease of which there are now several variants. An interesting case has appeared in the field of insect biochemistry. In 1930, von Frisch discovered that mannose was strongly toxic to honey bees, a fact which was confirmed in Spain. When Sols *et al.*[29] gave mannose to bees in the food, over 90 per cent died in 3 hr. In a beautiful series of experiments they traced this to a lack of balance between the enzymes hexokinase and phosphomannose isomerase. This would result in a competitive inhibition between mannose and the phosphorylation of glucose and fructose, thus giving rise to the accumulation of mannose-6-phosphate which could interfere with glycolysis. Hence it is an inhibitor of phosphoglucose isomerase. Honey bees are markedly dependent on a high level of sugar in their blood. The Spanish workers make the pertinent observation that the "mannose disease" is like the syndrome of galactosaemias, the biochemical lesion being produced by eating mannose because of an inborn error of metabolism.

Muscular Contraction

In pure physiology attempts have been made for at least 50 years to relate the chemical events in muscle tissue to the change in the mechanical state, "the contraction", dating from the lactic acid work especially of Fletcher and Hopkins. This work will always be associated with the names of O. Meyerhof and A. V. Hill, and G. Embden. The discovery by Harden and Young in 1904 of the part played in yeast fermentation by the hexosephosphates was also of much influence. Over the years, this work has become increasingly significant, leading to most important discoveries in biochemistry; for instance the finding of adenosine triphosphate, and of oxidative phosphorylation, Meyerhof[20] and D. M. Needham and Pillai[22]; in spite of this one cannot say even yet that it is certain how the chemical reactions in the muscle tissue give rise to the contraction. The initial lactic acid hypothesis has been replaced by several others.

Cholinesterase

On the borders of physiology and pharmacology, we have the classical researches which started with H. H. Dale's observations upon acetyl choline,[7] and which were followed by the discoveries of Loewi,[17] and his postulate of the chemical transmitter. The action of eserine was traced to the inhibition of the enzyme cholinesterase, an enzyme finally characterized by Stedman[30,31] (1933) as a specific enzyme. From this came a temporary cure for myasthenia gravis, a truly important clinical application. The researches have also opened up large programmes of work in neurophysiology in the attempt to settle the difficult questions, firstly of the reality of the chemical transmission, and secondly of the site of the enzyme and the factors concerned.

In 1940 a new turn was given to this field when it was discovered in Cambridge (by those working with Adrian and M. Dixon) that the toxic effect of the newly made diisopropylfluorophosphate was due to its most powerful inhibitory action upon cholinesterase. The research became an investigation of the detail of the biochemical lesion and of its possible reversal by logical approach. The application to chemical warfare was quickly reinforced by the wider application to new insecticides and the saving of life from dangerous casualties. In the hope of getting antidotes, there has probably been a more detailed examination of this biochemical lesion than of many others. Since excellent reviews are available,(40) (R. D. O'Brien(41) and D. Nachmansohn(40,42)), only a few points need be mentioned here. Confining ourselves to the true cholinesterase, as this is difficult to get in quantity even now, most of the research has been done with chymotrypsin, an enzyme which combines with the cholinesterase inhibitors, and which is easy to obtain in crystalline form. Most seem agreed now that it is not the –OH of tyrosine which is first attacked, but that there is an ultimate phosphorylation of a serine grouping, sited in the molecule of the enzyme next to a histidine residue. Based upon chemical arguments it was thought by Wilson (in Nachmansohn's laboratory) that the phosphorus group attached to the active centre and causing the inactivation, could be removed with an oxime. Independently workers in the U.S.A. and in Britain have been making and testing similar compounds which could act as reversing agents; they have had considerable success with some of the nerve poisons. The most potent compounds at present are those known as PAM and P_2S which powerfully reinforce the action of

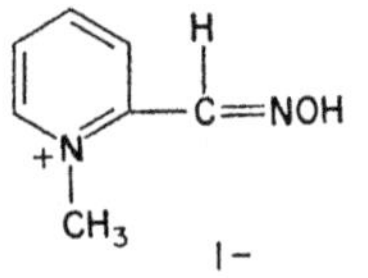

P_2S = Methane sulphonate salt of PAM

PAM. Pyridime–2–Aldoxime Methiodide

atropine. It has been found that some alkyl phosphates are not toxic *per se*; but that they must undergo some change in the liver to the active anticholinesterase; the details have been worked out well for Schradan. Bound up with this field is also the curious pathological demyelination, which can happen some weeks after administration of some of these insecticides and also of dinitrocresol; this demyelination may produce serious paralytic effects in animals such as the hen as well as in man. The factor in the chemical structure inducing this change is still not clear. The whole subject is of great interest, and it seems not too much to claim that the discovery of

BAL and of the antidote to cholinesterase poisoning are still the only instances where an antidote has been discovered by logical chemical reasoning, first the understanding of the biochemical lesion involved, and then the application of this to the finding of an antidote.

ANTIMETABOLITES

The biochemical lesion in anticholesterase agents is in principle irreversible, though this is not quite true in some cases. There is one large field of development which has not been referred to so far, that of the antimetabolites. Though we have not worked in this field, some reference should be made to it. The idea of the competitive inhibition of an enzyme *in vitro* has been in the literature in a concrete form ever since Quastel and Wooldridge(26) in 1928 showed that malonate inhibited the enzyme succinodehydrogenase competitively. Important as this was for the study of the kinetics of enzymes, the idea did not advance *in vivo* until Woods(36) found in Fildes's laboratory in 1940 that the effect of sulphonamides on bacteria could be reversed by para-aminobenzoic acid. Here it was shown that the biochemical lesion was due to the displacement of the toxic substance from some active enzyme centre. Based upon this in 1940 Fildes(9) with Woods proposed the "antimetabolite" approach to chemotherapy, an idea which was picked up rapidly by organic chemists, who synthesized large numbers of possible antimetabolites, based largely upon purely structural analogy. The approach is quite logical, but the lack of practical answers to therapy arising from this idea has led to some disillusionment on the part of many, and to the feeling that the older empirical approach is still the best hope, which I feel to be a counsel of despair. Surely one should try to understand why the success was not more marked; and I would put the question; Is it really surprising that purely structural analogy, as conceived largely on paper did not give new answers? Of course it would have been a short cut if it had done so; but surely a more certain approach is to centre the attack upon the biochemistry in the first instance, and to study the biochemical entity enzymic or otherwise which must be attacked *in vivo*. It is of course much slower; as can be seen by a perusal of some of the chapters in this book; but this mode of research has been used by D. D. Woods himself in Oxford, and it has led to definite success in the hands of D. W. Woolley.(37,38)

The poet Masefield has said that ideas come out of the East like locusts, and settle on the minds of men; and it is interesting how often several minds seem to come to the same point independently and about the same time. In this antimetabolite field, again about 1940, Jorgen Lehmann(16) in Sweden and Link in the U.S.A. were independently studying the principle in sweet clover responsible for the haemorrhage disease in cattle in researches which led to the development of dicoumarol as an anticoagulant in practical therapy. The ideas guiding this work were also based upon competitive inhibition of vitamin K; dicoumarin interferes with the synthesis of prothombin in the blood.

Some other biochemical lesions may now be considered.

The Vitamins

Looking back to our work upon the biochemical lesion in thiamine deficiency, we were lucky to have chosen practically the only case in which it was possible to take out tissue from an avitaminous animal and obtain in this way parallel effects *in vivo* and *in vitro*. This was partly due to the difficulty of getting the right pieces of tissue from animals deficient in other vitamins. For instance, in the B complex, one difficulty is that deficiencies of riboflavin show up as a recognizable pathological change first in the growing areas of the body, the corner of the mouth or some area of the skin or cornea, so that one can show up a deficiency of cofactors containing riboflavin, I know of no case where the tissue has been taken out and the lesion corrected *in vitro*.

There has been some approach to this type of study in the case of pyridoxine; in the absence of this vitamin M. Richards[27] (1938) found that some rats suffering from deficiency got fits, and this was also studied in pigs.[6] More recently in the U.S.A. children have been found to have "fits" as a result of pyridoxine deficiency. The first indications of this were due to Kline[15] as a result of his tracing to its cause an effect due to bad feeding in the children. Then Towers *et al.*[32] discovered that there were some children needing an intake of 10 mg of pyridoxine a day instead of the usual 0·3 mg. Withdrawal of the extra doses of vitamin B_6 led to an onset of the fits which were cured on increasing the dose of vitamin; the children were for some reason extravagant in their use of the pyridoxine. This condition is accompanied with an abnormal excretion of xanthurenic acid, the detection of which can be improved by giving a "loading" dose of tryptophane, of which it is a metabolite. It has been stated that some alcoholics also need large doses of pyridoxine to counteract a similar abnormal excretion. The condition in the children must be rated as an inborn error. One would like to know where the biochemical lesion is in the alcoholics.

TRYPTOPHANE XANTHURENIC ACID

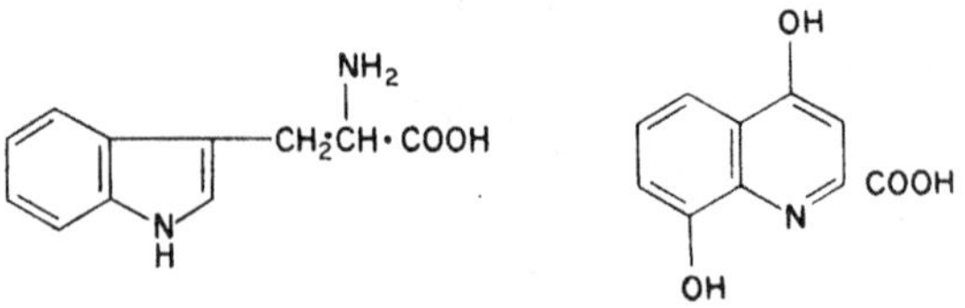

It appears that the fault is either in the transaminase systems or in some enzyme concerned with amine formation, and in 1948 it was shown that in rats deficient in pyridoxine there was a decrease in the capacity to form amines from the action of dopa carboxylase and cysteic decarboxylase. If the effect is a failure of the glutamate transaminase, one would like to think that it was the failure of the tissue to produce the substance, gamma-amino-butyric acid, nicknamed GABA, which would require pyridoxal phosphate for its formation from glutamate. If GABA is a substance which tends to damp down excitation and so keep cells in order, then the explanation would make a good story. Unfortunately there seems to be less interest in GABA as a substance playing such a role than there was a few years ago (see Hirsch and Robins, 1962).

In the vitamin B_6 field also lie some observations upon a substance called toxopyrimidine, which is the pyrimidine half of thiamine, and which R. Abderhalden discovered to be toxic to rats, mice and guinea pigs on a vitamin B_2 deficient diet. It has now been shown in Japan that toxopyrimidine phosphate blocks the action of pyridoxal phosphate *in vitro* in the tyrosine decarboxylase reaction, and that the toxic condition can be

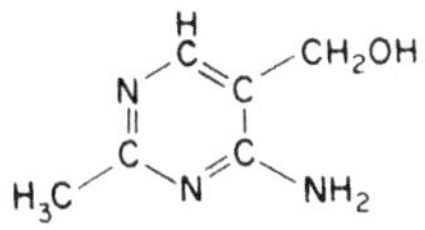

TOXOPYRIMIDINE

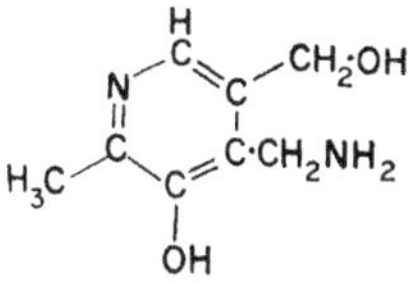

PYRIDOXAMINE

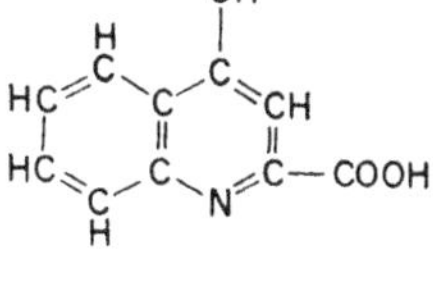

KYNURENIC ACID
(Compare with Xanthurenic acid.)

reversed by giving one of the vitamin B_6 group of substances. *In vivo*, toxopyrimidine inhibited glutamic decarboxylase of brain which may be significant in a reduction of γ-aminobutyric acid.[28]

Passing over effects due to lack of biotin or of pantothenic acid, it is interesting to note the position of research upon the cobalt-containing factor or factors in the vitamin B_{12} group, which are connected with the cure of pernicious anaemia. The astonishing effects of the addition of cobalt and molybdenum in small amounts to deficient pastures, are now classical; but it has only been found recently that there is a coenzyme containing cobalamine which is concerned in the metabolism of propionate.[19,4] The deficiency seems to hit the liver first. As indicated earlier, there is a puzzle here which does not only apply to cobalt. Cobalt salts have been proved to interfere with the lipoate in mitochonodrial preparations; they are less toxic to slices of tissue. But it is sensational to realize that injections of cobalt metal in finely divided state will induce rhabdosarcomas in rats.[13] This is another example of the importance of molecular regulation.

At the present time there is much study in progress of biochemical lesions, which I have not discussed in detail here, not because it is unimportant but because I do not feel that I have the same first-hand knowledge. For instance, the biochemical effects are being pursued of radiation[6a,39] of phloridzin,[18] of triethyltin[2,21] and of anti-inflammatory agents.[35] (See also ref. 11.)

With all this valuable physiological, pathological and pharmacological work and its most significant repercussions, it may be said again the significant fact was that our work on the biochemical lesion developed as an attempt to understand the action of a vitamin in the general metabolism of the brain, and that the study was made upon a condition which could be produced experimentally and was reversible. It has been said that, "It is the intelligent assembly of perhaps older ideas, which makes a new painting." I hope that this applies to what we have done. The comments of Z. Bacq[3] in the introduction to his book are relevant; his book is one of the very few devoted to the application of biochemistry to pharmacology.

> "The set idea of the 'biochemical lesion' is one of those concepts to which we have a weakness for feeling that we have lived through certain episodes which have contributed to its proposal. So far as biochemists and pharmacologists are concerned, it originates from a desire to understand the real cause of the lesions which have been described by clinicians and anatomopathologists, of knowing exactly what happens in the interplay of the reaction in the organism before the anatomical picture has developed." (Free translation)

In summary, we may say that biochemical alterations in metabolism can be divided into those that are permanent involving the genes, and those that are due to biochemical lesions; this is represented in Table 9.1.

TABLE 9.1. CLASSIFICATION OF BIOCHEMICAL ALTERATIONS

Biochemical lesions due to

(a) Nutritional deficiencies
(b) Toxic agents, metals, As, Hg, etc.
Vesicants and toxic gases
Temperature changes
Narcosis
Metabolic antagonists
Gases such as O_2, CO
Release of feedback control.
(Note the last is not yet known.)

Biochemical modifications due to alteration of the germ-plasm

(a) Inborn errors of metabolism such as alcaptonuria, porphyrinuria.
(b) Mutations due to
(1) Biological agents, cancer
(2) Chemical agents, e.g. mustard gas
(3) Radiation.

CONCLUDING REMARKS

In this book several biochemical lesions have been considered and the outlines of research upon them have been given. Briefly the biochemical lesion in thiamine deficiency was found to be a deficiency in the enzyme system metabolizing pyruvate, and that in arsenical poisoning of tervalent, mono-substituted compounds, to a deficiency in the same system, but introduced positively by toxic action on the lipoate component. The exploration of the biochemical lesion in mustard gas poisoning was unrewarding so far as it showed that it was virtually impossible to devise an antidote to this. At the same time the work led to a part of modern research upon nucleic acid and genetics, because mustard gas produces mutations. This again has been actively pursued as an alkylating agent against various types of malignant disease, with a large programme of synthetic nitrogen mustards.

There followed a chapter on the biochemical lesion induced by the cattle poison fluoroacetate, a carbon-fluorine compound. Exploration of this led to the concept that the biochemical lesion was due to a lethal synthesis to a fluorocitrate which then blocked the reactions of the citric acid cycle at the citrate stage. The facts are interesting in themselves, but they have been followed by remarkable developments again in the field of genetics, where the making of fraudulent nucleotides has had its applications again to malignant disease.

A chapter was inserted dealing with toxic agents and thiol substances in the attempt to underline the different reactions in the field and their importance. In the biochemical lesion in burns, the concept enabled us to find a new protease. This is likely to be of significance locally, but whether it bears upon the problem of toxaemia and shock in burns is still an open question. Unlike the other biochemical lesions studied, there is no evidence that the lesion in burns starts with damage to an enzyme.

Finally in the Linacre Lecture the possibility is considered that there may be more complex biochemical lesions, which involve the organization and chemical structure of the cell, a matter upon which the auther has had views for many years.

Each biochemical lesion has taken a long time to explore; but the writer feels that the results obtained by this method of approach have justified the efforts made, and that it is a question of the tortoise and the hare. By making certain that the biochemistry studied has direct relation to change in a physiological or pathological condition, any facts discovered are more likely to have their impact upon the machinery of life. I would urge that at this juncture, when attention is being turned to biochemistry in its relation to the organization of cells, some such approach is more than ever necessary for what is almost a philosophical reason. One cannot go astray with a study of the chemical compounds and of the enzymes present in living matter, because it is essentially an addition to pure chemistry, and remains this until it is carried into the attempt to relate this to the machinery of life. At this point one is obliged to study first organized entities of the cell like the mitochondria or the microsomes and to study them as a whole, their behaviour as an organized complex; since this behaviour is only of interest in the degree to which the reactions taking place in it contribute to the life of the whole cell and organism. Hence reactions outside the common conditions of pH, temperature and so on are of little real interest, in so far as they are neither additions to chemistry nor to the problem of life. This is how I come to think that relevance is ensured by work upon some biochemical lesion.

When we come to the action of drugs in the biochemical sense we are still woefully ignorant; but, as has been pointed out, there are several places where light is entering and where we can give a reasonable biochemical argument for the action. The attempt to enlarge our area of knowledge will take place quickest if we work upon the lines suggested here.

It is indeed likely that the complications introduced by the endoplasmic reticulum (page 187) will be best approached upon the basis of some form of biochemical lesions.

NOTES TO CHAPTER 9

1. "Über die Art und Weise, in welcher die Chemoreceptoren mit den entsprechenden Stoffen sich verbinden, ist es natürlich sehr schwer, präzise Ansichtēn zu äussern . . . Wenn wir aber denken, wie leicht bestimmte Gruppen der

Metalle imstande sind komplexe Verbindungen zu bilden, wenn wir sehen, dass die Oxyde des Arsens und Antimons mit Weinsäure, Gallussäure und einer grossen Reihe von orthoständingen Hydroxylgruppen Verbindungen eingehen, so wird man annehmen können, dass vielleicht solche orthoständingen Hydroxyl oder Sulfhydrylgruppen durch geeignete Bedingungen befähigt sein können, als Metallreceptoren zu fungieren." (Ehrlich)

References

1. Abderhalden, R. (1939), *Pflügers. Arch.* **242,** 199.
2. Aldridge, W. N., and Cremer, J. E. (1951), *Biochem. J.* **61,** 406.
3. Bacq, Z. M. (1950), *Sciences et Lettres*, Editeurs, Liège.
4. Barker, H. A., Smyth, R. D., Weisbach, H., Munch-Petersen, A., Tooley, J. L., Ladd, J. N., Volcani, B. E. and Wilson, R. M. (1960), *J. Biol. Chem.* **235,** 181.
5. Beadle, G. W. (1945), *Physiol. Rev.* **25,** 643.
6. Chick, H., Macrae, T. F. and Martin, A. J. P. (1938), *Biochem. J.* **32,** 2207.
6a. Creasey, W. A. and Stocken, L. A. (1959), *Biochem. J.* **72,** 519 and Ord, M. G. and Stocken, L. A. (1962), *Biochem. J.* **84,** 600.
7. Dale, H. H. (1934), *Proc. Roy. Soc. Med.* **28,** 319.
8. Ehrlich, P. (1909), *Ber.* **42,** 17.
9. Fildes, P. (1940), *Lancet* **1,** 955.
10. Florkin, M. (1948), *Exposés annuels de Biochemie Médicale*, 8 Serie, Masson, Paris, p. 97.
11. Gallagher, C. H. (1957), *Austr. Vet. J.* **33,** 70.
12. Garrod, A. E. (1923), *Inborn Errors of Metabolism*, 2nd ed., Henry Frowde, Ed., Hodder & Stoughton; (1931), *The Inborn Factors in Disease*, Oxford Univ. Press.
13. Heath, J. C., Daniel, J. T., Webb, M. and Lucy, J. A. (1960), Report of Strangeways Research Laboratory, Cambridge.
14. Hirsch, H. E. and Robbins, E. (1962), *J. Neurochem.* **9,** 63.
15. Kline, O. M. See Nelson, E. M. (1956), Publ. Health. Rep. U.S.A. **71,** 445.
16. Lehmann, J. (1959), Review in *Circulation* **19,** 122.
17. Loewi, O. and Navratil, E. (1926), *Pflügers Arch. Ges. Physiol.* **214,** 678.
18. Lotspeich, W. D. (1960), Harvey Lecture.
19. Marston, H. R., Allen, S. H. and Smith, R. H. (1961), *Nature,* **190,** 1085.
20. Meyerhof, O. (1937), *Naturw.* **25,** 443.
21. Moore, K. E. and Brody, T. M. (1961), *Biochem. Pharmacol.* **6,** 125.
22. Needham, D. M. and Pillai, R. K. (1937), *Biochem. J.* **22,** 689.
23. Nelson, E. M. (1956), Publ. Health Rep. U.S.A. **71,** 445.
24. Paget, G. E. (1962), *Chem. & Ind.* p. 776.
25. Peters, R. A. (1959), *Biochem. J.* **71,** 1 (Hopkins Memorial Lecture).
26. Quastel, J. H. and Wooldridge, W. R. (1928), *Biochem. J.* **22,** 689.
27. Richards, M. B. (1945), *Brit. Med. J.* **1,** 433
28. Rindi, G., Perri, V. and Ventura, U. (1959), *Ital. J. Biochem.* **8, 148.**
29. Sols, A. A., Cadenas, E. and Alvarado, F. (1960), *Science* **131,** 297/8.
30. Stedman, E. and Stedman, E. (1931), *Biochem. J.* **25,** 1147.
31. Stedman, E. and Easson, L. H. (1932), *Biochem. J.* **26,** 2056.
32. Towers, D. B. (1958), IV Inter. Congress Biochemistry (Vienna) Symp. III, 21.

33. VARETTI, E. (1902), See reprint *J. Biophys. Biochem. Cyt.* **10,** Suppl.
34. WARBURG, O. (1925), *Science* **61,** 576.
35. WHITEHOUSE, M. W. and BÖRSTROM, H. (1961), *Biochem. Pharmacol.* **7,** 135.
36. WOODS, D. D. (1940), *Brit. J. Exp. Path.* **21,** 74.
37. WOOLLEY. D. H. (1952), *A Study of Antimetabolites,* Wiley, New York.
38. WOOLLEY, D. H. (1958), *The Strategy of Chemotherapy,* p.139. 8th Symp. Soc. Gen. Microbiology, Eds. Cowan, S. T. and Rowatt, E., Camb. Univ. Press.

HISTORICAL REVIEWS

39. BACQ, Z. and ALEXANDER, P. (1961), *Fundamentals of Radiobiology,* Pergamon Press.
40. DAVIES, D. R. and GREEN, A. L. (1959), *Brit. J. Ind. Med.* **16,** 128.
41. O'BRIEN, R. D. (1960), *Toxic Phosphorus Esters.* Academic Press.
42. NACHMANSOHN, D. (1959), *The Chemical and Molecular Basis of Nerve Activity,* Academic Press.
43. ROSE, F. L. Ed. (1949), Symp. Soc. Exp. Biol. 111, *Selective Toxicity and Antibiotics,* p. 14. Camb. Univ. Press.

PART 2

CHAPTER 10

Cell Integration

LINACRE LECTURE: "THE CHALLENGE OF THE LIVING CELL"*

THIS Linacre lecture is part of the historic tradition of Cambridge; even after a lapse of 50 years I can recall the lecture to which I listened given by Sir Victor Horsley, the distinguished brain specialist. Thomas Linacre† gave his name also to the Professorship of Zoology in Oxford. He was a most distinguished scholar, having taken degrees in classics at Bologna and of medicine in Padua, where Dr. Caius and W. Harvey also studied medicine. No question here of two cultures. As is well known, Linacre was the first President of the Royal College of Physicians. The epitaph written to his memory by Dr. Caius, a later President, was as follows: "Fraudes dolosque mire perosus; fidus amicus; omnibus ordinibus juxta carus".

I have called my lecture "The Challenge of the Living Cell", and I propose to exclude as far as this is possible the cell in division and multiplication, also single cells like bacteria. Professor E. F. Gale dealt with bacteria recently.(35) I do not think that I need to stress the importance of the cell. In so far as it is in each case the fundamental unit of the tissues of our bodies, it is the unit upon which all the other phenomena of life depend. Even the corpuscles in our blood are single cells. Our organs such as the liver are multicellular. Our brains and nervous systems are a vast assembly of specialized cells. They have taken a long time to evolve, even from the earliest Primates some 70 million years.

The main purpose of the research in various fields of biology today, in a variety of disciplines, really is to understand in detail the constitution and workings of the living cells right down to the biochemistry and biophysics. One of our constant difficulties is that we cannot often observe changes when the cells are alive. It is true that Professor Keilin made a great advance by looking at the behaviour of the pigment cytochrome *in vivo*; but mostly it cannot be done, and we have to make inferences from killed tissue.

We can ask what are the main problems, and above all what should be

* Delivered 5 May 1962 to the Master and Fellows of St. John's College, Cambridge, and an audience of the Medical School and others.

† A portrait of Linacre of doubtful authenticity was shown at the lecture (Osler(58)).

our *strategy* in attacking these? How can we get a sufficiently comprehensive attack, so that the solution of the problems is not delayed too long; because a solution is important. Though we can look at this as an intellectual challenge with the rewards which come from increased knowledge alone, it is also of great importance for medicine to understand the living cells better, not only those of our liver, but more especially of our brains. So much of the modern world depends upon the capacity of our brains to comprehend and understand the vast problems on which mankind is now engaged; among which we may list the completely new field of exploration in outer space.

I can only give you a bird's-eye view of an enormous field, but I thought that by looking first at some dynamic reactions related to cell activity, and then considering them in relation to some modern structural advances, I might be able to show why I think that the time has come for a more integrated approach.

I will start first upon the dynamic side of biochemistry, which was so significantly emphasized by Gowland Hopkins in his 1913 address. I choose an illustration from our own field of work, and from the end effects of a toxic substance, which is found in plants both in South Africa and also in central Australia. These plants have the trick of combining together the fluorine and carbon atom to make an abnormal vinegar; an acetic acid with one atom substituted by fluorine ($FCH_2.COOH$). This simple chemical change converts the harmless vinegar into a most toxic substance causing severe convulsions, resembling tetanus in some animals, because it interferes with a particular part of the biochemical machinery of the brain; other cells are also poisoned in a similar way and have their activities cramped. The fluorine compound is synthetized in the body into a near relative of citric acid, which owing to the incorporation of the fluorine atom blocks a quite specific part of the apparatus concerned with what we call intermediary metabolism. Let us now look at the effect upon the cells of the kidney tissue of this substance fluorocitric acid:

FCH.COOH
|
HO.C.COOH
|
HCH.COOH

Working in Professor F. G. Young's Department, Dr. Cater of the Pathology Department and I found that it had a devastating action.(11) Figure L1 is taken from a section of normal rat kidney, and is a low power view, not much magnified.

For those who are not acquainted with the histological appearance of kidney tissue, I would draw your attention to two features, the initial filtering apparatus, called the glomerulus, and the tubules which lead from this; these are filled with a delicate layer of cells. It is easy to understand that if these cells are hit by some toxic agent, then the excreting capacity of the kidney will fail.

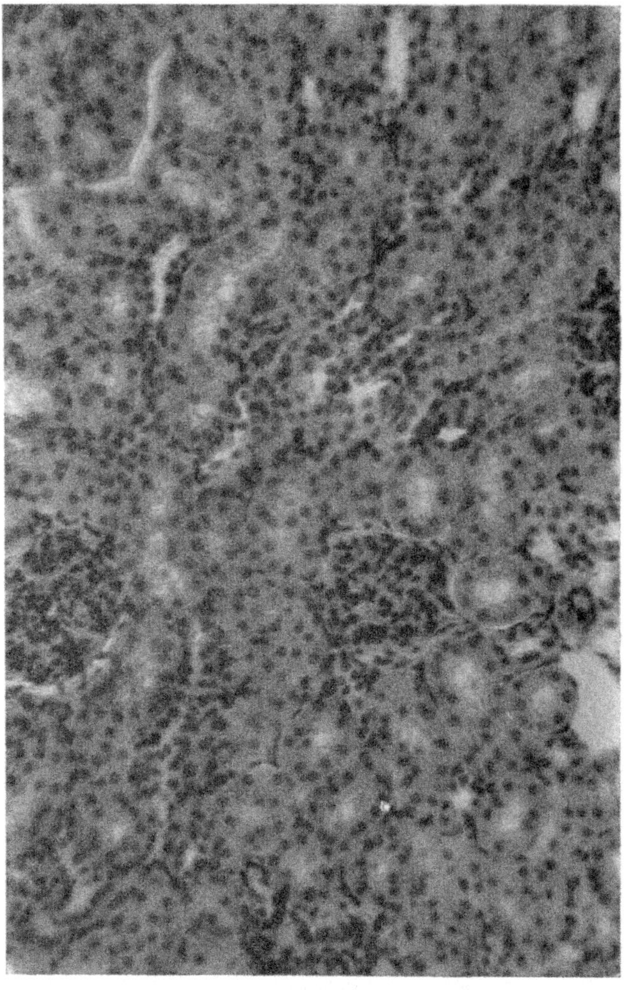

FIG. L1. Normal kidney.

Figure L2 shows a section with a higher magnification from the kidney of a rat which died 2 hr after receiving a dose of fluorocitrate. Even in this short time there is a large disturbance in the microscopic appearance of the section. The filtration from the glomerulus has been stopped by the extensive damage to the tubules and the whole surface of the tubule has been shed; it appears in the urine as the debris called casts by the clinician.

The experiment illustrates more than one point; we can be dogmatic and say that the disturbance is due to the blocking of one enzyme concerned with citric acid; because we can back this observation *in vivo*, with observations of the effects of the poison on pieces of brain and kidney tissue

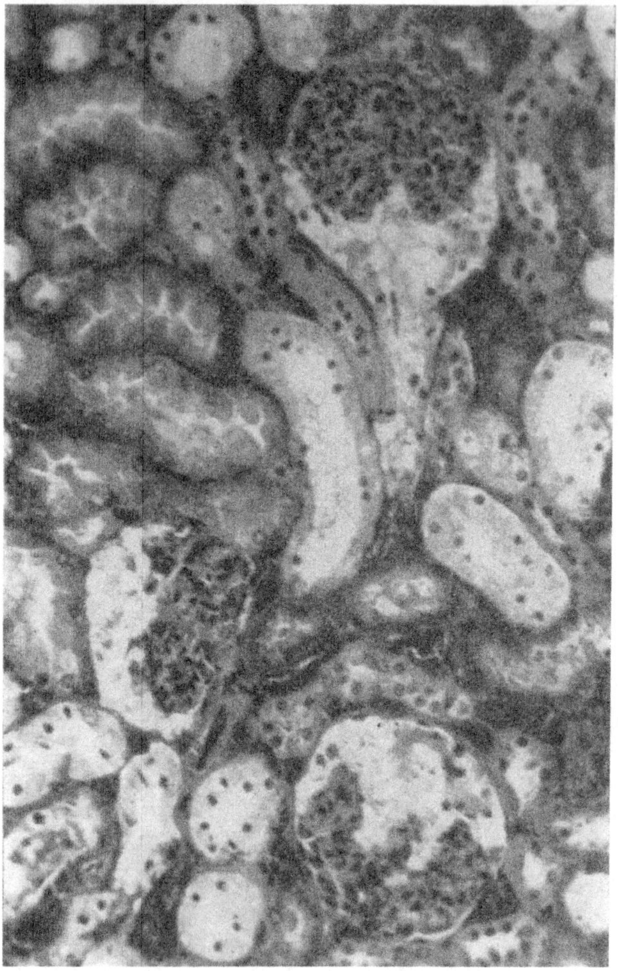

FIG. L2. Kidney from rat poisoned with fluorocitrate. (Cater and Peters, 1961.)

studied *in vitro*, where the same block in citric acid metabolism is found. Hence we can pinpoint a definite element in the biochemical machinery, and state that this pathological change starts with a fault in the biochemistry. It is fortunate that we can do this. All this can be summarized in the statement that fluoroacetate is toxic in the animal, because it is converted by the animal's biochemical machinery into fluorocitrate, which then blocks a specific catalyst in metabolism. We call the enzyme catalyst, aconitase, and the process "lethal synthesis".

This instance of a case where we can form a dogmatic conclusion as to the action of a toxin, could not have been reached without reliance upon a very large amount of careful biochemical work which has been done upon enzymes and enzyme systems isolated from tissues, a work of many years. In this case the enzyme blocked was discovered many years ago by Martius. The chart prepared by Dr. Nicholson of Leeds University (Fig. L3) and supplied by Messrs. Light & Co. represents the patient work of biochemists over at least 30 years, much of it since Gowland Hopkins gave his Linacre Lecture in 1938. It is a very good attempt to put on one sheet all the main biochemical changes found in cells; it is, of course, a very large undertaking and can be criticized here and there; it has also to be read with some knowledge, because it is a little like the map in the old Bradshaw railway timetable. In some places you will find bacterial or yeast reactions alongside those of the animal; for instance, the alcohol, ethanol is a terminus, which it is not in the animal. The stages for degradation of sugar and fat are written down. Each of these stages is catalysed by an enzyme or organic catalyst made by cells, which is mostly very specific for the stage concerned. Here you have the justification for Hopkins's continued encouragement of studies of specific enzymes. In their book, which is classical, M. Dixon and E. C. Webb[(28)] have now listed over 800 enzymes, and there are probably more to be found. Biochemists have not been idle!! The word *pyruvate* in Fig. L3, is a half-way house in the degradation of sugar, containing 3 carbon atoms ($CH_3.COC.OOH$) instead of the six ones in the sugar. The pyruvate oxidase system of enzymes is a very important one, and we owe to Martius and Knoop and to Krebs the generalization that the subsequent stages in breakdown go through a cycle of reactions called the citric acid cycle; it is shown on the chart entering this cycle through acetyl CoA. It is seen now why we get an accumulation of citric acid when we block this enzyme, and why we stop the subsequent course of energy production (Fig. L4). Actually I have found it to be a generalization that whenever one stops the course of reactions in the citric acid cycle, or as we used to say the pyruvate oxidase system, one gets *fits* in animals. A good example of this is the loss of thiamine (vitamin B_1) at the early stage, in vitamin deficiency; another instance, where we can relate the biochemistry *in vitro* to the condition *in vivo*. I think that it is an important fact that biochemists can take out pieces of brain tissue and work at some aspects of their biochemical metabolism; Professor McIlwain[(53)] and his colleagues have been able to get electrical responses from brain tissues *in vitro*. No one would be unwise

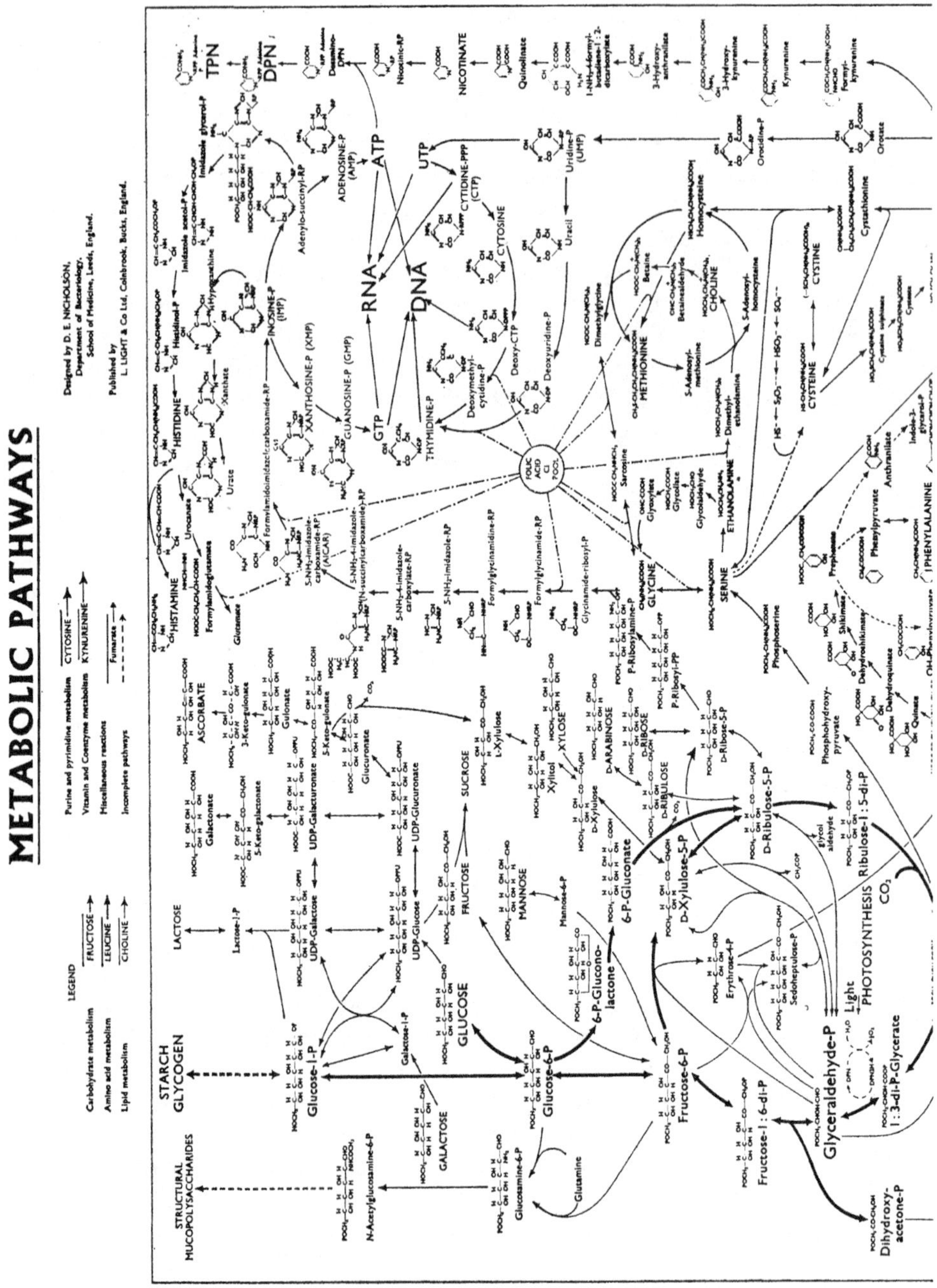
METABOLIC PATHWAYS
LEGEND
Carbohydrate metabolism
Amino acid metabolism
Lipid metabolism
Purine and pyrimidine metabolism
Vitamin and Coenzyme metabolism
Miscellaneous reactions
Incomplete pathways
Designed by D. E. NICHOLSON, Department of Bacteriology, School of Medicine, Leeds, England.
Published by L. LIGHT & Co Ltd, Colnbrook, Bucks, England.
STRUCTURAL MUCOPOLYSACCHARIDES
STARCH GLYCOGEN
LACTOSE
Lactose-1-P
Glucose-1-P
Galactose-1-P
GALACTOSE
UDP-Galactose
UDP-Glucose
GLUCOSE
Glucose-6-P
Glucosamine-6-P
N-Acetylglucosamine-6-P
Glutamine
FRUCTOSE
SUCROSE
MANNOSE
Mannose-6-P
6-P-Glucono-lactone
6-P-Gluconate
Fructose-6-P
Fructose-1:6-di-P
Dihydroxy-acetone-P
Glyceraldehyde-P
1:3-di-P-Glycerate
Erythrose-4-P
Sedoheptulose-P
PHOTOSYNTHESIS
Light
D-Xylulose-5-P
D-Ribulose-5-P
Ribulose-1:5-di-P
CO_2
Galactonate
5-Keto-galactonate
UDP-Galacturonate
UDP-Glucuronate
Glucuronate
Gulonate
3-Keto-gulonate
ASCORBATE
5-Keto-gulonate
L-Xylulose
Xylitol
XYLOSE
D-Xylulose
D-ARABINOSE
D-RIBOSE
D-RIBULOSE
D-Ribose-5-P
P-Ribosyl-PP
P-Ribosylamine-P
Glycinamide-ribosyl-P
Formylglycinamide-RP
Formylglycinamidine-RP
5-NH₂-imidazole-RP
5-NH₂-4-imidazole-carboxylate-RP
HISTAMINE
Glutamate
Formylamidoglutamate
Urocanate
HISTIDINE
Histidinol-P
Imidazole acetol-P
Imidazole glycerol-P
Urate
Xanthate
Hypoxanthine
INOSINE-P (IMP)
XANTHOSINE-P (XMP)
GUANOSINE-P (GMP)
Adenylo-succinyl-RP
ADENOSINE-P (AMP)
GTP
ATP
RNA
DNA
UTP
CYTIDINE-PPP (CTP)
CYTOSINE
Uracil
Uridine-P (UMP)
THYMIDINE-P
Deoxymethyl-cytidine-P
Deoxy-CTP
Deoxyuridine-P
FOLIC ACID C1 POOL
TPN
DPN
NICOTINATE
Nicotinic-RP
Desamido-DPN
Quinolinate
3-Hydroxy-anthranilate
3-Hydroxy-kynurenine
Kynurenine
Formyl-kynurenine
Orotidine-P
Orotate
SERINE
GLYCINE
Sarcosine
Dimethylglycine
Betaine
Betainaldehyde
CHOLINE
ETHANOLAMINE
Glycolaldehyde
Glycollate
Glyoxylate
METHIONINE
S-Adenosyl-methionine
S-Adenosyl-homocysteine
Homocysteine
Cystathionine
CYSTINE
CYSTEINE
Phosphoserine
Phosphohydroxy-pyruvate
Dehydroquinate
Quinate
Dehydroshikimate
Shikimate
Prephenate
Phenylpyruvate
PHENYLALANINE
Anthranilate
Indole-3-glycerol-P

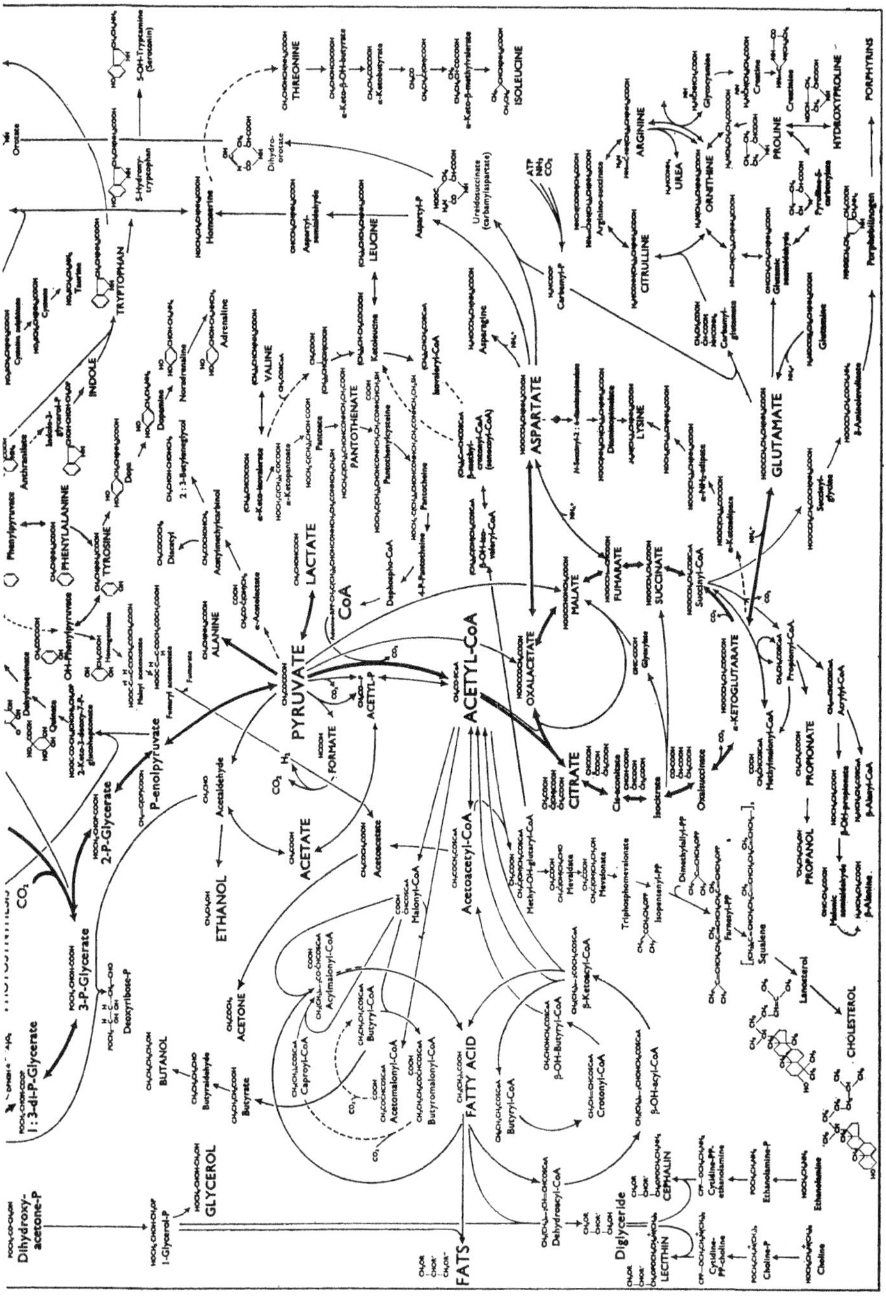
PYRUVATE
ACETYL-CoA
CoA
LACTATE
ALANINE
FORMATE
ACETYL-P
ACETATE
ETHANOL
Acetaldehyde
P-enolpyruvate
2-P-Glycerate
3-P-Glycerate
1:3-di-P-Glycerate
Dihydroxy-acetone-P
1-Glycerol-P
GLYCEROL
FATS
FATTY ACID
Diglyceride
LECITHIN
CEPHALIN
Choline
Ethanolamine
BUTANOL
Butyrate
ACETONE
Acetoacetate
Acetoacetyl-CoA
Malonyl-CoA
Mevalonate
Squalene
Lanosterol
CHOLESTEROL
CITRATE
Cis-aconitate
Isocitrate
Oxalosuccinate
α-KETOGLUTARATE
Succinyl-CoA
SUCCINATE
FUMARATE
MALATE
OXALACETATE
Glyoxylate
Propionyl-CoA
PROPIONATE
PROPANOL
Methylmalonyl-CoA
Acrylyl-CoA
ASPARTATE
Asparagine
GLUTAMATE
Glutamine
LYSINE
CITRULLINE
ARGININE
ORNITHINE
UREA
PROLINE
HYDROXYPROLINE
Creatine
Creatinine
PORPHYRINS
Porphobilinogen
LEUCINE
VALINE
ISOLEUCINE
THREONINE
Homoserine
TRYPTOPHAN
INDOLE
5-OH-Tryptamine (Serotonin)
PHENYLALANINE
TYROSINE
Dopamine
Noradrenaline
Adrenaline
PANTOTHENATE
Pantetheine
Deoxyribose-P

FIG. L3.

enough to claim that much of importance has not dropped out in getting the biochemical preparations; but much that is valuable still remains.

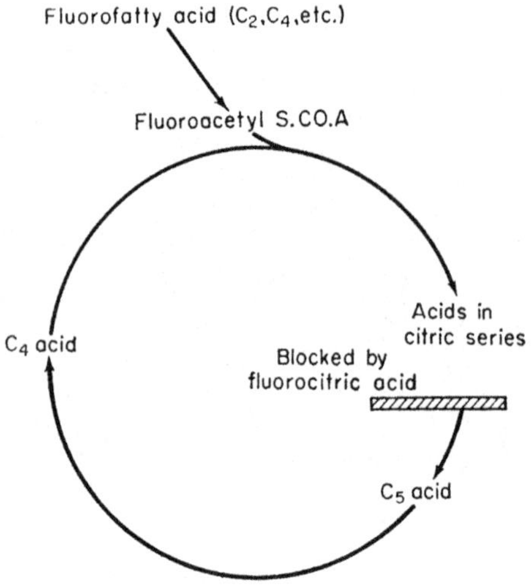

FIG. L4.

With this introduction of more classical biochemistry, and I have chosen only one example of several that would illustrate the points equally well, I will now turn to the problem as to how far the structural organization of cells may intrude upon our understanding of the chemistry of the reactions taking place in cells.* I will confess that when I first thought of lecturing upon this, some months ago, I was held back by the feeling that it might be thought to be too speculative. When I have tried to deal with function and structure in the past, the subject has been caught in a cross-fire between zoologists who, quite naturally, felt that they looked at more cells than we did; I refer especially to the brilliant researches on the eggs of the sea urchin. More chemically minded persons seemed to think that the introduction of structural considerations implied travelling too far from normal ideas of chemical equilibria. I am bound to say that I had the feeling that people thought it rather like talking about flying saucers or the Erewhonians.† Now, however, a remarkable change has happened. There has been a sudden swing of opinion in the last 4 months, so that I find myself talking about an immediately topical subject. I do not think

* K. C. Dixon[27] is one of the few who early emphasized the importance of structure in considering the biochemistry of brain tissue.

† Samuel Butler who wrote "Erewhon" was a Fellow of St. John's College, Cambridge.

that this is only the Scientific American! I got a reprint from Dr. Paul Weiss on the subject very recently indeed.*

Anything below 0·2μ (200 mμ) is too small to be seen even under the highest power of the light microscope. All the viruses, for instance, must be examined by the electron microscope.

When we examine a cell by ordinary light microscopy, we all know that we have the striking cell membrane, which has been the subject of intensive study in the case of muscle and nerve, its electric charge, and semi-permeable properties for various ions, especially Na and K; researches well known in Cambridge and elsewhere by the group in the Physiology Department (Professor Hodgkin and his colleagues).

Inside cells we have the nucleus in most cases, in which there may be a nucleolus, thought now to consist of RNA. The nucleolus rolls about in the nucleus as Sir J. Gray found many years ago, but J. E. Harris in 1939 proved in the egg of the sea urchin that the axis of rolling showed polarity there being viscosity in the nucleus.(43)

In the cells of blood and secreting organs there are often granules.

Then there are the mitochondria, on which I should like to concentrate now. Their importance in the life of the cell was first appreciated by biochemists 28 years ago, when Bensley and Hoerr(7) started to centrifuge fragmented cells of liver and so on at different speeds. At low speeds of the centrifuge, they spun out the nuclei, at intermediate speeds the mitochondria, and at very high speeds smaller particles called the microsomes. All this is very well-known to biochemists in this audience, and the work has been developed especially by Claude,(14,15) Hogeboom and Schneider and Palade.(46)

To return to the mitochondria, the interesting point arose that the pyruvate oxidase system of which I have spoken, containing the citric acid cycle is found in the mitochondrion, which thus becomes a kind of power house for synthesizing adenosine triphosphate (ATP). Hence we now have a location in the cell for some of the enzymes in the diagram. Before this connection was tied up I had been trying to get some evidence that the enzymes of this system were organized; and in 1938 I published with Wakelin some very simple experiments using long chain fatty acids, or more colloquially soaps, with a special length of carbon chain.(61) The experiments showed that the pyruvate oxidase system in brain, which we were using was wrecked completely by adding comparatively small amounts of the C_{10} acid called caprate (neutralized to pH 7·0). It can be seen that

* Dr. Weiss has condemned rightly the static idea of a cytoskeleton; but towards the end of his chapter makes many points which we share in common; see *The Molecular Control of Cellular Activity*, Academic Press, 1961. See also *The Cell* by J. Brachet and A. E. Mirsky.

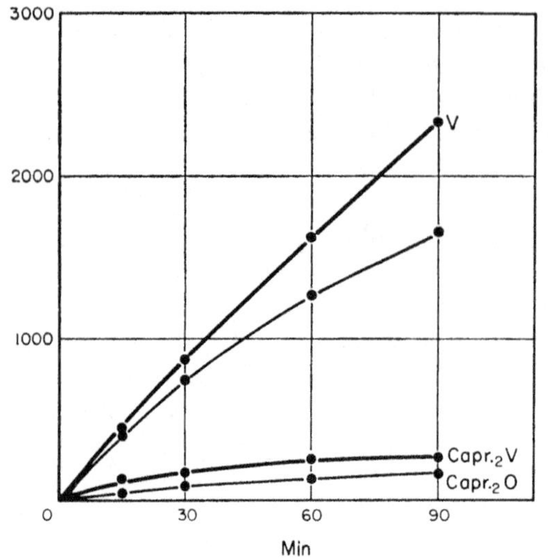

FIG. L5. The destructive effect of small amounts of Na caprate (decanoate), 0·7 mg, upon the pyruvate oxidase system in pigeon brain tissue. This was done upon a brei from a pigeon with thiamine deficiency, and shows that the fatty acid inhibited the extra O_2 uptake induced by the addition of vitamin. Hence it inhibited the pyruvate oxidase system. (Peters and Wakelin, 1938.)

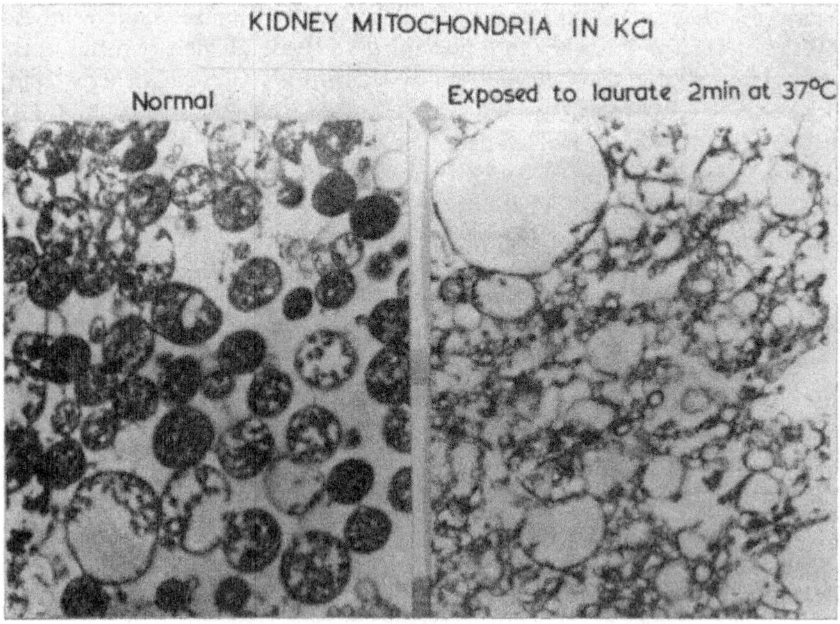

FIG. L6.

no oxidation of pyruvate takes place. What then has the fatty acid done? As I have mentioned, we know now, a fact that we did not know 24 years ago, that this complex of enzymes in the pyruvate oxidase system is in the mitochondria. We also know from studies of Palade especially that there is a structure in the mitochondrion, which is revealed by the electron microscope. Examination of a highly magnified photograph of mitochondria from the kidney of a rat, shows that there is a double membrane on the outside, and that there are partitions inside called cristae (Palade).

Recently I thought that it would be interesting to see whether a fatty acid of 12 C atoms would change the appearance in the electron microscope of kidney mitochondria, with which we work and which give the reactions expected with the citric acid cycle. We confirmed that these are wrecked by addition of the fatty acid. Mrs. Legge and I exposed some of our mitochondria to laurate ($CH_3(CH_2)_{10}COOH$). They were then embedded by Dr. D. H. Northcote, and cut by Dr. P. F. Baker, to both of whom I am most grateful. The results are interesting. The figures show the normal mitochondria with which we work, which have good biochemical properties. In the other part of the figure the effect of 2 min exposure at 37°C to laurate is seen. All the structure (Fig. L6) of the inside has disappeared, though curiously enough there is still evidence of some membrane outside. The correlation of structure with function, however, seems to be clear in this rather simple experiment. (I should explain that the mitochondria are rounded, because they were prepared in salt solution (KCl)).

Study of the mitochondrion at the moment is being very rewarding. There has been much work in the U.S.A. and also here on the biochemical activities of this organized complex of enzymes. One can mention the name of D. E. Green;* A. Lehninger has taken up the question of structural changes, especially the swelling which has been further studied here by Dr. G. D. Greville and Dr. J. B. Chappell. Lehninger considers that the mitochondria not only are a complex of enzymes, but in view of their organization also take part in osmotic activities and specific ion separation, an activity which is wrecked by certain enzyme uncouplers; they also can change their shape, can swell and contract. When seen in a *tissue culture*, mitochondria are elongated and show rather rapid changes; they will not be the same for a minute at a time; they may divide and then coalesce.† About a month ago I saw a film of a tissue culture from chick heart made in ultraviolet light by Professor Buffa of Modena with Professor Godina and Dr. Barasa of Turin. They found that treatment with fluoroacetate

* D. E. Green and T. Oda[38] (1961) have calculated dimensions in the mitochondrial unit; a single mitochondrion has some 17,000 repeating enzyme units.

† It is not certain that mitochondria are so elongated in a tissue like liver or so motile.

showed a curious bulging after a short delay.* I am indebted to them for these unpublished results.

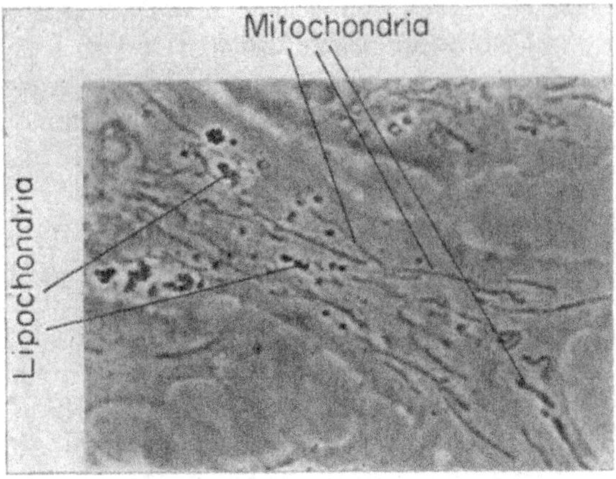

FIG. L7. Normal filamentous mitochondria in tissue culture from a chick heart, lipochondria present. Phase contrast time-lapse cinematography. Only part of the cytoplasm of three cells visible. (Barasa, 1960.)

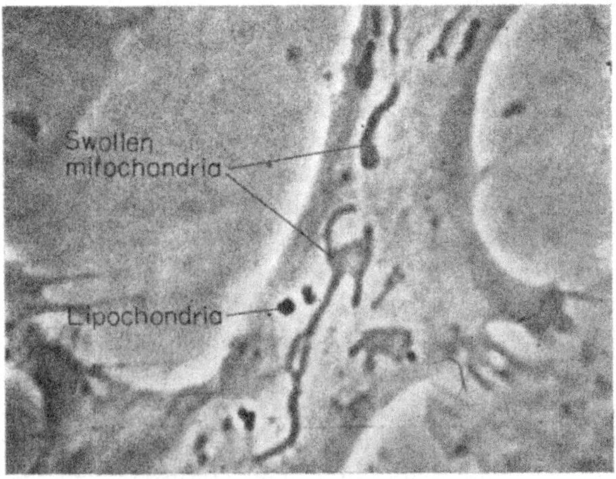

FIG. L8. Swollen mitochondria due to treatment with fluoroacetate (103 M). Picture taken after 3·5 hr. A 7 hr hanging-drop culture of heart tissue from a 5-days-old chick embryo. Tyrode solution. (Buffa, Godina and Barasa, 1961.)

* This is consistent with the idea that the fluoroacetate is synthesized to fluorocitrate, which then induces an increase in citrate, and so an increased osmotic pressure.

I am vividly interested in this subject of organized structure in organelles of the cell like the mitochondria, because over 30 years ago, in 1929, I was so convinced of the strategic importance in research of considering the organization of the cell as well as of studying the enzymes and their specificity, that I ventured to postulate the presence of an organized structure inside cells which could be responsible for coordinating their reactions.[63] To quote from my remarks in 1929: "The fact that the cell contents are often quite fluid can be reconciled with the fact that the living cell shows a continuous directive power by the view that protein surfaces in the cell constitute a mosaic from which radiate chains of molecules, controlled by their attachment to the central mosaic. These constitute a fluid anatomy in the cell, and the central mosaic behaves as a kind of central nervous system". The need for some such view was also felt a little later in 1935 by Joseph Needham[56] to explain the action of organizers in development in which he was so much interested; he termed it a "cytoskeleton", a term which has stuck, and which is all right provided that is it not confused with the older and rather rigid cytosquelette of the histologists.

It is relevant to mention that Wigglesworth[76] has always felt the need for a cytoskeletal conception to explain some of the effects which he found in the integument of an insect *Rhodnia*, with a transplanted cuticle; the piece which was transplanted still grew according to the original orientation. What I proposed I felt to be a matter of logic, needed to explain in chemical terms the facts that the cell must have compartments which were separated from one another and also that there was coordination to surface reactions; and yet the system was fluid. I could have made no proposals at all in terms of chemistry without the conception of orientation of long chain molecules in the interfaces of separation between aqueous and fatty solutions.* In this University, I feel that we should be proud that experimental evidence of this orientation in surfaces was first produced by W. B. Hardy in 1913,[42] 50 years ago. Sir William Hardy was a great figure as a Tutor in Gonville and Caius College; but he was also a physiologist, a biological secretary of the Royal Society, in his spare time a Master Mariner, and, an unusual honour, gave both the Croonian and the Bakerian lectures, the Biological and Physical Lectures of the Royal Society. Hardy's "interfacial tension" work was extended by Harkins in 1917. Another slant

* I think that those who attacked this hypothesis should have been prepared themselves to suggest a better, which would have logical validity in chemical terms; for this was the real point. No pupil of Gowland Hopkins could ever again be content with any scheme for integration which did not carry the chemistry right inside the cell. So far as I am aware no one else did suggest any other hypothesis. In fact A. R. Moore[55] (1935) found a similar hypothesis necessary to explain his observations upon *Plasmodium*, which would pass through pores of an average size 1 μ; but were disrupted readily if forced through pores of less than 0·2 mm.

was given to the subject by Langmuir in 1917 with his famous trough, on which he could plant layers of fatty acids one molecule thick, and which has become a schoolboy experiment. All this work has been consolidated and much extended by N. K. Adam,[1] E. K. Rideal,[21] and their schools. But the point which I wish to emphasize is that without this conception of

W. B. HARDY.

orientation, it was not possible to make any proposal of an internal fluid cytoskeleton in chemical terms. It meant that with only a chain of some 10 to 12 $-CH_2$ groups in between the ends of the molecules one could have on one side a layer of paraffin wax, CH_3 groups, and on the other a group such as $-COOH$ taking part in aqueous reactions.

I can recall that when W. B. Hardy first showed me the paper of Langmuir about 1920, I went over to see a distinguished organic chemist whom I knew, and whose early demise was a big loss to chemistry. When I asked him about these exciting results of orientation, he was most worried, thought that it was going much beyond the facts, and said that the formulae which they drew on paper helped chemists in their synthesis, but should not be taken more seriously. I mention this only to show that the conceptions were new at the time. At the present time the dominating importance of surface structure in the life of the cell is generally accepted.[77]

Though I have continued to think it important to investigate problems connected with interfaces in the cell, the proper application of these ideas has had to wait a long time; the waiting period has been due in part to the difficulty of finding methods of proof, and partly to the other more concrete and straightforward problems waiting to be investigated. In the interim, I was able to prove that there was a shift of apparent pH at an interface for a long chain fatty acid (see p. 210).

It is a generalization that molecules in an interface tend to be less dissociated, as the well-known experiment of Deutsch,[24] with bromothymol blue shows.* The soap experiment which I have mentioned already convinced me that there was organization in the pyruvate oxidase system, and some work published in 1940 by Keilin and Hartree[50] showed that the cytochrome, succinate oxidase system from heart was less efficient when put together artificially than the original, again suggesting structure.

In the meanwhile in 1936, Danielli and Davson[18] proposed a structure for membranes of two opposed layers of long chain fatty acids, to which is attached on each side a layer of protein. This structure for membranes may be said to have received proof from two directions. Electron microphotographs show a kind of double surface for the membrane with about the right dimensions. Quite recently, Professor Lehninger has told us of some experiments which he has in progress, in which they have made artificially membranes from long chain fatty acids, which become stable when they reach a dimension of two molecules thick. It is possible to pass a pin through them without disruption. There are other successful attempts in progress, so that we may soon look forward to knowledge of the behaviour of mixtures of the simple fatty acids with phospholipids and proteins.

Passing now from the membranes and mitochondria, to the other contents of the cell. Is it just a loose bag of enzymes and other constituents such as proteins, which orient themselves slowly in relation to one another and find stable positions by simple equilibrium alone?—or have we to consider something more organized? In the cell there certainly seem to be partitions; otherwise one does not see how various chemical reactions can be fenced off from one another; there seems to be an incredibly accurate organization even of the fats. I have seen electron microscope pictures in which there are globules of fat next door to the mitochondria; yet from the pictures shown on page 182 any release of fatty acids in excess might be expected to wreck the mitochondrial structure. How are the lipases controlled?

The microsomes have not yet been considered; they are that part of the cell which it is difficult to bring down in the centrifuge; many of these are the ribosomes, particles containing ribonucleic acid and actively concerned in the synthesis of protein, as thought originally by Brachet and Casperson. It is now clear that the microsomes are in their turn part of a structure inside the cell which is called the Endoplasmic Reticulum, or briefly the E.R. This discovery has been developed at the Rockefeller Institute, especially by Palade and K. R. Porter.[60] This recognition of a continuous

* Deutsch's experiment was shown at the lecture. A buffer solution of about pH 7·2 containing bromothymol blue is shaken with benzene. The droplets of aqueous phase change colour from blue to yellow, returning to blue when shaking is discontinued.

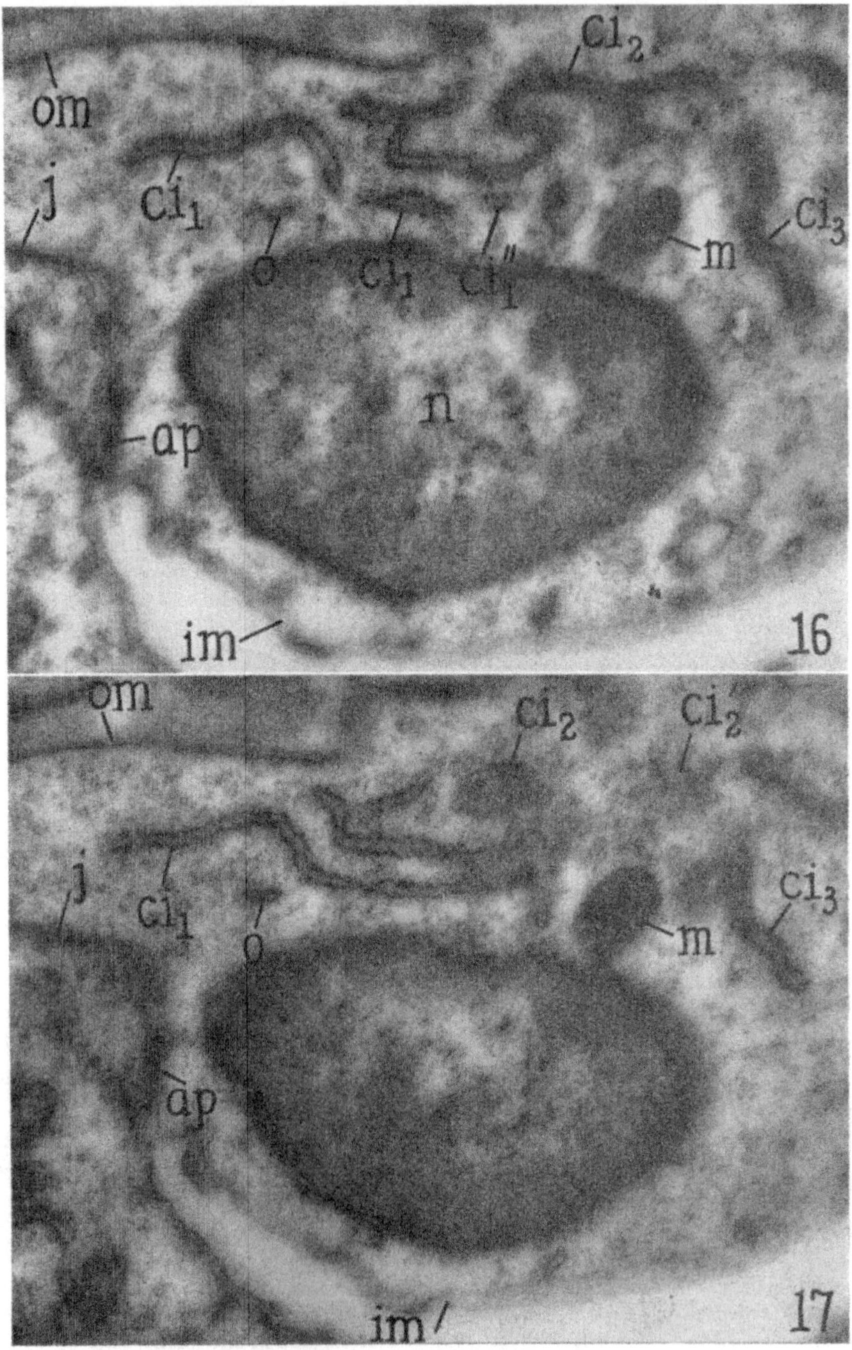

FIG. L9. Endothelial cell in the wall of a blood capillary of the myocardium. Two serial pictures, etc. (Palade and Porter, 1954.)

n = nucleus; *m* = mitochondrion; *O*, Cl_1, Cl_2, Cl_3 = ER.

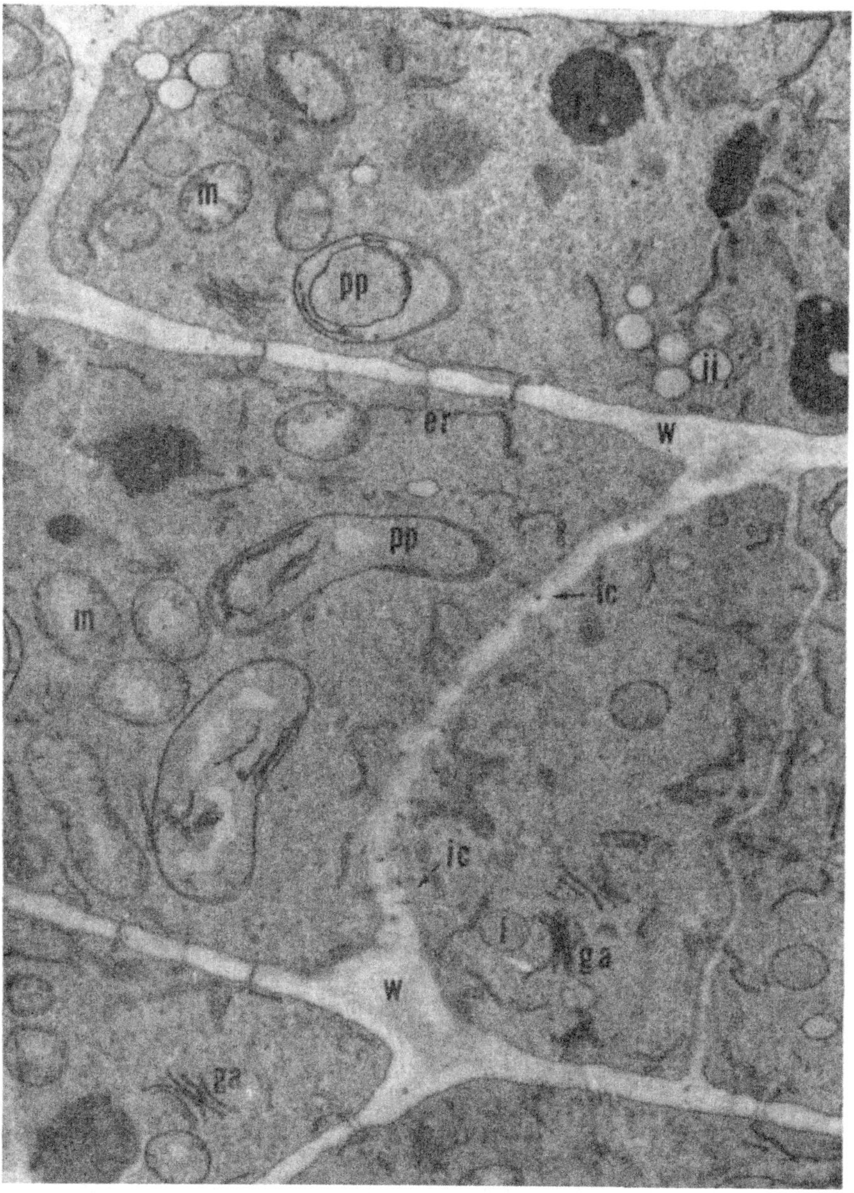

FIG. L10. Electron microscope photograph of an onion root tip, showing the manner in which the endoplasmic reticulum passes from one cell to another across the cell wall. *m*, mitochondria; *pp*, proplastid; *i* and *ii*, unidentified; *er*, endoplasmic reticulum; *w*, wall; *ic*, intercellular cement; *ga*, golgi apparatus. (From, Whaley, Mollenhauer and Leech, 1960.)

E.R. seems to me to be one of the great achievements of electron microscopy. In Fig. L9 you have two contiguous sections, and you will see that taking the two together there is good evidence of tubes in connection with the oval bodies in the top picture and the tubules in the lower. Of course the first reaction of anyone is to ask the question whether the structures are artefacts of fixation, a matter which electron microscopists have much in mind. There are many ways to argue about this; what I find convincing is that by different techniques the same structures appear. It is also impressive that as the technique has improved over the last 10 years, the fine detail has become more evident, and even yeast shows mitochondria in the latest work,[79] and bacteria, cytoplasmic membranes.[36] Of course no one would like to say that fixation and coagulation of proteins have made no change.

This double membraned structure, the E.R., does come rather close to the fluid cytoskeleton, which I thought must be present. In so far as enzymes and ribosomes may be attached to it, it will be in control of some of the actions of some enzymes; this was a point which I always felt to be a logical extrapolation. Movements of this structure would be even possible which could bring enzymes to substrates, so that there would be an additional control, in addition that is to say to any control by specificity.

The question arises whether the surface membrane of the cell is attached to this E.R. In the case of the plant cell, Dr. Northcote takes the view that there is no doubt about the continuity of the E.R. with the nuclear membrane, but tells me that there is still argument about the degree of attachment to the external surface. I am indebted to him for introducing me to figures of the E.R. in plant cells, root tips from onion and maize, one of which is shown in Fig. L10 by Whaley *et al.* The plant evidently has the same structures in principle.[68,75]

There are some criticisms in detail about the E.R., a healthy sign, I think. Birbeck and Mercer[8] for instance, think that the E.R. is only present when the cells are making proteins for export so to speak; and that it is not present in cells which do not have to excrete; but I find that this is not universally accepted. The matter should perhaps be settled soon.

We have travelled a long way from the position years ago that the effect of internal surfaces was not believed in at all to a position where we have too many, the biochemical influence of which we have still to understand. Certainly surface charge can matter in enzyme action. Bangham and Dawson have shown[5] (Fig. L11) that the action of a lecithinase, phospholipase C from *Clostridium Welchii*, upon lecithin at an interface is entirely controlled by the electric charge upon it. It is of very great interest that Varetti should have described the reticular structure in muscle as early as 1902. In fact the whole trend of much exciting present work is in line with the conception that internal and external surface structures can be controlling enzymes, for instance the influence of DNA upon RNA, and of RNA upon the synthesis of protein.

Clearly the matter of how enzymes act in interfaces is raised, and I am glad that P. Mitchell(54) is working upon the theory of this, on chemiosmotic equilibrium.

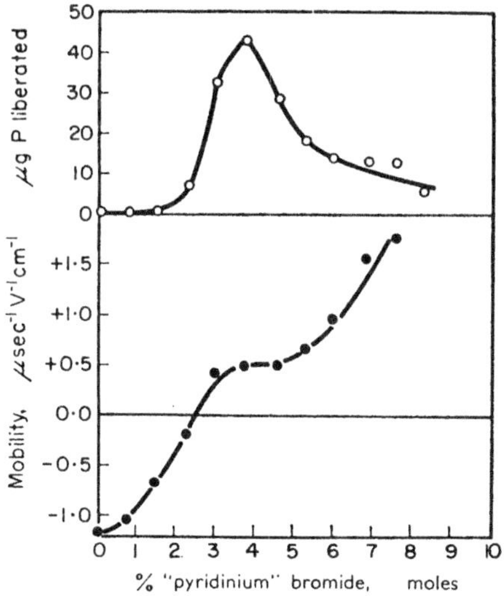

FIG. L11. Action of phospholipase C upon lecithin at an interface. Relationship between the percentage of the detergent docosanylpyridinium bromide added to lecithin containing 2·3 per cent (mole) diacetylphosphoric acid; o—o the release of water-soluble P by α-toxin; ●—● the electrophoretic mobility of the micelles. 25°C, pH 7·6. (Bangham and Dawson, 1962.)

The electron microscopists can distinguish in liver a smooth and rough endoplasmic reticulum, depending upon the presence of more or less ribosomes attached to it. Emmelot(31) states that by feeding with the amino acid cysteine, the amount of the rough E.R. can be increased. It is a challenge to us all to work out techniques which will enable us to integrate our present knowledge of enzymes with the action of these remarkable interfacial structures.

I would now like to enlarge a little upon the vistas which lie before us.

I do not need to speak of the work on the gene, that is to say upon the nucleic acids and the means by which the genetic substance stamps its code afresh at each new division of the cell. Here in Cambridge we are lucky to be near the high priests of this work. But it is interesting to see how the cell pushes aside its endoplasmic reticulum when it is beginning to divide, in Fig. 12 for which again I thank Dr. Northcote.(68,10)

It is better for me to spend time in speaking to you of the applications

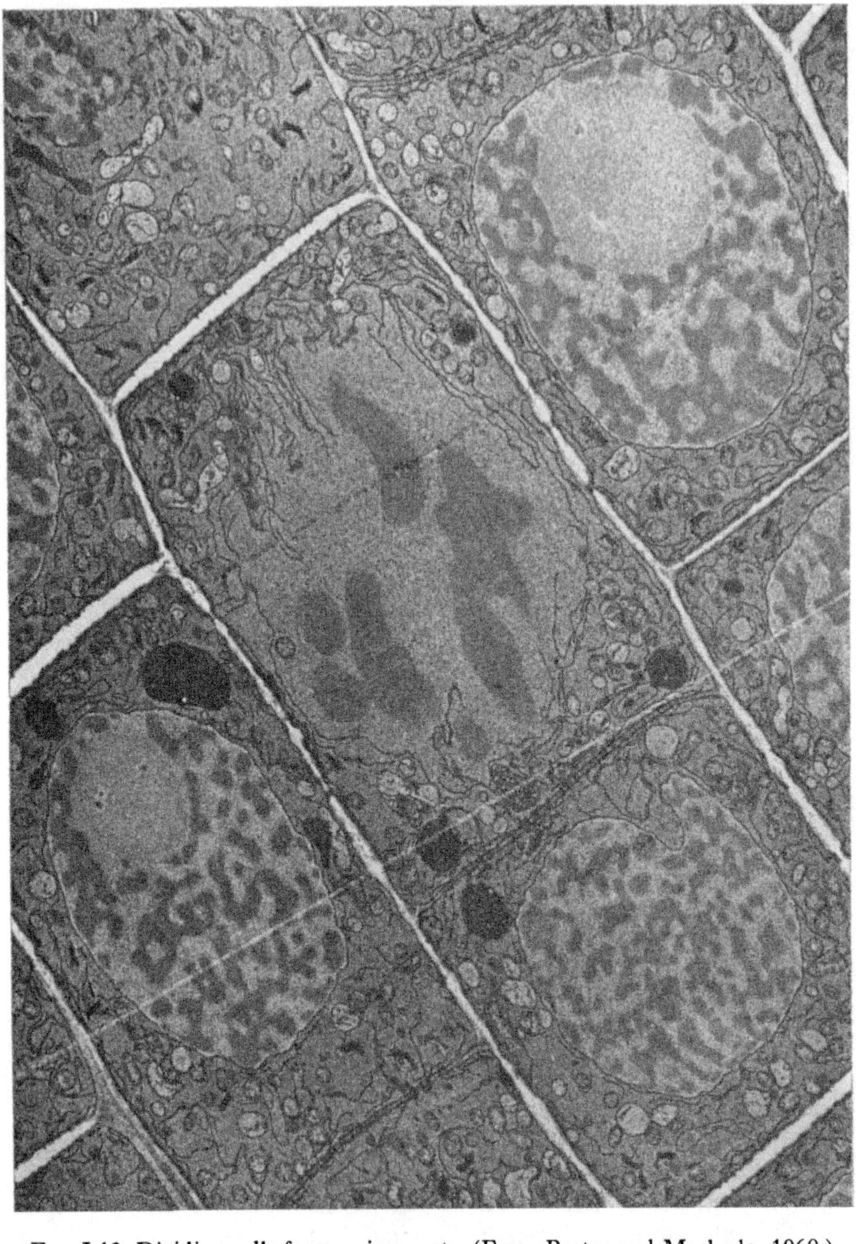

FIG. L12. Dividing cells from onion roots. (From Porter and Machado, 1960.)

of the fluid cytoskeletal ideas. By this term it is convenient to include all those structures forming part of a more permanent fluid organization. I mean that even among the various movements inside cells, these structures retain a certain amount of organization in relation to one another. (This is not very different from saying that in a benzene ring there is a permanent organization, even though the individual parts of this are in rapid motion about a mean position.) It was always rather clear to me, and would be to anyone I think, that once one admitted the existence of the cytoskeleton, much followed in consequence. For instance, it then became likely that parts of the internal structure might move about and be accurately directed, so that reaction could be accelerated or the reverse. Further if the surface membrane was concerned, a stimulus to this might well spread along some internal channel quicker than by diffusion alone, a circumstance virtually proved for muscle by A. F. Huxley and colleagues.[49]

One consequence, therefore, would be that a complicated molecule like a hormone such as a steroid or protein could take part in a reorientation of the cytoskeleton, and in this way might alter the action of several enzymes. For some years I did not dare to suggest this: but I proposed it

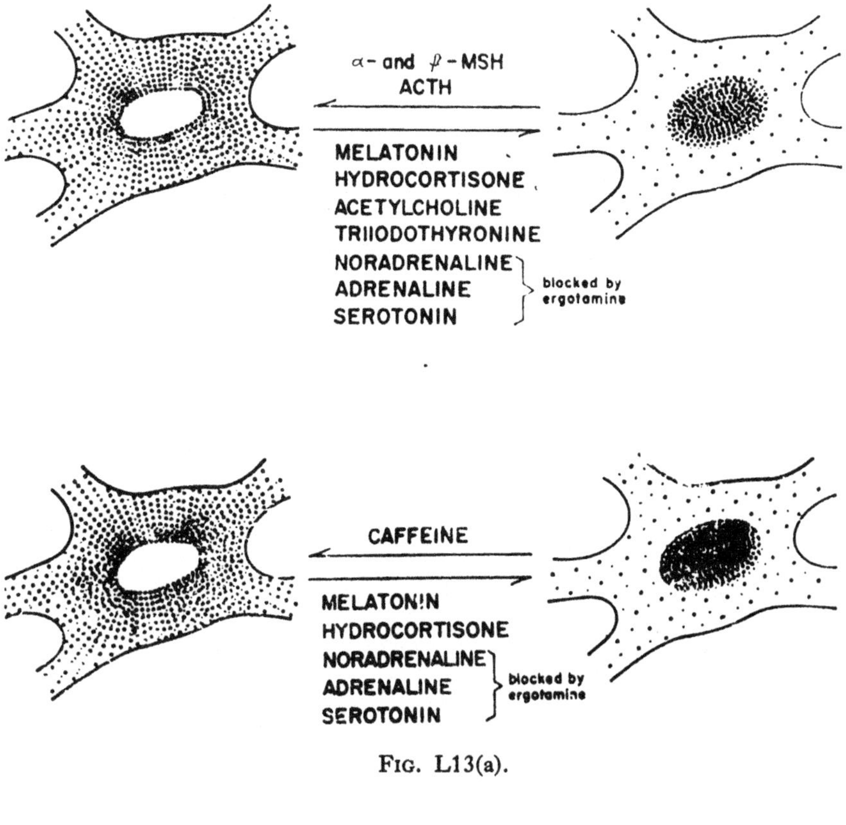

FIG. L13(a).

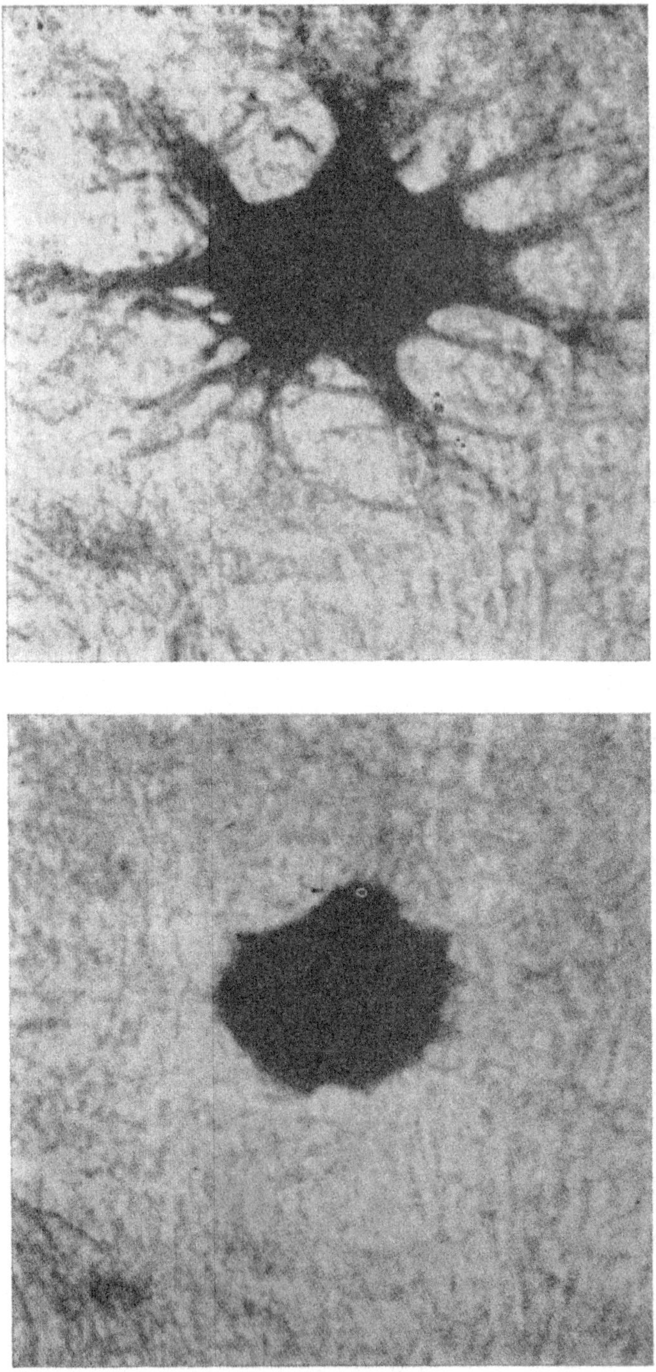

FIG. L13(b). Frog melanocyte with melanin granules clumped around the nucleus (left) becomes completely visible (right) after MSH is added, causing melanin granules to spread through all the branched processes of the cell. Granule dispersal makes frog's skin darken. (Lerner, 1959.)

finally in print in 1956, 6 years ago. About the same time Bacq and Alexander[3] proposed that this might be one of the main effects of damage by radiation, a damage which occurs with very great speed, within 1 min, though the effects may show up only slowly. So far as a hormone is concerned these ideas have been brought to the decisive stage of proof by

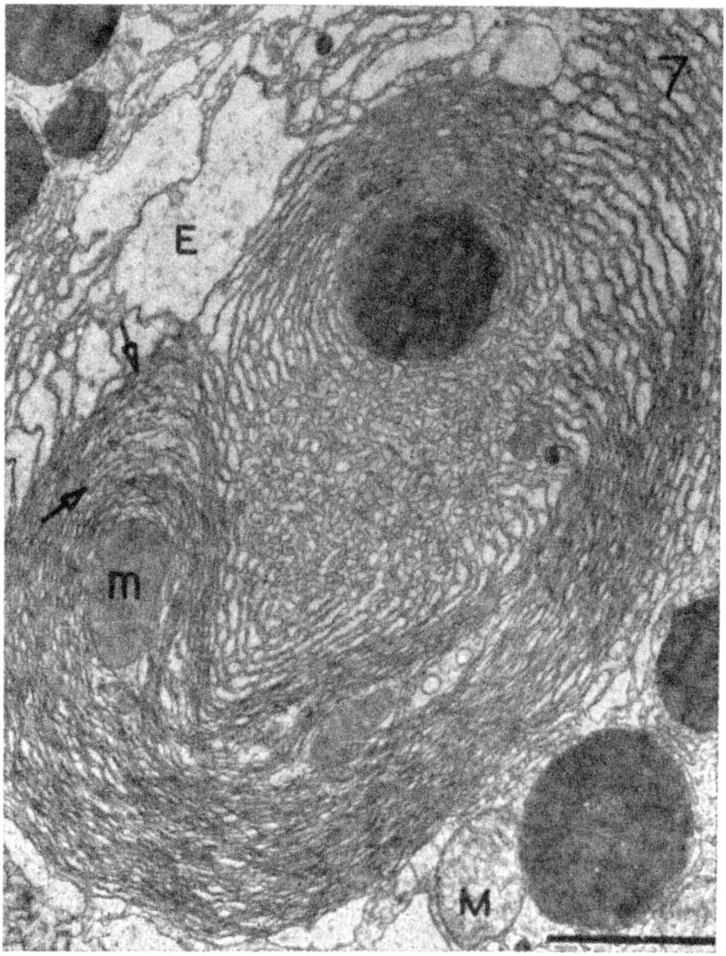

FIG. L14. Effect of the carcinogen DMNA, 20 mg, administered to a rat upon the fine structure of liver cells taken 20 hr after giving the dose. A large ellipsoidal body consisting of an elaborate mass of double membranes very closely packed together (bottom) communicating with a complicated tubular system (centre) and some ER membranes with attached (——→) particles. A microbody (*m*) and a large lipid inclusion are contained in these structures which seem to be continuous with the ER vacuoles (*E*). (Emmelot and Benedetti, 1960.)
(From Plate 218, 7, *J. Biophys. and Biochem. Cytol.*)

Dr. A. B. Lerner of Yale University, whom we have had the privilege of meeting in Cambridge this year.[51] Working with frog's skin, he has been able to show that the granules containing the dark pigment melanin are moved about in the cell by exposing the whole cells to very small amounts (10^{-11} M, 0·17 μg/ml) of the melanocyte stimulating hormone of the pituitary gland; similar changes can be produced with some salts, but much higher concentrations are needed than of the hormone. In this case, we have a proof that a hormone acting upon the surface of the cell can induce a rearrangement of the cell contents (Fig. L13 a and b).

If one is thinking upon these lines it is an obvious idea to extend this to the possibility that there may be a change in the cytoskeleton in cancer, and I did in fact point this out two years ago,[65] not knowing that Porter and Bruni,[67] and Emmelot and colleagues[29,30] had work already on the way. Emmelot and Benedetti, for instance, have found that treatment of rats with dimethylnitrosamine, a compound which induces cancer of the liver, even within 3 hr of injection causes changes in the appearance of the Endoplasmic Reticulum, and after 24 hr the changes are as great as appear in Fig. L14.

It has been said that one knew when one was on the right lines in research when the evidence begins to fall in rather than falling out. On the matter of surfaces, there is now much evidence falling in. Here in our midst, Dr. Willmer has been investigating the behaviour of a strange unicellular organism (*Naegleria gruberi*), which is like the genie of old[78] (Fig. L15). If placed in a medium of lower osmotic pressure, it can change from an amoeboid form rolling along the ground, to a free-swimming flagellate; and—the important point—this can be retarded by the action of certain steroids, such as deoxycorticosterone and aldosterone.

Then again, Dr. Honor Fell and her colleagues at the Strangeways Research laboratory have been following the work (Fig. L15) done with Sir E. Mellanby,[32] in which they found that excess of fat soluble vitamin A added to tissue cultures of ectoderm from the chick suppress keratinization and induced the formation of mucus secreting cells; it had a plastic effect. Now the Strangeways group[33] find that one marked and specific effect of the vitamin A in excess is to cause the liberation of some proteases, enzymes which can take to pieces proteins; this is evidently an effect on the surface of the cell, and influences growing bones (Fig. L16).

The problem of membranes carrying an electric charge inside the cells as well as on their surface may be more than a scientific abstraction because an application on these lines has been made to the problem of poisoning of the liver by carbon tetrachloride. In Sir Roy Cameron's department at University College, Judah and Rees[13] with colleagues have been working for some time upon the cause of liver damage which shows up as a leakage of enzymes into the blood stream, of fatty accumulations, and mitochondrial

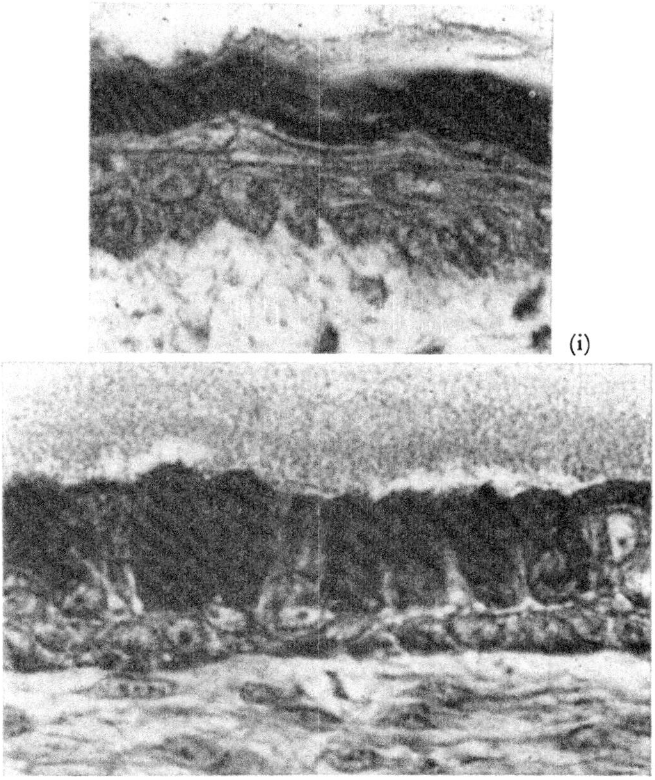

(i)

(ii)

FIG. L15. In tissue culture of chick ectoderm, vitamin A keratinization (i). *Normal,* (ii) +Vitamin A, being grown in a medium enriched with vitamin A. Tissue resembles nasal mucosa; no sign of keratinization.

damage; the effect upon the mitochondria now seems to be later than the other actions, and Recknagel and Lombardi(69) think that the initial effect is on the E.R. which shows evidence of early change. It was found that a variety of substance like the tranquillizer chlorpromazine could prevent this kind of liver damage. Now Bangham, Rees and Shotlander(6) (1962) have found that by thinking of the problem in terms of substances which induce charges on membranes, much more powerful antidotes can be found; in especial stearylamine, and CTAB (cetyl trimethyl ammonium bromide). Surface charge may therefore be of medical importance!

One is glad to know that research is in progress which may well extend our knowledge in these directions. For instance, Dr. Northcote is now able to grow sycamore cells in culture, and there has been an advance in centrifuging, the gradient technique, in which a centrifuge tube is layered with sucrose of different strengths, and so a further differential centrifuging can be used to separate fractions. Gray and Whittaker have in this way separated nerve ending in brain tissue(37) (Figs. L17 and L18).

The experiments which I have outlined briefly raise points of much subtlety. From many sides we have evidence of a regulation which seems well nigh miraculous. I have referred already to the fatty acids, but the experiments on vitamin A also underline this. Essential in the diet, it can produce hypervitaminosis even at tissue level.

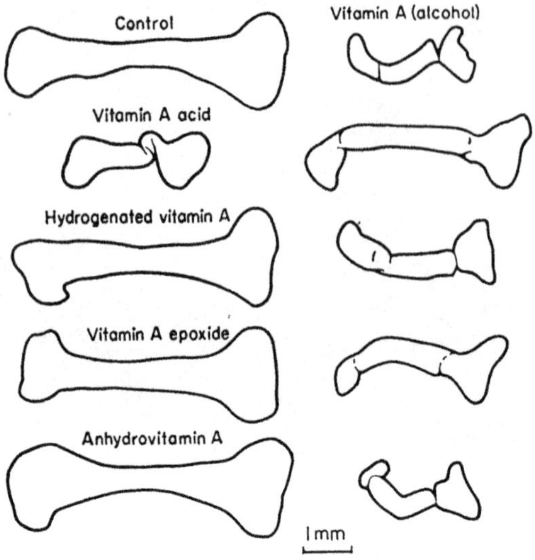

FIG. L16. Effect of vitamin A (alcohol) and other analogues on development of bone in chick embryos. Camera lucida drawings of five pairs of living femora after cultivation for 8 days. All the explants on the right were grown in medium containing 0·01 mole of vitamin A. Corresponding rudiments on left grown either in normal medium (top) or in the presence of 0·01 μ mole of an analogue of vitamin A/ml. Vitamin A acid alone has the same effect as vitamin A. (From Fell, Dingle and Webb.[33])

Then we have the trace metals. We must have copper in our diets; yet when a copper salt is injected into the brain in small amounts (10 μg in the pigeon), it is rapidly toxic. Dr. Walshe and I are now working on this,[20] in connection with the clinical condition of Wilson's disease, in which the patient accumulates copper. The metal cobalt is another case. It is an absolute essential in the form of vitamin B_{12} which cures pernicious anaemia; without it soils are not fertile. Yet the "Strangeways" group here are working upon the phenomenon that a few milligrams of cobalt metal injected into the muscles of the rat will form rhabdosarcomas. Again we are up against the problem of regulation—and I expect of segregation of substances within the cells.

It is only by courtesy that I have been able to speak today about "cells" because there are so many different kinds of cells. Nevertheless much of

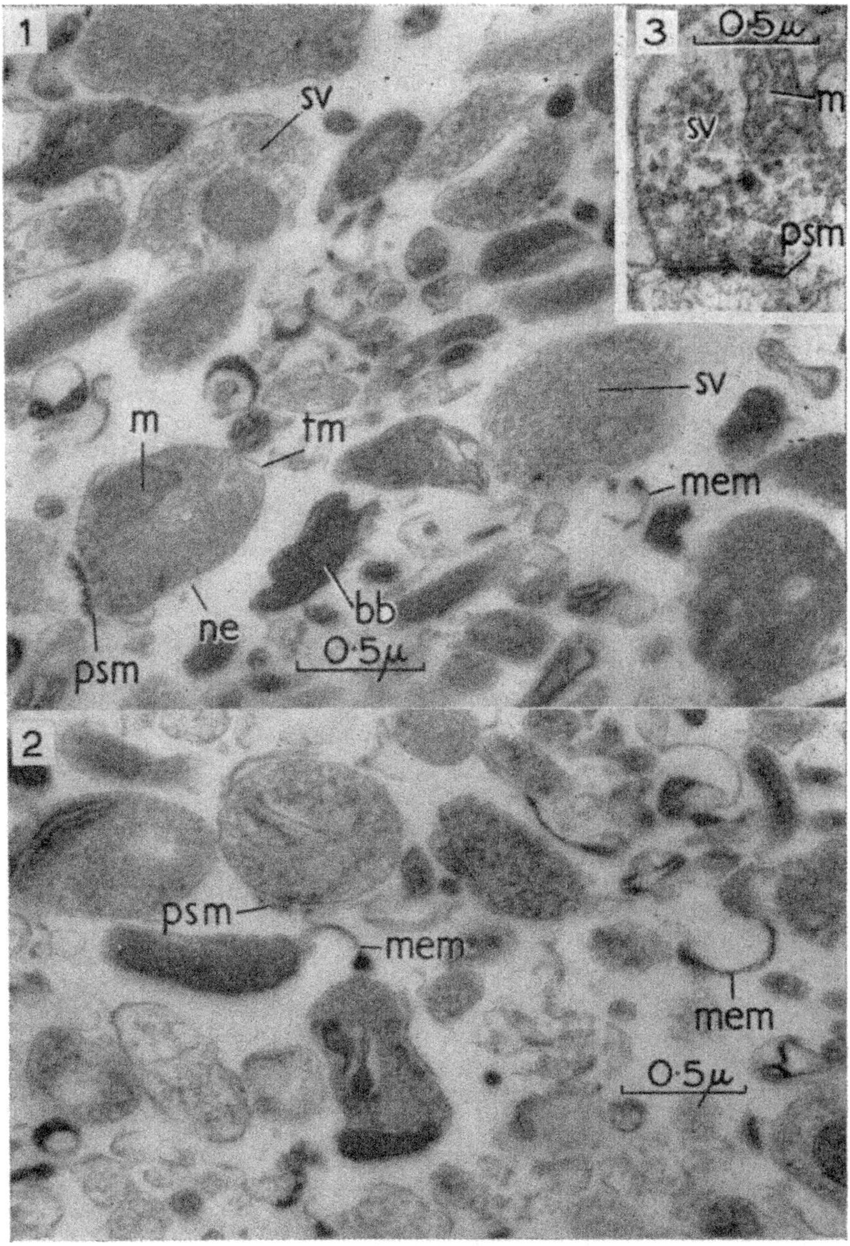

Fig. L17, PLATE 1.

FIG. L17. Electron microscope photographs from brain tissue of fractions containing nerve endings prepared by centrifuging by the gradient density, and relatively free from mitochondria. Two plates.

PLATE 1. 1 and 2. Note synaptic vesicles (*sv*) and mitochondria (*m*) enclosed within membrane (*tm*) to form a nerve ending particle (*ne*) which may have a thickened portion of the post synaptic membrane (*psm*) adherent to it. Isolated membranes (*mem*) may represent psm's.

3. Whole tissue section showing nerve ending for comparison.

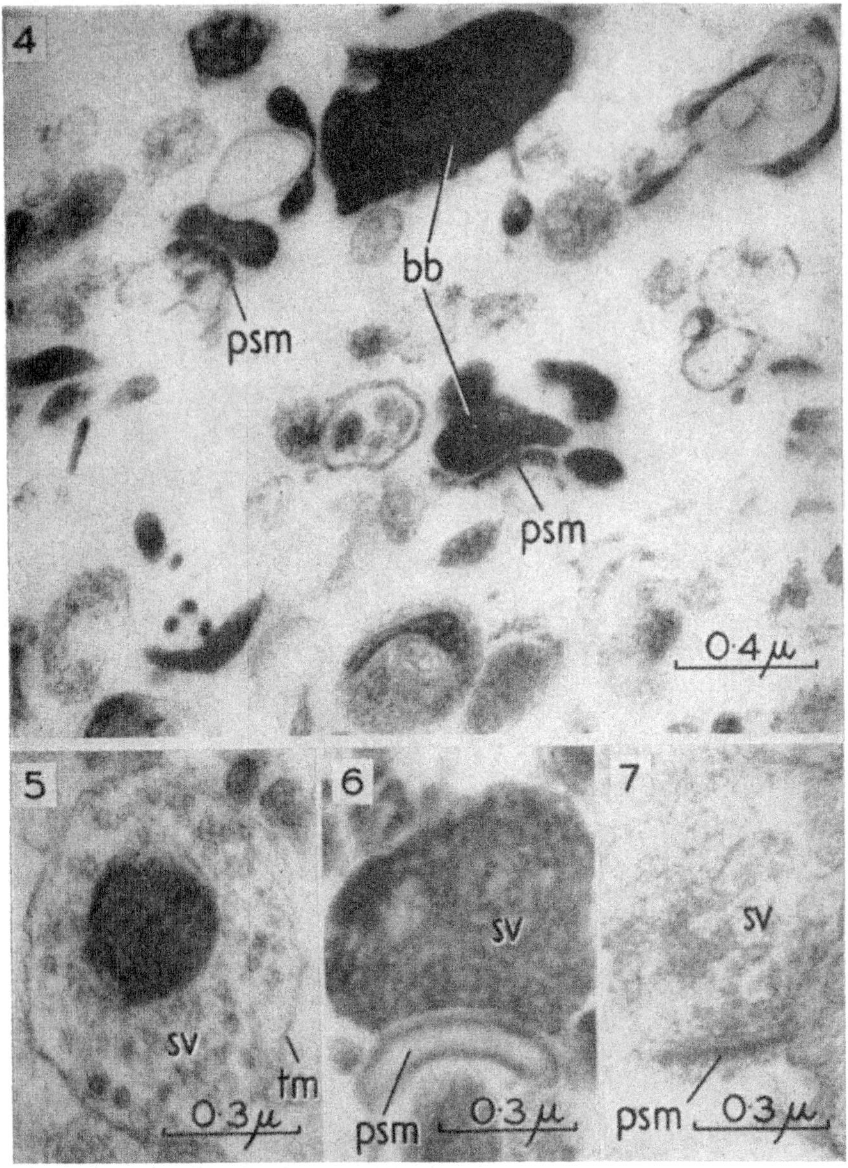

Fig. L17, PLATE 2.

PLATE 2. 4. Portion of fraction B (spun out of layer from density gradient without prior dilution) showing numerous "black bodies" with attached psm's.
5, 6, 7. Nerve ending particles.
Gray and Whittaker (1962)[37]

which I have spoken seems to be general to many types of cell. But all the experiments which I have mentioned (which are only a selection) raise some of the most challenging problems in biology. If we summarize, I think that we have evidence of a fluid cytoskeleton in cells of various types. The nucleus has a structure; the various semi-permeable membranes are organized; the mitochondrial structure is very clear; in fact it is regarded almost as a solid. The endoplasmic reticulum runs throughout the cell as a tubular structure; there is also a structural Golgi apparatus to be considered. It is very probable that more of the cell background may turn out to be organized. But what we want now is solid information on the place of enzymes in these structures; this is a most difficult task. If you isolate the structures, you have to decide whether what you find was originally in the particle or whether it has become stuck on to it during the centrifuging process. Then again the E.R., what are the convolutions which matter, and what is trivial, and are the tubes excreting or acting as a circulating system? Is there evidence for some way of conducting information more quickly about the cell, than could occur through diffusion alone? Furthermore, accepting that the junctions between the E.R. and the surfaces of the cell are like the thixotropic gel, how far can the influence of such a delicately organized structure extend; that is how deep can it be in terms of molecules? W. B. Hardy thought it could be some distance (and I think that some Russian workers still think this), but Dr. Bowden tells me that in his laboratory they can get no evidence for it.[(9)] I feel that the time has come when we may with advantage devote more of our time to a study of the behaviour of biochemical entities inside the cells and to their intracellular organization, which should lead to an understanding of the orientation of these structures, and to the exact place of enzyme specificity in this organized cytoskeleton. It seems to me that progress will come best by an integration of scientists of different disciplines working together, and that it would be as well if we could train students to this end.

I find it interesting that Gowland Hopkins at the end of his Linacre Lecture made a plea that "biochemists should strive as far as was humanly possible to be biologists as well as chemists", a plea not far off that of my last remarks.

May I end with a translation of a statement by Aristotle, and by doing so I hope strike the right note of humility, in case I have seemed to be too arrogantly dogmatic.

"The investigation of the truth is in one way hard, and in another easy. An indication of this is found in the fact that no one is able to attain the truth adequately while, on the other hand no one fails entirely, but everyone adds something true about the nature of things". (Aristotle *Metaphysics*, Book II α, Chap. 1, Vol. [VIII, Oxford Translation of Aristotle, W. D. Ross].)

NOTES ON THE LINACRE LECTURE

(1) The cytoskeleton.
(2) Localization of enzymes.
(3) Enzymes in surfaces.
(4) Mitchell's theory
(5) The action of hormones.
(6) Cancer and the cytoskeleton.
(7) The amoeba and the packing of steroid films.
(8) Fat soluble vitamins.
(9) Ionization at an interface between a lipid and an aqueous phase.
(10) Hydrogen ion concentration in cells.

1. *The Cytoskeleton*

As the original Harben lecture upon this and the paper presented in 1930 to the Faraday Society are difficult to find, these are reprinted together with the other two Harben lectures (pages 216 *et seq.*) of the series which formed a starting-point of my work on the biochemical lesion. Some comments are added and references (see Appendix).

Some expansion of the short statement in my lecture is wanted. The cytoskeletal hypothesis was based, as I thought, upon a logical examination of the facts then well known about cells, and upon personal observations of the behaviour of ciliate protozoa under the microscope. It was essentially an attempt to visualize the co-ordination of the cell in terms of chemistry and biochemistry. One way of putting it is to say that I suggested the existence of an organized heterogeneity superimposed upon what Hopkins termed the "dynamic equilibrium", and that this heterogeneity ran as a continuum throughout the cell, taking part in the organization of biochemical events and acting in the cell as a nervous system does in the whole animal. As I thought then, and think still, this went beyond Hopkins's qualification of his phrase "dynamic equilibrium" by his words "in a polyphasic system", as my hypothesis suggested that this was in turn organized in a special way. An indication was then given by me as to how this integration could be accomplished by surface films. An essential part of my thesis was devoted to the significance of surfaces and to the importance of studying these. The term "cytoskeleton" is a reasonable compromise, provided that it is understood not to imply a rigid bone-like structure; I meant something which was functional, which could alter its shape and—an important point—in the last limit control enzymatic activity. The attachment of enzymes in some cases to the internal parts of the surface of the mosaic could decide whether a particular enzyme was brought into contact with a substrate. Looked at from the point of view of coordination in the cell, this means that though the specific enzyme will often be a deciding factor in the direction of the metabolic reactions, it will not always be the final arbiter. That I was thinking on these lines is made clear from my later suggestion (1937) that the change from respiration to fermentation, the Pasteur reaction, in yeast might come about through a change of orientation.

At the time, and since then, I have felt it to be a matter of *strategy* in biochemical research to encourage studies of surface phenomena alongside those of enzymes and their specificity; and that this justified my speculations. It is little more than happens in theoretical physics. It is nicely put by Mazia[52] (1961): "The difficulties (of analysis) . . . arise chiefly from our lack of a biochemistry of structure and of multimolecular phenomena. This could hardly have been asked of the infancy or adolescence of biochemical science, but we may expect it now from its maturity."

The point of view has been justified in the march of events. I understand that my Harben lecture was responsible for the entry of at least one prominent organic chemist into the nucleoprotein field. *Inter alia*, the cytoskeleton implied how compartments could arise in the cell; so it provided a basis for understanding how reactions could be proceeding simultaneously which required different environments of pH or of —SH groups. We know that some enzymic reactions need the —SH form, whereas others must need this in the —S.S.— condition. Also it can explain the old puzzle how heart tissue can respond to small amounts of acetylcholine, which could cover only a small part of the surface (Clark[12]).

At the time that my hypothesis was advanced, it was most difficult to prove. I used the gene theory, as then known, to speculate upon the kind of chemical groups which might be concerned in interfaces. Subsequently I investigated the behaviour of —COOH and —NH_2 groups in interfaces, in work which is summarized briefly below (p. 209). Ever since 1929, I have been watching for instances where one might get a compelling reason for thinking that a fraction from the cell was behaving differently from the whole cell itself in some enzymic reaction. In addition to the points mentioned there are a few others. With the enzyme aconitase, for instance, there appears to be some difference, perhaps structural, between the enzyme containing the Fe^{2+} atom in the natural state, and that where the Fe is added back artificially, because there is much greater sensitivity to the inhibiting effect of fluorocitrate in the former. Further, a different optimum pH for the enzyme in the intact mitochondrion has been found as distinct from that in a disrupted preparation (Dickman and Speyer[25]). They also found differences in catalysis between the three substrates in intact mitochondria.

In line with my original thesis, the Endoplasmic Reticulum is considered to control some kinds of enzyme activity, and this is consistent with the known fact that there are enzymes like β-glucuronidase (Levvy), which may be inactive until they are liberated from the mitochondria or the like. Such an idea too is the essence of the way in which the de Duve lysosomes behave; they are inactive bags of enzymes until liberated. For a most interesting review on present structural views, the Symposium on Tissue fine structure in 1956 can be studied,[59] and especially the paper by Palade, a leader in this field, which gives the pros and cons for using the term E.R. In a recent summary,[66] K. R. Porter writes: "It does mean however that the form the ER adopts is determined in part at least by the molecular architecture of the system itself and more remotely by genic information. In determining through its morphology the spatial intracellular disposition of biochemical reactions it becomes an instrument for the nuclear control of these phenomena."

The dividing egg of the sea urchin has also provided information of great importance. M. M. Swann[72] studied mitosis in detail making a quantitative research upon the birefringence in spindles and asters of the dividing cell. During anaphase the birefringence decreased while the structures were growing in size. As the mitotic figure grows, there is an increase in birefringence, which Swann thinks is probably due to an orientation of new material. Then we have the work of Mazia and his colleagues carried out over recent years of which he has produced some summaries.[52] The significant point in the present connection is that they can stabilize the spindle structure by the use of the —S.S.— reagent, dithiodiglycol,

$$\begin{array}{l} S.CH_2.CH_2.OH \\ | \\ S.CH_2.CH_2.OH \end{array}$$

and that though this cannot be seen during the intervals that the cell is not dividing, the fact that it can be formed is a strong argument for the arrangements of some structures in the cell which precedes the appearance of the visible spindle.

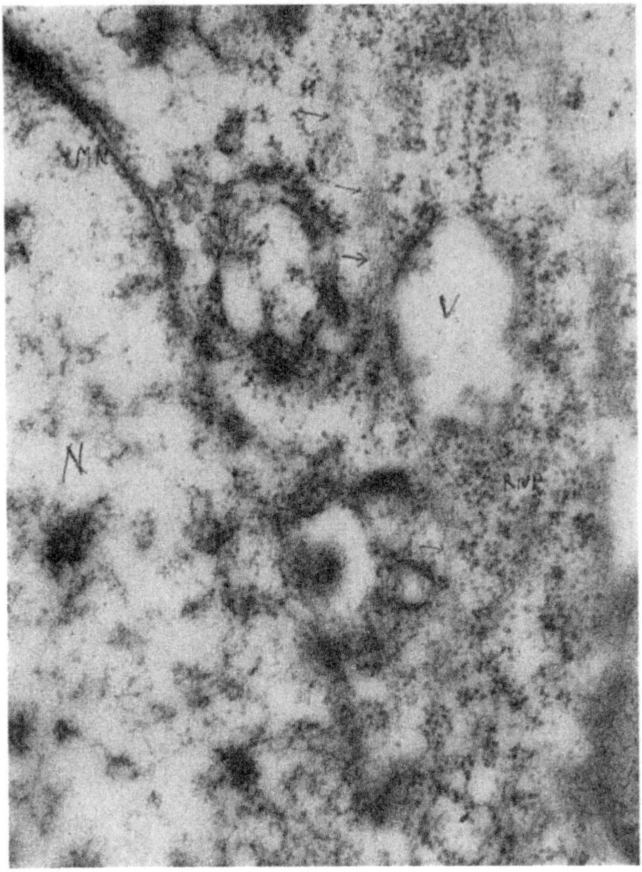

FIG. L17(a)

In extension of the remarks about the E.R., Dr. E. Hagenau has sent me very kindly photos taken by the electron microscope (Fig. L17) of the cytoplasm from cells infected with the Rous virus. The magnification is large (about × 43,000) and the small lines show an extremely fine detail suggesting an organization which is thinner than the E.R.[40,41]

2. *Localization of Enzymes*

Attention should be drawn to the work of Holt,[48] who has gone deeply into the theory of staining, reaching the conclusion that it is not possible to locate the action of an enzyme more closely that to 10μ; he has devised new stains for the localization of esterases, and is already getting valuable results for their location in liver, kidney, and muscle. Coloured complexes of an indigo type are formed.

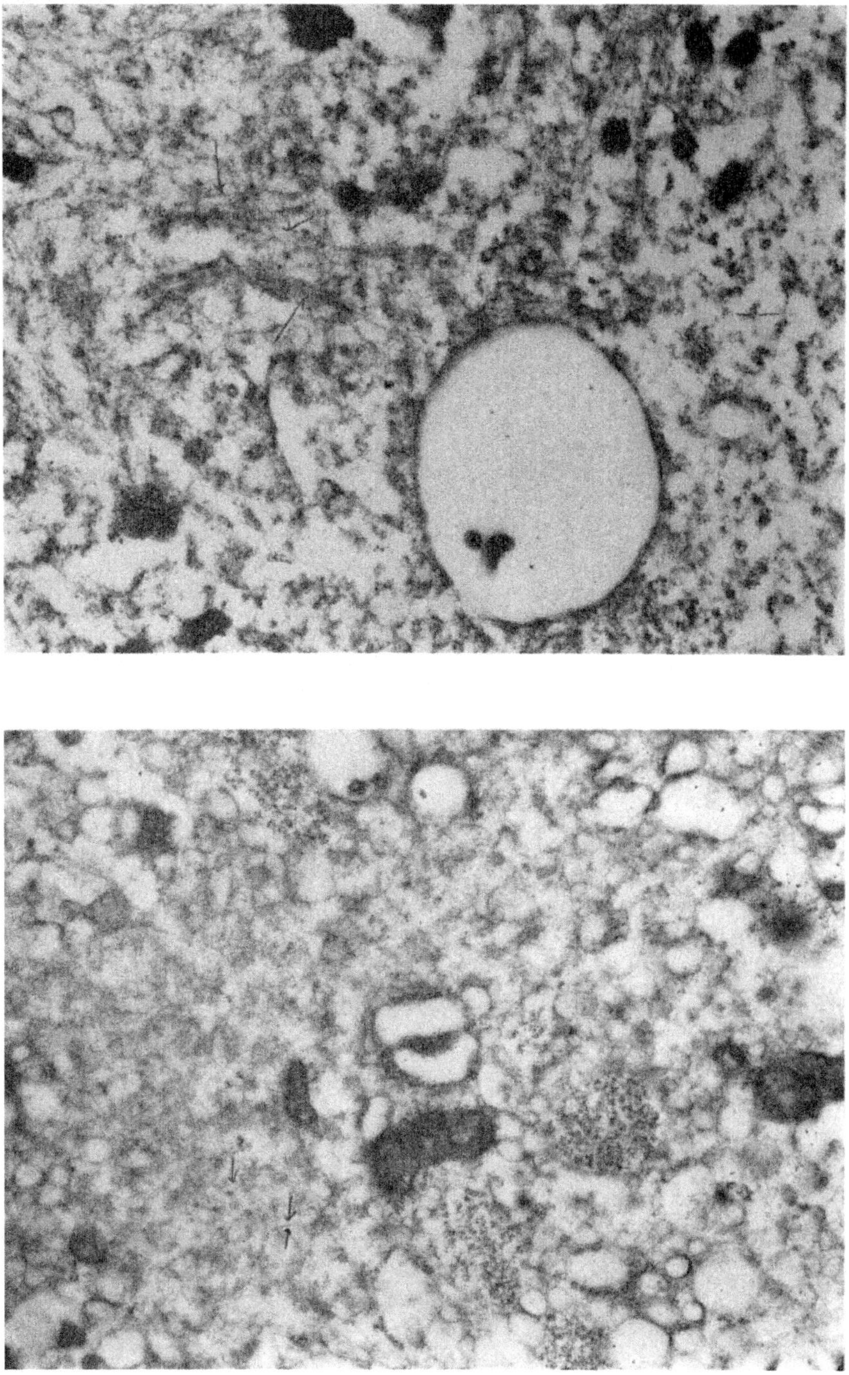

FIG. L17 (b), (c).

3. *Enzymes in Surfaces*

In addition to the researches mentioned, Bangham and Dawson have investigated the attack made by phospholipase A upon a lecithin film spread on a surface as a monolayer, using lecithin labelled radioactively, and estimating the rate of enzyme action by the decrease of radioactivity on the surface. The activity of this enzyme depends entirely upon the charge (the potential); it must be overall negative, which can be accomplished either by adding cardiolipid or by adding synthetic molecules with a hydrophobic and a polar group, such as diacetylphosphoric acid or sodiun dodecyl sulphate (the detergent); but in the latter case observations must be made quickly because the detergent destroys the enzyme. When ions were added, the hydrolysis was stopped when the negative mobility was reduced to zero; this varied for the cation, e.g. UO_2 0·19 mN, Th 2·2 mN, Ca 7·0 mN. They point out that variation of the charge could be a trigger mechanism in the cell.

4. *Mitchell's Theory* (1960)

One of the essential features which he postulates is an enzymic ATP-ase fixed permanently in the semi-permeable membrane together with an oxido-reduction system; together they catalyse a translocation of electrons, so that one gets conduction of protons through the membrane by adding uncouplers of oxidation.

5. *Action of Hormones*

Lerner makes the point that in the case of the action of the melanocyte-stimulating hormone, there is a "change in the location of enzymatically active particles which is followed by a change in the rate of enzymic reactions forming melanin". The action definitely shows that the hormone acts upon the whole cell, as nothing occurs with homogenates; and it constitutes, therefore, cogent evidence for the cytoskeleton. It has been suggested in Lund that the mechanism of the change of distribution of the pigment is operated by a series of contractile fibrils placed rather close to the surface membrane of the cell. Mercer's group have also studied melanin granules, and Birbeck *et al.*[(4)] find that the tyrosinase particles which they call melanosomes are different from the mitochondria, but they also have a well-organized structure (Fig. L18).

It is to be noted that Hechter and Lester[(45)] (1960), after a thorough examination of the hormone problem, agreed with these conceptions, concluding that the hormone insulin may act by "altering units of an ordered lattice system which represents a continuum throughout the cells" (see also Siekevitz[(71)] (1959)).

Van Overbeek[(73)] (1961) has now become dissatisfied in looking for the primary mode of action of the plant hormone auxin "in the direction of covalent bonding, activation of a single enzyme system", and is exploring in some detail the consequences of thinking that the hormone affects the cytoskeleton. This is leading him into the question how far the auxin can move into possible cavities in the membrane of the cytoskeleton.

The conception that there are definite tubes connecting the E.R. to the outside of the cell, such as are shown in the diagram of Robertson,[(70)] seems to have been given up. In conversation I found that Dr. Palay regards the integration of the E.R. with the surface of the cell as a dynamic affair. This would come closer to the suggestions often mooted of a thixotropic gel structure for this part of the cytoskeleton. It is also the kind of structure which could be deduced from the experiments on the viscosity of cells made by Crick and Hughes[(17)] using iron particles imbedded in the cultures and magnets.

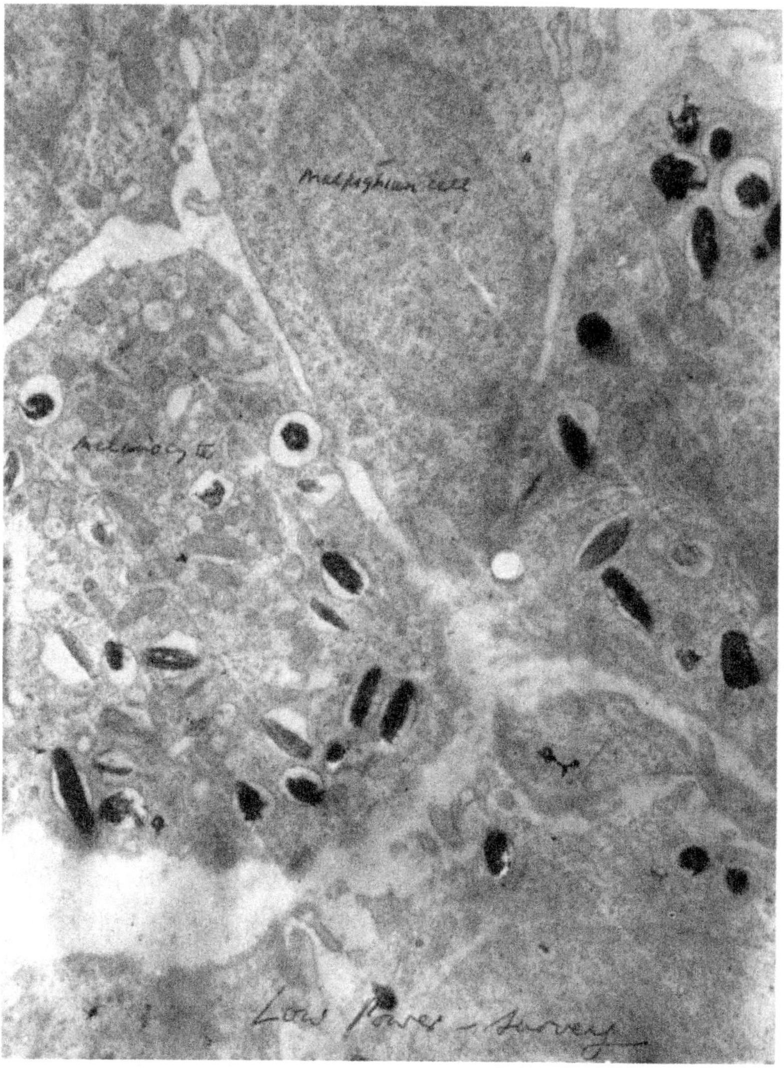

FIG. L18. Illustrations from the work of Mercer and colleagues.
(a) Shows a low-power photograph of a melanocyte next to a Malpighian cell.

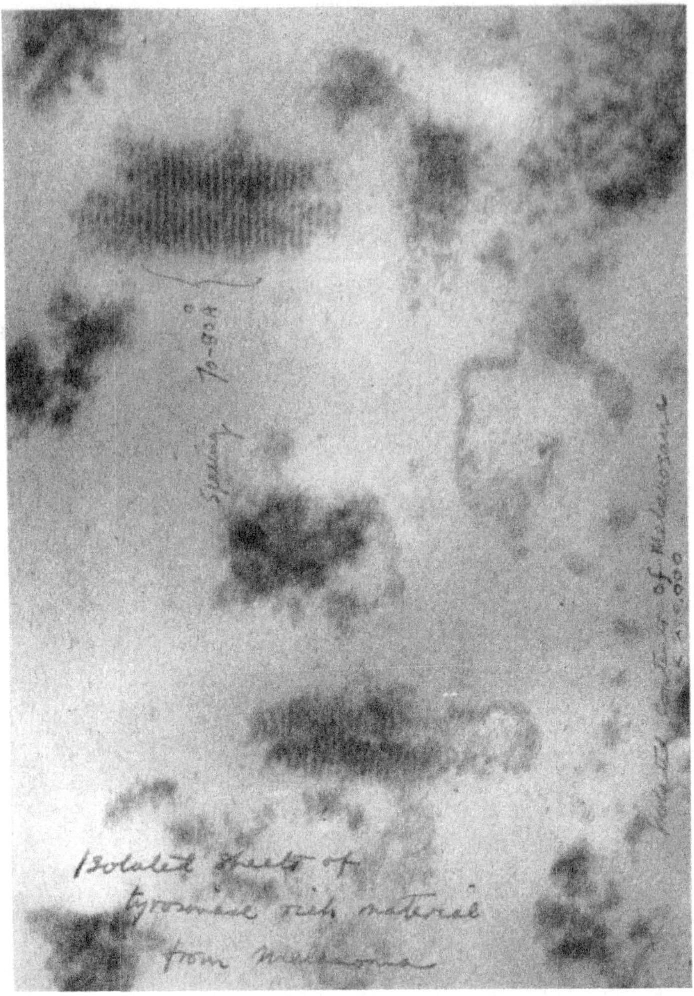

FIG. L18

(b) is a high-power picture of isolated sheets of tyrosinase-rich material from a melanoma, showing a cross-hatched appearance, which is also seen in melanosomes. (Birbeck and Barnicott (1959), *Pigment Cell Biology*, p. 549, Academic Press.)

6. *Cancer and the Cytoskeleton*

In a recent symposium on cancer held in Louvain (1960), at which we were asked for bold ideas, I pointed out that there was a subtle way in which another form of lethal synthesis might have to be considered in the cancer field; this could come about through the entry into the cell of some substance which could alter the cytoskeleton and give a new direction to the enzymes. Here the distinction between the genetic effect and that of the biochemical lesion would break down, because the cancer cell would be altered genetically. A similar state may happen with the action of the "mustards"; Alexander and Lett[2] suggested that there could be some combination of these with the phospholipid components of the submicroscopic barriers in the cells, and so there could arise modification of the cytoskeletal functioning of the enzymes attached to it. There was, however, then no evidence that there was any change induced in the E.R. by substances which could cause malignant disease, a position which has now altered. The observations already mentioned (p. 195) do not stand alone as Hagenau and Hollman[41] (1960) describe changes in the ergastoplasm of the mammary gland in carcinoma (the older term is the equivalent of the E.R.). If the hypothesis is correct, then there should be a search for substances which would reverse the cytoskeletal change.

If it is realized that the E.R. and the cell surface in bacteria, which may be held now to include internal membranes, each perform an integrating function in the cell, and that they are in that sense a real part of the physiology, the ground already explored points the way to future research. There are certainly great complexities still to be unravelled.

7. *The Amoeba and the Packing of Steroid Films*

Willmer[77] has studied with models the possible way in which the steroids can pack their molecules into the phospholipin cholesterol monolayers. It is a significant point that the 5 β-steroids are not able to pack into these layers so effectively as the α-steroids, and they are inert physiologically; they possess bent molecules. Dr. Willmer thinks that the stimulus for the growth of the flagellum is in the surface membrane. In this connection the work of Clowes *et al.*[16] is pertinent, who studied the effects of mixing polycyclic hydrocarbons with sterols in surface films; the intensities of surface pressures so formed could be sometimes increased by adding fatty acids; their work included the carcinogenic compound, benzpyrene. They finished with the following statement: "The strongest binding of polycyclic hydrocarbons so far observed is exhibited by mixtures of certain sterols with certain fatty acids or phosphatides. These attractive forces which link together hydrocarbons and cell lipids, or hydrocarbons and the non-polar parts of other cellular constituents are apparently of the same order of magnitude as those which maintain cell structure."

8. *Fat-soluble Vitamins*

Vitamin A has also a remarkable effect in changing the shape of red blood corpuscles, in very small concentrations and very quickly, Dingle *et al.*[26] (1962). It is interesting to note that changes have been noticed with vitamin D. Mitochondria isolated from rats deficient in vitamin D were found to differ morphologically from the normal.[23] Under the electron microscope, they showed large intercristal spaces and disrupted cristal systems.

9. *Ionization at an Interface between a Lipid and an Aqueous Phase*

The two curves (a) and (b) in Fig. L19 represent the observations which I have made[64] upon the apparent pK of —COOH groups and —NH_2 groups of long chain fatty acids suspended in benzene against aqueous buffered solutions.* It shows data which can only be interpreted upon the idea that the fall in interfacial tension is due to some form of ionization; it shows that such changes are taking place some 3 units in each case towards the neutral point as compared with dissociation in a bulk solution. It was a genuine pH change as one could get the same effect with different buffers of the same pH. The question is whether this is to be described as a change in surface pH. I was never quite certain that it was fair to go further than to call it a change in the apparent p*K* and received some support for this (Ogston[57]). But Danielli[19] thinks it is a genuine change in surface pH in a small layer and has produced equations fitting the facts reasonably well, with application to proteins in solution. G. S. Hartley and Roe[44] have found the p*K* differences at an interface approaching the matter from another angle, getting the equation $pH_s = pH_b + E\Psi/2{\cdot}3\,kT$, where pH_s is surface and pH_b bulk pH.

It seems probable that the fall in interfacial tension is related to the gradual acquisition of a charge by the —COOH or other group in the interface, with its consequent tendency to repel other charged groups in the head of the chain. This will be very similar to what Adam found for surface films on acid. In extension of this it seems that the change must also be happening in the cell at some of its interfaces, and that it may help to explain the large variation in optimum pH of enzymes in their relation to any bulk pH of the cell. But there are certain experimental facts about this change which require emphasis. If the buffer solution is diluted, or alternatively in the absence of added electrolyte, such as NaCl, the fall in interfacial tension does not occur. These facts of mine were confirmed by Danielli using brombenzene. Hence the change in ionization depends upon some state of the water molecules in the interface. Furthermore, a similar fall is found even with pure ethyl palmitate starting at a more alkaline point, pH 7·0. It is natural to suppose at first that this is due to the formation of soap, for instance the change to Na palmitate. But this simple explanation does not hold as it stands, because the fall also takes place with a glycine buffer containing much less Na, and in any case any Na palmitate would not remain in the interface. A similar fall was observed by Hartridge and Peters with thoroughly washed triolein (1922). If we could interpret the phenomena to mean that the first change is to the ionized $—COO^-$, and that this is succeeded by an actual formation of soap, when the interfacial tension has fallen to a low value at about pH 9·0 then we have a logical explanation. In any case I find it difficult to think that these changes are not of significance in the various interfaces of the cell, where we have compounds like the phospholipids and phosphoproteins, the reactions of which have been shown by Bangham and Dawson (p. 191) to be affected by surface charge.

10. *Hydrogen Ion Concentrations in Cells*

The calculation which I published in my first Harben lecture (pp. 216–239) has still to be properly assessed, I feel; and it is related to the question whether there is an actual change in pH at an interface, where one interpretation is that molecules in the interface are less dissociated, but that nevertheless this lesser dissociation indicates that the interface is more acid than the bulk solution when it is a —COOH group and more alkaline when it is an $—NH_2$ group in the interface.

* Similar curves have been obtained for the acids, Stearic (C_{10}), Lauric (C_{12}), and Capric (C_{10}) and for the base Octadecylamine (C_{18}). Peters.[64]

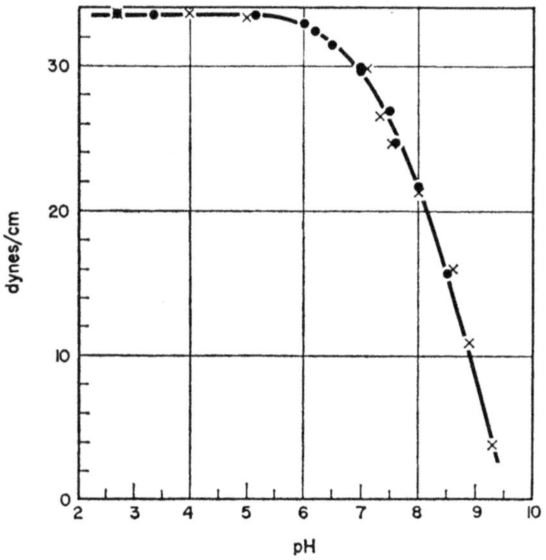

FIG. L19 (a). Palmitic acid 0·39 g in 100 g benzene. Temp. 20°C ± 0·1. ●, Phosphate buffers. ×, Glycine buffers.

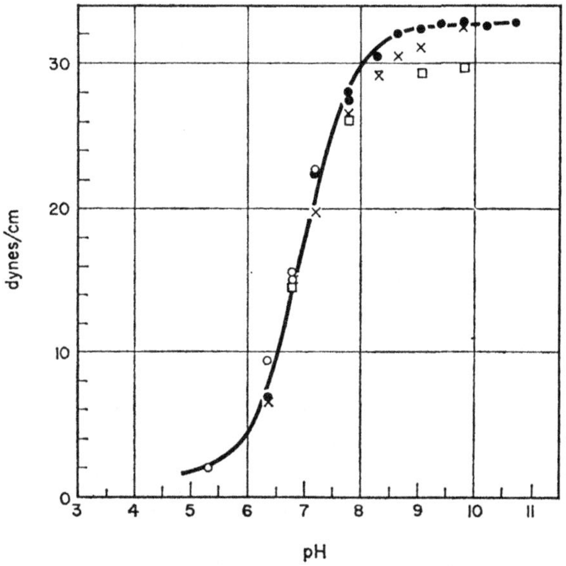

FIG. L19 (b). Hexadecylamine, 0·14 g in 100 g benzene. Expt. 1, ×. Expt. 2: ●, outer rim of tip; ○, inner rim of tip. Hexadecyamine carbonate, □. (Peters.)

The calculation showed that there was on average only one water molecule dissociated at the neutral point in an organism with a volume of 0·2 μ^3. This means that all those structures and microorganisms which must be seen with the electron microscope are in a condition in which on average they will have less than one hydrogen ion present at the neutral point. Clearly living organisms do not disobey the laws of thermodynamics; otherwise we should not have to eat! We may say that the equilibria involved in the dissociation of water are relatively so much faster than any other reactions taking place that the average equation has validity; but such an argument is based upon the idea that there is a free exchange of electrons at least between the inside of the living organism and the bulk solution. Yet the evidence is that several of the membrane barriers are rather impermeable. Even in an aqueous solution, Webb and Danielli[(74)] have shown that Ca is much more concentrated in the surface of a solution; the ratio Ca : Na is much greater than in the bulk phase. Hence the bulk pH, which has been so well determined by Chambers and others for the cytoplasm, cannot be taken to mean that it is the same as the pH on individual interfaces in the cell, as, for instance, on the surfaces of the Endoplasmic Reticulum. If the internal parts of the cell are in fact comparatively insulated from the more external parts, then it is not correct to predict the pH of the inside parts from an examination of the bulk pH. It seems that the local pH will depend upon the ionization of the groups immediately in contact with the small compartment. These are complicated matters, but I feel that they need an answer as so much of our interpretation of cell behaviour will depend upon them. For instance, the local pH of a micelle concerned with the action of an enzyme could depend upon a local rise of pH to a point which would be known to be sufficient to hydrolyse an ester in a bulk solution. Since the ionization at the interface also depends upon the salt concentration of the solution, so far as simple interfacial tension measurements are concerned, there is another complicating factor; and all these may operate in addition to any catalytic effect of metals. For further references see Peters, *Perspectives in Biochemistry*, Eds. Needham and Green (1937), p. 36, Camb. Univ. Press: *Proc. Roy. Soc.* (1937) B, 121, 587; Address to Section I British Assoc. (1949), in *Advances in Science*.

Further Note

Since writing the above, my attention has been drawn by Dr. J. B. Davenport to the most interesting work of McLaren and colleagues (see especially McLaren, A. D. (1961), *Enzymologia* **21,** 356). This is the first time that I have had the pleasure of reading a paper from someone who has understood some of the difficulties of applying observations upon enzymes in bulk solution to the behaviour of the same enzymes in the interfaces inside cells, and also the curious problems raised by small compartments and hydrogen ion concentration.

Summary to Chapter

There is now a considerable body of evidence in favour of the view that there is in the cell a fluid cytoskeleton, which takes part in the direction of biochemical reactions occurring in the cell by exercising some control even of the enzymes. This direction is in addition to that provided by the specificity of the enzymes themselves. The cytoskeleton is an organization based upon surface structures, and is in some way integrated with the surfaces of the cells. Upon the assumption that there is a cytoskeleton, it follows that a hormone acting upon the whole cell, rather than upon the enzymes inside it, should act by controlling the arrangements of the cytoskeleton. This has been proved in the case of the melanocyte-stimulating hormone acting upon the melanocyte of the frog. The idea of the presence of a cytoskeleton leads naturally to the view that the action of substances inducing cancer (and of radiation), the biochemical lesion, may be to alter the cytoskeleton in such a way that a permanent change may result. There is some support for this from the work of electron microscopists; and it is clear from recent most important researches that there exists an endoplasmic reticulum in most cells, which seems to be the major part of any cytoskeleton.

References

1. ADAM, N. K. (1938), *Physics and Chemistry of Surfaces*, Oxford Univ. Press.
2. ALEXANDER, P. and LETT, J. T. (1960), *Biochem. Pharmacol.* **4,** 341.
3. BACQ, Z. and ALEXANDER, P. (1961), *Fundamentals of Radiology*. Pergamon Press.
4. BAKER, R. V., BIRBECK, M. S. G., BLASCHKO, H., FITZPATRICK, T. B. and SEIJE, M. (1960), *Nature*, **187,** 392.
5. BANGHAM, A. D. and DAWSON, R. M. C. (1962), *Biochem. Biophys. Acta* **59,** 103.
6. BANGHAM, A. D., REES, K. R. and SHOTLANDER, V. (1962), *Nature*, **193,** 754.
7. BENSLEY, R. R. and HOERR, N. L. (1934), *Anat. Rec.* **60,** 251; (1937), *Anat. Rec.* **69,** 341.
8. BIRBECK, M. S. C. and MERCER, E. H. (1961), *Nature*, **189,** 558.
9. BOWDEN, F. and BASTOW, S. H. (1935), *Nature*, **135,** 828.
10. BUCK, R. C. (1961), *J. Biophys. Biochem. Cyt.* **11,** 227.
11. CATER, D. B. and PETERS, R. A. (1961), *Brit. J. Exp. Path.* **42,** 278.
12. CLARK, A. J. *et al.* (1936), *Proc. Roy. Soc.* B, **121,** 580.
13. CHRISTIE, G. S. and JUDAH, J. D. (1954), *Proc. Roy. Soc.* B **142,** 241.
14. CLAUDE, A. (1946), *J. Exp. Med.* **84,** 51.
15. CLAUDE, A. (1954), *Proc. Roy. Soc.* B **142,** 177.
16. CLOWES, G. H. A. (1943), *Amer. Soc. Adv. Sci.* No. 21, p. 1.
17. CRICK, F. H. C. and HUGHES, A. F. W. (1950), *Exp. Cell. Res.* **1,** 37 and 505.
18. DANIELLI, J. F. (1949), *Surface Chemistry* (discussion) p. 87. Butterworths Scient. Publications, London.
19. DANIELLI, J. F. (1951), *Advances in Enzymol.* **11,** 35.
20. DAVIES, A., PETERS, R. A. and WALSHE, J. M. (1961), *J. Physiol.* **159,** 32P.
21. DAVIES, J. T. and RIDEAL, E. R. (1961), *Interfacial Phenomena*, Academic Press.
22. DAWSON, R. M. C. and BANGHAM, A. D. (1958), *Biochem. J.* **72,** 493.
23. DELUCA, H. F., REISER, S., STEENBOCK, H. and KAESBERG, P. (1960), *Biochem. Biophys. Acta*, **40,** 526.
24. DEUTSCH, D. (1927), *Ber.* **60B,** 1036.
25. DICKMAN, S. R. and SPEYER, J. F. (1954), *J.B.C.* **206,** 67.
26. DINGLE, J. T., GLAUERT, A. M., DANIEL, M. and LUCY, J. A. (1962), *Biochem. J.* **84,** 76P.
27. DIXON, K. C. (1953), *J. Physiol.* **120,** 267.
28. DIXON, M. and WEBB, E. C. (1958), *Enzymes*, Longmans Green and Co.
29. EMMELOT, P. and BENEDETTI, E. L. (1961), *Protein Biosynthesis*, p. 99. Ed. R. T. C. Harris, Academic Press.
30. EMMELOT, P. and BENEDETTI, E. L. (1960), *J. Biophys. Biochem. Cyt.* **7,** 393.

31. EMMELOT. P., MIZRAIL, I. J. NACCARATO, R. and BENEDETTI, E. L. (1962), *J. Cell. Biol.* **12,** 177.
32. FELL, H. B. and MELLANBY, E. (1952), *J. Physiol.* (*London*) **116,** 320.
33. FELL, H. B., DINGLE, J. T. and WEBB, M. (1962), *Biochem. J.* **83,** 65.
34. FREY-WYSSLING, (1949), *Research* **2,** 300.
35. GALE, E. F. (1957), *Proc. Roy. Soc.* B **146,** 166. Leeuwenhoek Lecture.
36. GLAUERT, A. M. and HOPWOOD, D. A. (1961), *J. Biophys. Biochem. Cytol.* **7,** 479.
37. GRAY, E. G. and WHITTAKER, V. P. (1962), *J. Anat.* **96,** 79.
38. GREEN, D. E. and ODA, T. (1961), *J. Biochem.* (*Japan*) **49,** 742.
39. GREEN, D. E. (1957), *Symposium Soc. Exp. Biol.* **10,** 30.
40. HAGENAU, F. and BERNHARD, W. (1952), *Exp. Cell. Research* **3,** 629.
41. HAGENAU, F. and HOLLMAN, K. H. (1961), *Biological Structure & Function.* Ed. T. W. Goodwin and Lindberg, vol. 1, 169. Academic Press.
42. HARDY, W. B. (1913), *Proc. Roy. Soc.* A, **88,** 311.
43. HARRIS, J. E. (1939), *J. Exp. Biol.* **16,** 258.
44. HARTLEY, G. S. and ROE, J. N. (1940), *Trans. Faraday Soc.* **36,** 101.
45. HECHTER, O. and LESTER, G. (1960), *Recent Progress in Hormone Research* p. 121. Ed. G. Pincus, Academic Press.
46. HOGEBOOM, H. G., SCHNEIDER, W. C. and PALADE, G. E. (1948), *J. Biol. Chem.* **172,** 619.
47. HOGEBOOM, G. H. and SCHNEIDER, W. C. (1951), *J. Biol. Chem.* **186,** 417.
48. HOLT, S. J. (1959), *Experimental Cell Research,* Suppl. 7, 1–27; (1958), *Proc. Roy. Soc.* B **148,** 465; Biochem. Soc. Symposium No. 16, p. 44 (1959).
49. HUXLEY, A. F. (1959), *Ann. New York Acad. Sci.* **81,** 446.
50. KEILIN, D. and HARTREE, E. F. (1940), *Proc. Roy. Soc.* B **129,** 277.
51. LERNER, A. B. (1959), *Nature* **184,** 674.
52. MAZIA, D. (1961), *Biological Structure and Function,* p. 475. Ed. T. W. Goodwin and Lindberg, Academic Press.
53. MCILWAIN, H. (1959), *Biochemistry and the Nervous System,* J. & A. Churchill Ltd.
54. MITCHELL, P. (1961), *Nature* **191,** 145; *Symp. Biological Structure and Function,* Vol. II, Ed. T. W. Goodwin and O. Lindberg, Academic Press, and earlier papers.
55. MOORE, A. R. (1935), *J. Cell. and Comp. Physiol.* **7,** 113.
56. NEEDHAM, J. (1935), Terry Lectures. Yale Univ. Press.
57. OGSTON, A. G. (1947), *Physiol. Rev.* **27,** 228.
58. OSLER, W., Linacre Lecture, 1909.
59. PALADE, G. E. (1956), *J. Biophys. Biochem. Cyt.* **2,** Suppl. 85 (Review).
60. PALADE, G. E. and PORTER, K. R. (1954), *J. Exp. Med.* **100,** 641.
61. PETERS, R. A. and WAKELIN, R. W. (1938), *Biochem. J.* **32,** 2290.
62. PETERS, R. A. (1956), *Nature* (*Lond.*) **177,** 426.
63. PETERS, R. A. (1929), Harben Lectures: see Appendix.
64. PETERS, R. A. (1931), *Proc. Roy. Soc.* A **133,** 140; (1938), *Trans. Faraday Soc.* **34,** 537.
65. PETERS, R. A. (1960), *Biological Approaches to Cancer Chemotherapy.* Ed. R. J. C. Harris, Academic Press.
66. PORTER, K. R. (1961), *Biological Structure & Function,* vol. 1, p. 127. Ed. T. W. Goodwin and O. Lindberg (Review).
67. PORTER, K. R. and BRUNI, C. (1959), *Cancer Research* **19,** 997.
68. PORTER, K. R. and MACHADO, R. D. (1960), *Biophys. Biochem. Cyt.* **7,** 167.
69. RECKNAGEL, R. O. and LOMBARDI, B. (1961), *J. Biol. Chem.* **236,** 564.

70. ROBERTSON, J. D. (1959), Biochem. Soc. Symp. No. 16, *The Structure and Function of Subcellular Components*, p. 3.
71. SIEKEVITZ, P. (1959), Ciba Symposium. Ed. Gen. Wolstenholme and C. M. Connor (J. & A. Churchill), pp. 17–45.
72. SWANN, M. M. (1957), *Cancer Res.* **17,** 727; (1958), *ibid.* **18,** 1118 (Review).
73. VAN OVERBEEK, J. (1961), From *Plant Growth Regulation*, p. 449.
74. WEBB, D. A. and DANIELLI, J. F. (1940), *Nature* **146,** 197.
75. WHALEY, W. G., MOLLENHAVER, H. H. and LEECH, J. H. (1960), *Amer. J. Bot.* **47,** 401.
76. WIGGLESWORTH, V. B. (1953), *J. Embryol & Exper. Morphol.* **1,** 269.
77. WILLMER, E. N. (1961), *Biol. Rev.* **36,** 368.
78. WILLMER, E. N. (1961), *Exper. Cell Research & Suppl.* **8,** 32–46.
79. VITOLS, B., NORTH, R. J. and LINNANE, A. W. (1961), *J. Biophys. Biochem. Cyt.* **9,** 689 and 707.

(*MS. was completed in Sept. 1962*)

APPENDIX 1

THE HARBEN LECTURES, 1929

Lecture i. Cell Surfaces

WITH REFERENCE TO THE PART PLAYED IN THE INTEGRATION OF CELL PROCESSES BY ADSORPTION

The object of these discourses is to advance the cause of the public health. With some aspects of biochemistry the contact of this is clear; the subject of nutrition in all its phases is obviously fundamental to the health of the community. There are, however, other sides of biochemistry where the appeal is less direct, but which in the end are likely to have an even wider application. My first lecture is devoted to a theoretical problem in which I have been interested over a period of many years, and in which I believe even slight progress to be important to the human race.

When we review broadly the progress of biochemistry, we see that descriptive biochemistry has led to dynamical biochemistry. This statement requires some expansion. Much of the earlier work in our subject was occupied with a determination of the constitution of the chemical substances isolated from the body. This was a necessary stage in the growth of knowledge, and one which must receive the skilled attention of the best chemists for many years to come. Even now we know but little of the constitution of many compounds of great importance. The need for more research in descriptive biochemistry is urgent. From those earlier studies naturally followed the attempt to trace the chemical changes induced in living matter. This led to the conception of the living system as a centre of dynamic chemical change, to "dynamic biochemistry". We may instance the emphasis given to this point of view in the masterly address of Hopkins to the British Association in 1913. Stress was laid by him upon the fact that the transformation of quite simple substances, such as lactic acid, are chemical events of prime importance in living matter. His text was taken from the fascinating work upon the function of lactic acid in muscle by himself and Fletcher in 1907. We have seen a long period in which biochemists have studied the various changes induced by isolated perfused organs, and in which they have followed the behaviour of oxidation systems and various ferment systems in the test tube. In this they have paid but scant attention to the immediate application of such work to the whole organized animal, and have received much criticism for the fact that their studies have been upon artificial systems. Such criticism is unfair; firstly, because there is no reason why they should not study phenomena which are reproducible for their own sake, just as other chemists and scientists do; secondly—and this is more important—because without this knowledge it would be impossible to proceed towards a synthetic conception of the living system. I think that anyone who studies the admirable treatise by Dakin, in which knowledge of this type has been summarized, must realize that there has been collected now a noteworthy generalized conception of the chemical limits within which life is made manifest.

Now that we have this knowledge the time has come for us to try to apply it. We want a working conception of the cell itself. We must try to paint a picture of events inside the cell, impressionist though this be in the present stage of our work. As Leathes has said, "Such an analysis of the chemical meaning of material life viewed in the light of scientific facts has to be largely an exercise of the imagination, but it may present itself as an intellectual necessity." In order to focus attention upon this attempt to coordinate our knowledge. I propose to call it "Coordinative Biochemistry", and it is the aim of these lectures to sketch some aspects of this problem. In using this term, I wish at the outset to disclaim any implication that the majority of workers in this field are not endeavouring to apply their investigations to the cell. I merely think that it is convenient to group such attempts under a definite title.

In this first lecture I shall deal with general conceptions of the microchemical structure of cells, and with the kind of reactions which may proceed there. In this I wish to indicate as clearly as I can how I think that the surfaces of cells in relation to absorptive processes are of importance in the integration of cell activity. I should state at once that I distinguish two types of surfaces, the external surface or the one which is presented by the cell to its environment, and various internal surfaces or interfaces within the cell itself. Let me first indicate the reason why I came to think that physical views of adsorption are not sufficient to give a conception of what is happening in living matter.

In 1911 in my first research, Professor Barcroft directed my attention to a problem in adsorption. In those days there was a considerable belief that the combination of oxygen with haemoglobin was chemical; but this was not settled, because others held that haemoglobin was a colloid, and by its surface merely adsorbed oxygen. Working with Barcroft's new methods of oxygen determination, and with the new titanium methods for iron analysis, I was able to show that the specific oxygen capacity, the ratio O_2 : Fe never exceeded the definite value of 401.* This settled the

* This value, the S.O.C., is of some interest in disease. Engelkes (1929) has recently shown than in apparently healthy rabbits intraglobular sulphaemoglobinaemia may cause a fall in S.O.C. In cases of secondary anaemia the S.O.C. is unchanged, but in haemolytic processes such as pernicious anaemia the S.O.C. may fall in periods of blood destruction.

matter so far as adsorption was concerned, but I was never quite satisfied in my own mind that this was all. Somehow I felt that there was some deeper significance in regard to life to be extracted from this fact. It was only slowly that I took in the full significance of it. Laidlaw had shown in 1904 that the union of iron with haemoglobin was intimately connected with its oxidation. These facts meant that a colloid such as haemoglobin attaches itself reversibly in chemical combination with a gas at a definite part of its molecule, which part is a small one compared with the volume of the whole. The beautiful experiments by Barcroft and his collaborators on the effect of salts upon the reversibility of this combination emphasized the importance of the immediate ionic environment to this union, and Christiansen, Douglas and Haldane, 1913, showed that union with oxygen made haemoglobin a more acid compound. This meant that it altered some —H ion grouping proximate to the Fe. Such an effect of oxygen is seen in the case of the simple acids; lactic acid is, for instance, a stronger acid than propionic acid.

These simple but very significant facts really taught us that a large molecule such as a chromoprotein may have a small part of its surface in a specialized state, that a colloid can interact at its interface chemically, and that such a combination can be influenced by the reaction of the aqueous environment. But even in spite of this I still continued to look at colloids and surfaces from a physical point of view, and it was not until 1918 that the inadequacy of physical views of adsorption to explain living events was really brought home to me in connection with gas poisoning A short description of some experimental work will make this clear.

The effect of the war gases such as phosgene and chlorine upon the permeability of lung tissue showed how a poisonous gas could upset the surface of a living membrane. It will be remembered that the prominent feature was the appearance of a massive oedema at a variable period after exposure to the gas, several hours in the case of phosgene. The first theory which presented itself to explain this was the effect that was produced by the intracellular liberation of acid. Phosgene readily splits up in the presence of water according to the following equation:

$$COCl_2 + H_2O = 2\,HCl + CO_2$$

Upon physical views the liberation of acid would produce some general change in the ionization of the surface proteins and certain lipoid substances, so affecting their colloidal behaviour. But upon examination, the simple acid theory breaks down for several reasons (Marshall, 1920). I may indicate briefly a few of the reasons.

There is the general fact that the constitution of living matter tends to protect it against changes in reaction, owing to its buffering capacity. Hydrochloric acid in reasonable doses does not produce these serious toxic effects. More cogent evidence against this view has been obtained both by an examination of irritant arsenic derivatives, and from research upon mustard gas, dichlorodiethyl sulphide.

A comparison of the toxic effect of various arsenic derivatives upon quite simple organisms is clear proof against the acid theory. For instance, in 1917 McCleland and I compared the toxicity of various arsenic derivatives upon crude cultures of hay infusion protozoa, and concluded that the chemical constitution of the compound determined in large measure the toxicity. Primary compounds were less toxic than secondary and aliphatic than aromatic. Further in the case of the arsenic compounds, the oxides were more toxic than the chlorides. These preliminary conclusions upon impure culture material were placed upon a firm basis and considerably extended in careful work by Dr. E. Walker (1928). He worked with pure strains of protozoa, which I had isolated, viz. *Colpidium colpoda* and *Glaucoma scintillans*, and the figures in Table 1 are taken from his work. This is an exceptionally valuable research upon a single uniform strain of organism. He added the conclusion that tervalent arsenic derivatives were more toxic than quinquevalent derivatives.

TABLE 1. TOXICITY OF CERTAIN ARSENIC DERIVATIVES TO THE PROTOZOA COLPIDIUM COLPODA AND GLAUCOMA SCINTILLANS

(From E. Walker, *Biochem. J.* **22**).

	Compound Tervalent	Toxicity relative to standard (Dephenylchloroarsine) Colpidium	Glaucoma
Primary	Phenylarsenious oxide	5	4
	Methyldichloroarsine	1	1
Secondary	Diphenylchloroarsine	100 →	← 100
	Diphenylarsenious oxide ..	124	94*
	10.cyanophenoxyarsine	145	117
	10.chloro 5 : 10 dihydrophenarsazine	57	42
	QUINQUEVALENT		
Primary	O.-benzarsinic acid	0·1	—
Secondary	Diphenylarsinic acid	1	—
	α-naphthylarsinic acid	0·5	0·5
Tertiary	Triphenylarsinic dichloride ..	0·5	—

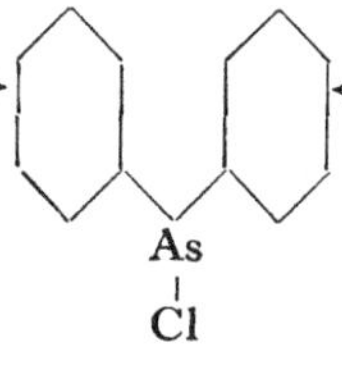

The toxicity of the standard diphenylchloroarsine to *Colpidium* = M/180,000 and to *Glaucoma* = M/420,000.

The standard time of death = 3 min.

* Unpublished figure, kindly supplied by Dr. Walker.

It is clear that toxicity is a question of chemical constitution and the fact that the oxides are as toxic as the chlorides shows that toxicity cannot be due to the liberation of intracellular acid.

The theory of local hydrolysis of acid at some interface has also been used to explain the vesicant action of "mustard gas". Here too this cannot be the main cause of the effect, as was shown by experiments carried out by Walker and myself in 1921 upon the rate of hydrolysis of "mustard" and its derivatives. This in no wise followed the vesicant power. To illustrate this, dichlor diethyl sulphone is not much less vesicant than dichlor diethyl sulphide, yet the latter hydrolyses 100 times more quickly. Dichlor dimethyl sulphide again is non-vesicant and yet hydrolyses 5 times faster than dichlor diethyl sulphide. These rates of acid production were measured at a constant reaction of pH 7·0 to imitate the reaction of the cell. The acid theory has broken down upon test, and an examination of some of these data, as for instance, the fact that αα' dichlor diethyl sulphide is not vesicant at all makes it evident that we are dealing with something much more subtle than gross physical change.

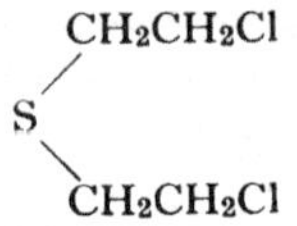

ββ' dichlor diethyl sulphide

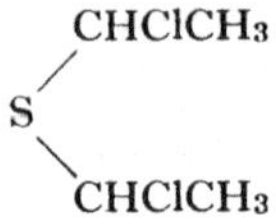

αα' dichlor diethyl sulphide

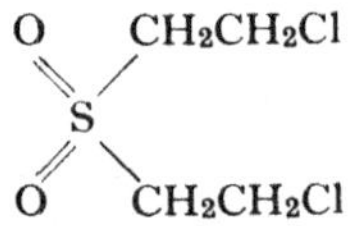

dichlor diethyl sulphone

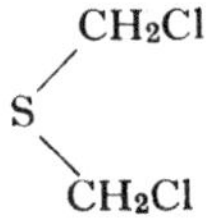

dichlor dimethyl sulphide

Though the explanation of the action of these substances is not yet clear, the point I wish to emphasize is that their toxic effect cannot be explained as due to a mere physical interference with the degree of dispersion of certain colloids. Some structural factor intervenes, and that almost certainly upon a surface. As the result of war work upon the toxic gases in 1918, I realized therefore that physical conceptions of colloidal behaviour in living tissue were not enough to explain the problems of permeability, but as yet did not see what to substitute for them. Up till then and for some time later, I think it is fair to say that it was orthodox to keep physical and chemical views strictly apart, so far as surfaces were concerned. No help could be obtained from a consideration of theories of the constitution of the cell, which postulated life as proceeding in a many phase heterogeneous colloid system. It was when groping my way towards some explanation of the action of the toxic gases that I first began to appreciate the importance of Hardy's observations (1913 and following) upon the relation between interfacial tension and chemical constitution. This work showed that surface force depended upon organic chemical structure.

Hardy's work was developed upon similar lines by Harkins and his colleagues (1917), and upon rather different lines by Langmuir (1917). With the difficulties of physical theories of adsorption confronting me, I can hardly describe now how I rejoiced when I read this work. Langmuir's work which has been consolidated and considerably extended by N. K. Adam (1925) gave a peculiarly concrete form to the conception. I do not think that we have even yet sufficiently appreciated the importance of these researches for a proper understanding of events in the living cell.

In brief, it has been proved that orientation of organic compounds exists at an interface. In the case of a long chain fatty acid, such as palmitic acid spread upon water, the molecules are oriented so that the —COOH group lies towards the water, whereas the methyl group lies towards the air. Hence the surface presented by a monomolecular film of palmitic acid upon water is a surface of methyl groups which are notoriously unreactive. To all intents and purposes it is a surface of paraffin wax. At a distance therefore of only some 10^{-7} cm apart at the interface, we have on one side non-reactive groups, and upon the other —COOH groups, anchored apart from one another in a structural arrangement. It would take too long here to discuss the other implications of this interesting work. The matter has been expanded by Leathes (1923) in his Croonian lectures, where he has added some observations of his own on the behaviour of lipoid substances such as lecithin. The important point is that we have here a chemical view of adsorption, and with this evidence behind us we can picture with reasonable probability that a cell surface whether internal or external may be made of molecules so anchored that they constitute a chemical mosaic. This conception makes coordinative biochemistry possible.

In order to indicate what I mean I wish now to direct your attention to a much wider problem than the one which I have used to introduce these remarks, and the investigation of which has been made possible by this physicochemical work upon surface orientation. It is necessary for me to spend a short time upon certain theoretical speculations in order to develop a basis for some experimental work which I wish to bring to your notice.

Let us consider the phenomena involved in the development of the full-sized individual from the egg, perhaps the supreme problem of all. Recent research by Morgan has shown that the constitution of the individual depends in some way upon the presence of definite genes carried in the chromosomes. Very probably chromosomes and their immediate molecular environment interact as development takes place; but at rock bottom we reach the fact that a series of particular chemical substances (for so they seem to be) influence in some way the subsequent history of each individual. This tremendous fact has as yet been hardly realized.

How can this possibly happen in chemical and physical terms? The problem is staggering when we examine it more closely. It extends beyond the formation of definite structure to the maintenance of a chemical "sett" in the whole organism.

A good example of this was given by Onslow (1923). The Dalmation hound unlike other animals excretes uric acid. When this was crossed with a terrier, the first generation destroyed uric acid. Among their offspring, however, were again found some which excreted uric acid. The chemical trait is inherited upon Mendelian principles.

The matter is much complicated by consideration of the physical state of living matter. During life the cell contents are often fluid. This is shown by the fact that centrifuging the cell drives the nucleus to one side (Lyon, 1906). (See also Leathes Croonian Lecture.) Underlying this fluidity, nevertheless, there must be structure. Parts of the cell at times become a gel, and there are seen in the cell radiant strands of more solid material. Some cells of a highly organized nature such as the protozoa have a complicated semi-rigid structure, capable of bearing stresses. It is fascinating to watch a large paramoecium bend to pass an obstruction and then return to its original shape immediately afterwards.

Let us examine then views as to the constitution of living matter and see whether they tell us how the whole is integrated. We often hear that the essential point about life is its colloid state. This carries us a certain distance but it is not enough. It is quite clear that reactions are directed in the cell. I can see no way in which a mere heterogeneous collection of colloids can do this. As Leathes has put it, "This faculty (of life) implies the power of introducing order into the chaotic movements of adjacent molecules (matter) in conformity with patterns which it possesses." It is of no avail to fall back upon the environment and to say that this is causing the directed stimulus. The cell is greater than its environment. It has been customary sometimes to picture the direction of reactions in the cell as due to the action of specific enzymes, and it may be conceived that these are grouped into systems. As illustrations of such systematized grouping of enzymes, we may instance the "cycle" of Meyerhof, formed as a coordinate picture of various facts known about the carbohydrate metabolism of isolated tissues. Keilin (1929) has also related together in one whole scheme the facts about oxidation systems in the cell.

These conceptions are of the greatest value in their own sphere, and I shall have occasion to refer to them later in the lecture. They do not help us in the problem which I have outlined, because it is impossible to get out of the conception of the enzyme more than has been put in. What has been put in is that there are specific catalysts attached to different parts of the cell, and able to operate possibly in stages? The conception of the enzyme involves the idea of something which accelerates a process proceeding by mass action effects. The presence of a specific enzyme may determine that out of a given number of possible reactions, this one will take place rather than that; it may act as it were as a chemical efferent nerve, to borrow a phrase from neurology. I do not see how it can continuously direct or maintain integrity of structure in the sense that must exist. A distinction of course must be made between the external surface of the cell and surfaces inside. In a brilliant piece of work it was shown by Quastel and Whetham that the surfaces of washed bacteria would catalyse an equilibrium between methylene blue and succinic and fumaric acids. This is important because it shows that enzymes even when fixed upon the surface of a cell can still exert their effects, and that, if given the chance by being supplied with enough molecules, they will behave as other catalysts do. It is valuable because it indicates that the cell has agents which behave as proper chemical substances should, and conform to ordinary chemical laws of equilibrium. Inside the cell, however, reactions appear to run to completion or sometimes not to take place when they should do so. What is the reason for this? The conditions in the cell must be different? On the external surface reactions may sate all their desires for chemical equilibrium. Inside the cell they meet control. I think that it is a mistake to believe that ordinary mass action laws have much significance inside cells. When once a molecule enters the surface film which it may do merely by the forces of diffusion, it becomes subjected mainly to surface forces and therefore to forces which are different in kind. I need give only one illustration of the enormous difference which must exist inside the cell as regards its chemistry. An organism which lies upon the limit of microscopic vision is approximately some 0·2μ in diameter.

According to the laws of mass action as applied to the dissociation of water, we are accustomed to write the equation, $C_H \times C_{OH} = k_w$ and to believe that approximately one water molecule in some 555,000,000 molecules of water are dissociated at ordinary temperature. Now the volume occupied by this amount of water is approximately a cube with a side of 0·25μ length.

Therefore such a small organism upon our ordinary physicochemical conceptions will contain on an average only one hydrogen ion at pH 7·0. Upon all the implications of this curious calculation, I do not wish to dwell here, but it emphasizes the point that the smaller divisions of the cell are working with too few molecules for ordinary statistical laws to be of direct service. The enzymes are valuable servants in the cell; they are not masters. Organization implies a continued and orderly stream of reactions.

Bacteriophage and filter passers (Boycott, 1928), for instance, must lie outside the ordinary laws of H_2O dissociation and so far as different small parts of the cell are isolated from one another by lipoids and such like substances, so must they.

The idea of the cell as a heterogeneous colloid system with specific enzymes, which has been useful during the development of biochemistry to this stage, cannot be of use much longer.

What then can be suggested to take its place? In a recent lecture, A. V. Hill has said "The living cell is a complex organized system of enzymes, interfaces, potential and osmotic differences, chemical substances; infinitely improbable in the thermodynamic sense, and yet existing in a steady state so long as free energy is available to maintain the organization." In this conception we are travelling towards the idea of an organized equilibrium, but I do not feel that from this we can get a view as to how the direction takes place. The question can be answered better by returning to our developing egg. Since the chromosomes carry the genes, and the genes determine the characters, we must examine first the chemistry of the chromosomes substance. If we take the simplest case of this, in fishes such as the salmon, the nuclear substance consists of protamines combined with nucleic acid. The nucleic acids for animals appear on the whole to be essentially similar. Not so the protamines.

Leathes has referred to the calculation of Steudel that nucleic acid = 40 per cent of the chromosome material and has calculated that as many as half a million molecules of nucleic acid exist in a single sperm cell. It may well be taken as almost certain that protamines build with nucleic acids an edifice of great complexity. To dwell upon the visions raised by such calculations makes any attempt to progress appear not merely puny, but impossible. If we, however, confine out attention merely to points which are known to differ, the prospect becomes brighter. We should become equally miserable if we tried to picture all the cell molecules in a complex aqueous solution of buffer substances. In the latter case, we have learnt that the hydrogen ion concentration is of importance, and the chance arrangement of the remainder of secondary interest. In fact, we are still content to be ignorant of the composition of water.

Great variety is shown even among those which have been investigated. In this variety then is the place for us to look for our answer. We can put the question into the form, how do these protein-like compounds exert a continuous effect upon their immediate environment in the cell? That we have to look to protein-like compounds for an explanation receives support from the fact that it is the proteins which are species and cell specific. It may be taken as almost certain that any influence is exerted at the interface between the molecules of protein and the fluid nuclear contents. These internal surfaces or interfaces hold the key to the problem. Now the exercise of a continuous influence implies some continuity of design. Translated to terms of a surface, this can only mean surface pattern. We may ask therefore whether it is likely that such a pattern or chemical mosaic exists, and if so what is its probable constitution?

For the sake of clarity we shall set aside the complications presented by films of other substances with which the protamines may be in immediate contact in the cell. The fats, for instance, are general inhabitants of surfaces; they have not been shown to be species specific, nor do they modify the antigenic properties of the proteins, as has been shown for some carbohydrates, the specific polysaccharides. In spite of the important role assigned to the fats by an authority such as Leathes, and of the evidence which he has adduced to show how they may readily assume cell-like structure, there is no real indication that the lipoid substances are the primary source of pattern. So far as we know they do not differ from one cell to another. What do we know of the structure of the protamines? Considering their significance for heredity, it is surprising how few workers have turned their attention to this field. In fact most of what we know is due to the work of Kossel and his associates. It has been learnt that these interesting compounds consist of varying groups of the basic amino-acids histidine, arginine and lysine together with some mono-amino acids. Both nucleic acids and protamines consist of compounds very rich in nitrogen. The protamine salmine for instance has the general composition a_2m, where a is arginine and other amino acid molecules are represented by m. It is to be noted especially that the predominate units in the protamines are the basic amino acids arginine, histidine and lysine. Some of the protamines, the triprotamines, contain all the three; sturine $(ahl)_2m$ is a good example. Little enough is known about these compounds, but still less about histones. This is in part because it is difficult to get uniform material for analysis, but I think in part also to the failure to realize the peculiar interest of such knowledge. A consideration of the probable groups which such protamine molecules present to their immediate surroundings cannot be wide of the mark, and should lead to pertinent facts. Now we know enough about the constitution of these compounds to say what are the main groups concerned; these will be many —CONH— groups, many basic groups and perhaps a few acidic —COOH. These will be anchored at various distances from one another, varying from say 10^{-8} cm to a limit outside the influence of one group for another, perhaps 10^{-6} cm. It is possible that such groups may have a limited amount of movement towards or away from one another, owing to varying hydration of the protein molecule. A mosaic of this kind is in some way responsible for the control which we know to exist and we have not gone much outside known facts in postulating it. The conclusion that directive pattern is especially concerned with proteins has experimental support from the phenomenon of anaphylaxis. In the classical research of Dakin and Dale (1919), it was shown by the racemization method that chemical differences exist in the terminal amino acid groups in the case of two albumens, the hen and the duck, which showed specific actions upon the guinea pig's uterus.* I contend therefore that we know in dim outline already

* See also D. Jordan Lloyd, 1928.

some of the main groups which constitute these surface mosaics of the cell. It is these which in my opinion must be immediately responsible for the continuous control of the chemical changes, and which therefore are acting in this sense as the central nervous system of the cell. It is evident that if this view is right, much more knowledge is wanted of the exact nature of these surface patterns; that is to say in the first instance much more descriptive biochemistry of the nuclear compounds. So only can we achieve a full understanding and control of heredity.

Meanwhile, we may apply what I feel that we know and perform experiments with the object of ascertaining how the influences of such a surface pattern could be made effective. We may first enquire what forces can be set up by ionizable groups such as —COOH groups at an interface. Can such a —COOH group when welded into a surface film respond to the ionizing influence of an aqueous medium of appropriate reaction and if so what quantitative relationships exist? In order to simplify

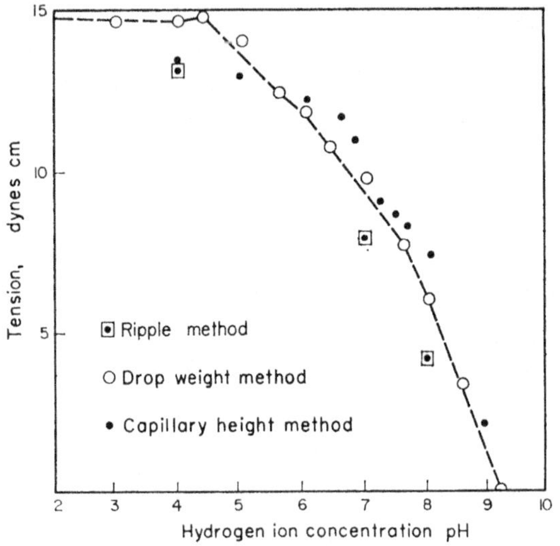

FIG. H1.* Per cent fall in interfacial tension of a benzene-palmitic acid solution against aqueous buffered solutions of different pH.

the experiments simple fatty acids can be used. Some work which Professor Hartridge and I carried out in 1920–22 throws some light upon this problem. It had been known in a general way that alkali lowered the surface tension of fatty substances (Reinders, 1910). More precise investigations showed that quite remarkable changes in interfacial tension can be

* By permission of The Royal Society. Hartridge and Peters (1922). *Proc. Roy., Soc. A.* **101,** 353

produced by merely changing the pH of an interface between an oil or fatty acid dissolved in benzene and an aqueous buffer solution. The accompanying figure shows the change in interfacial tension produced in an olive oil–benzene mixture upon passing from an acid to an alkaline pH. The general character of the curve suggests its relation to ionization.

Points here were determined by different methods with consistent results as may be seen. With a series of pure substances of which typical examples are given in Table 2, it can be noticed that only those substances possessing a —COOH grouping, or having a —COOH grouping in ester formation, e.g. the fats showed this change.

TABLE 2.* SUBSTANCE DROPPED THROUGH AQUEOUS PHASE

Substances	Concentration of buffer	Density	Temperature °C	Interfacial tension in dynes per cm against an acid buffered phosphate solution pH 5·6	Interfacial tension in dynes per cm against an alkaline buffered phosphate solution pH 7·4
(1) Benzene	1	0·8787	20·2	33·4	33·4
(2) Toluene	1	0·8680	18·2	34·3	34·4
(3) Octane	1	0·7060	17·5	44·6	43·7
(4) Octylene	$\frac{1}{3}$	0·7226	16·7	23·1	23·9
(5) Caprylic acid	$\frac{1}{3}$	0·9118	17·0	7·31	6·54
(6) Octyl alcohol	1	0·8290	16·3	8·69	8·76
(7) Hexane	$\frac{1}{3}$	0·6601	18·0	50·97	49·33
(8) Cyclohexane	$\frac{1}{3}$	0·779	16·3	44·75	45·74
(9) Cyclohexanol	$\frac{1}{3}$	0·946	18·0	3·81	3·85
(10) Oleic acid	1	0·898	16·1	12·6	1·45
(11) Ricinoleic	$\frac{1}{3}$	0·9538	17·0	6·85	4·41
(12) Triolein	—	0·915	16·3	20·4	12·1

Our results showed that interaction between the group —COOH at a surface and the fluid around it can occur in such a way that the change in tension produced depends upon the ionization. We found that the fall in interfacial tension in the case of tripalmitin took place at a rather more alkaline reaction than in the case of triolein. This suggested that the nature of other groups in the compound influences the ionization of the —COOH.

* Hartridge and Peters (1922), *Proc. Roy. Soc. A.* **101,** 363.

even when this is set in an interface. In some preliminary experiments* which I carried out some years ago. I found that the break in the curve occurs very sharply. In palmitic acid it is approximately at pH 6·0 (see Fig. H2).

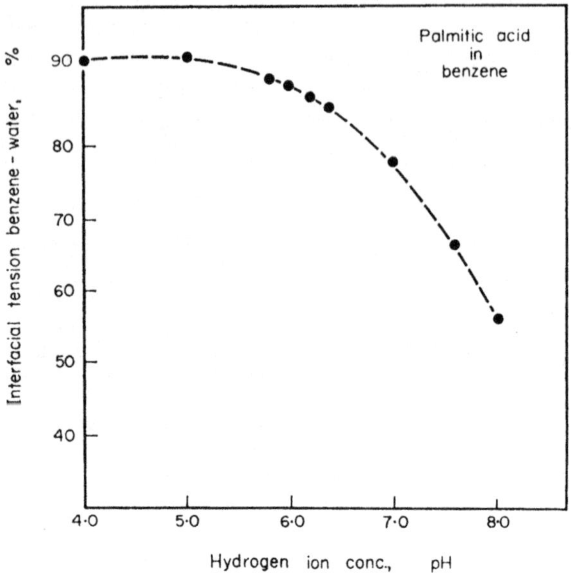

FIG. H2

Experiment. The results were obtained by a drop weight method. The palmitic acid solution in benzene contained 0·1465 g palmitic acid in 29·513 g benzene. The points pH 4·0 and pH 5·0 were determined against phthalate buffers and the remainder against phosphate buffers (Clark and Lubs). At each point in the curve the size of the drop formed was compared against the size of that formed by pure benzene under the same conditions. The same tip was used throughout. With capric acid the fall in interfacial tension begins at a rather similar point. With erucic acid (an unsaturated compound), the fall commences at a rather more acid pH.

Since the group —COOH when it is attached to a simple acid such as propionic acid, and therefore entering into true solution in water, commences to ionize at pH 3·0 approximately, it can be seen that the conditions of ionization at the interface are quite different. We have also here what I believe to be the first attempt to measure directly the strength of ionization at the interface. According to the results, the —COOH groups at interfaces of this type in the cell will be approximately half ionized at C_H which would correspond to a neutral reaction upon the cell surfaces. These

* Previously unpublished results.

results have been obtained perforce by the use of fatty acids, but the experiments indicate how such groups may be affected by pH. It is doubtless also true for the bases, though experiments upon this point are wanted. We also need to know urgently how far in terms of molecular distance the ionizing force of such a group radiating from the surface can extend. It is interesting to note that soluble fatty acids must be completely present as salts at the reaction of the cell. By attachment of —COOH to a surface, a device comes into play which makes it possible to use the ionizing forces of an aqueous solution at the reaction of the cell.

It is possible to calculate (from these results) how many —COOH groups are in contact with one hydrogen ion at pH 7·0. At this reaction 1 H ion is present on an average in $5{\cdot}55 \times 10^8$ molecules H_2O, or in a cube of which the area of one face is $6{\cdot}45 \times 10^{-10}$ cm² ($6{\cdot}45\ \mu^2/10$). The face of this cube would accommodate $2{\cdot}57 \times 10^5$ or 257,000 —COOH groups. The face of this cube containing on an average 1 H ion at the neutral point has 257,000 COOH groups. If we increase the C_H by 1000 (change of pH 7–4) the number of H ions in contact with our interface increase by 100, or we have now 1 H ion in contact with 2570 —COOH groups. In the case of palmitic acid in benzene a change in pH from pH 6·0–pH 9·0 makes a difference of say, 60 per cent in the interfacial tension. Hence, if we assume that in the case of muscle, acid production occurs in a box-like structure at some interface, it appears that much less lactic acid than is required to cover the surface may produce quite large effects, perhaps not more than 1/2000 of the amount required to cover the surface. Since the particular argument of A. V. Hill (1925) against the surface tension hypothesis of muscular contraction depends upon the assumption that a continuous mono-molecular layer of lactic acid is necessary, the evidence above invalidates his argument. Where he calculates that the rise of surface tension must be 4800 dynes/cm, it need not perhaps, be more than some 2–10 dynes cm, assuming that the active interface consists largely of —COOH groups.

Let us now consider another aspect from which these coordinating influences may be made effective. According to the views here put forward, organization as a directive force radiates from the mosaic patterns of proteins and perhaps other surfaces. Local modifications of this radiating pattern will take place if substances are adsorbed at special points. The researches of Warburg (1926) and of Meyerhof (1926) have led us to realize the importance of the surface factor in oxidation. According to Warburg, cell surfaces are the active agents; the extent of their living activity varies with the iron which they contain, and this must be bound to the surface in a special relation to nitrogen in order to reach its maximum activity. Ashed haemin is especially potent. Urethanes and alcohol inhibit oxidation according to the degree to which they are adsorbed. They block the surface. My interest in the problem of selective adsorption received some help from another angle. In the course of our attempts to isolate vitamin B, Mr. Kinnersley and I (1927) found that the adsorption of these compounds by charcoal was much influenced by the reaction of the solution from which adsorption took place. That reaction mattered for the vitamin had been clearly shown in the case of Fuller's earth by Seidell (1916) some time previously. This is not unexpected because Fuller's earth is an acidic substance. The interest of our observation lay in the fact that it was obtained with charcoal, a substance of more or less neutral properties. The matter warranted an independent investigation, in which I was joined by Mr. H. J. Phelps (1929) as a colleague. Not only have the results proved of particular interest, but they have opened out a promising field of research. The problem upon which we embarked was to find out which chemical substances showed differences in adsorption by charcoal as the pH of their solutions was varied.

Owing to the fact that amino-acids show an optimal precipitation at the so-called isoelectric point (for alanine this would be about pH 5·0),

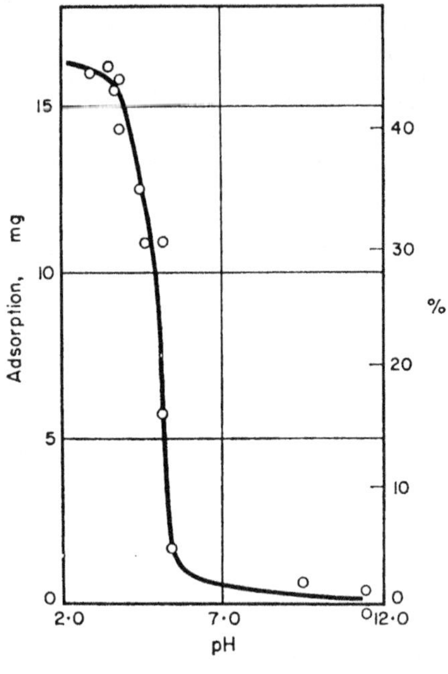

FIG. H3.* Showing the influence of hydrogen ion concentration on the adsorption of propionic acid. The adsorption is expressed in milligrams per 200 mg of charcoal from 20 ml of solution containing originally 1·8 mg per ml. The hydrogen ion concentration is expressed in pH units.

The scale on the right-hand side of Figs. H1 and H5 inclusive represents the adsorption as a percentage of the total amount of adsorbate originally present.

* By permission of The Royal Society. Phelps and Peters, *Proc. Roy. Soc. A.* **124,** 559.

it seemed to be reasonable to imagine that at pH 5·0, an amino acid would tend to leave the solution and cling to the charcoal. At reactions more acid or more alkaline than this, it might have been expected to reappear in solution. We therefore commenced with the amino-acids, and to our surprise obtained no differences in adsorption with change in pH at all for various monoamino-acids. This confirmed an observation of Michaelis, which we had hitherto been unable to believe. I should mention that in the experiments which we carried out, we worked with Norite charcoal highly purified by Miller's method (1912). A surface such as charcoal does not adsorb amino-acids. This perhaps explains why they pass through the bodily tissues so easily, but makes it difficult to understand how they are captured by any growing surface for synthesis of protein. It means, I suppose, that the arms which catch them must be charged. After some

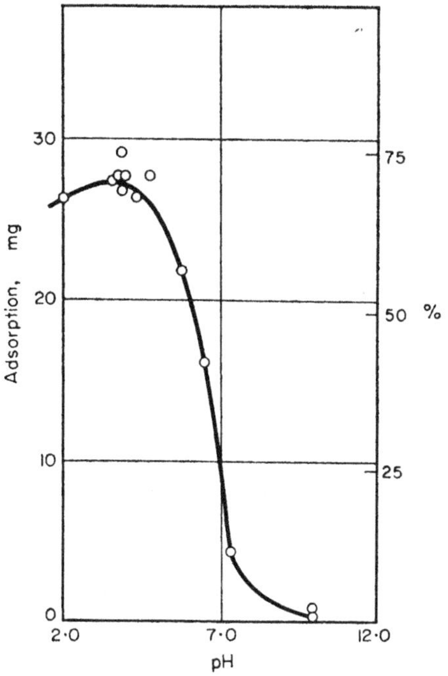

FIG. H4.* Showing the influence of hydrogen ion concentration on the adsorption of caproic acid. The adsorption is expressed in milligrams per 200 mg of charcoal from 20 ml of solution containing originally 1·94 mg of the acid per ml. The hydrogen ion concentration is expressed in pH units.

* By permission of The Royal Society. Phelps and Peters, *Proc. Roy. Soc. A.* **124,** 559.

searching, we found that simple weak acids and bases were the compounds for which we were looking. When the degree of adsorption of caproic and propionic acids was plotted against the pH, the curves shown are obtained (Figs. H3 and H4). Adsorption upon charcoal is a property of the unionized molecule. It may be noted in passing that the curves follow approximately the ionization curve. The point, however, for our present purpose is that the simpler acids at the reaction of living tissues do not tend to be adsorbed upon a neutral surface at all. Bases on the other hand show increasing adsorption with increased alkalinity. In this case the adsorption does not follow the curve of dissociation. The results obtained by us for the acids confirmed and extended some pioneer researches of Fromageot and Wurmser (1924), which seem to have escaped attention in the literature.

Mr. Phelps (1929) has continued this work with an interesting problem. He has found that the two isomers maleic and fumaric acids give different curves of adsorption, from it which can be concluded that the two isomers are differently orientated at the surface of the charcoal. This physico-chemical fact gains in importance when it is remembered that a mould like *Aspergillus niger* will utilize fumaric acid for growth and not maleic.

These experiments throw light upon some problems in the realm of colloid chemistry, and have their bearing upon immunological and other reactions in which precipitation play its part. In the present connection, they show how changes in reaction alone could modify the influence of groups in such a mosaic pattern as we have been postulating, by altering the degree of adsorption of substances already present.

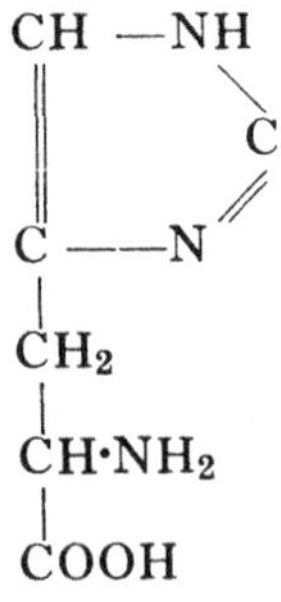

Histidine

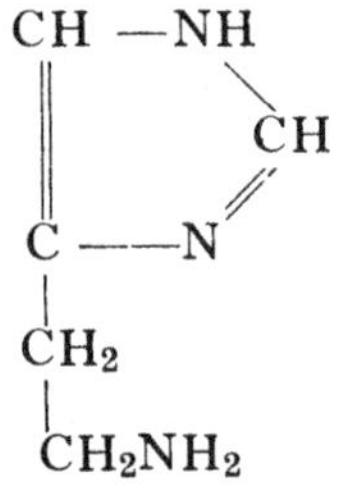

Histamine

The cases of histidine and histamine deserve special comment. Though the simpler amino-acids show no tendency to adhere to charcoal at any pH, this is not true of the more complex. Histidine has a complex dissociation curve. It is a substance possessing as a speciality the glyoxaline ring, and shows changes in adsorption over the limits of pH 5–8·0 approximately; it is interesting that this is the range over which the vitamin shows a differential adsorption (Fig. H5). When we compare histidine with histamine, it is evident that whereas the adsorption of the two compounds is identical within the limits of experimental error to the acid side of the reaction of the body, to the alkaline side it is not so. As it will be seen from the curve, small changes in alkalinity largely increase the adsorption of histamine but do not increase that of histidine. We have therefore a means by which the surface concentration of such a substance can be changed; a very small alteration in pH at the reaction of the body would suffice to settle how much of the base was concentrated upon a surface.

Since it has lately been shown by Dale and his colleagues (1929) that the reactive substance histamine is a drug normally used by the body in its adjustment of circulatory activity, the result at once becomes of biological interest. If the biological activity of histamine depends upon its

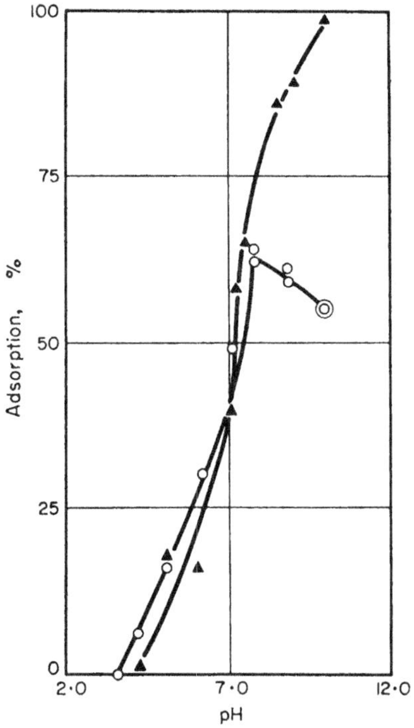

FIG. H5.* Showing the influence of hydrogen ion concentration on the adsorption of histidine (circles), and histamine (triangles). The adsorptions are expressed as percentages of the total amounts originally present in solution which are adsorbed by 50 mg of charcoal. The hydrogen ion concentrations are expressed in pH units.

* By permission of The Royal Society. Phelps and Peters, *Proc. Roy. Soc. A.* **124,** 565.

adsorption, then the effects of histamine upon tissues should vary with H ion concentration, and the responses of the tissues be larger at a more alkaline pH. Mr. Phelps and I have tried to obtain some evidence upon this question working with the guinea pig's uterus. Using the technique elaborated by Dale and Burn (1922), we have exposed guinea pig's uteri to Ringer's solution of varying pH. Our results can only be regarded as preliminary at present, and have not given as decisive results as we had wished. In the majority of experiments which we have done, histamine acts more powerfully in the more alkaline solution. With phosphate buffers, we have never obtained a case in which the response was increased in acid solution.* It should be emphasized that in alkaline solution the histamine is less ionized, so that its action in these cases may well be bound up with adsorption upon some surface in the unionized state. Such a conclusion receives some support in the literature. Andrus (1924) found that both adrenaline and tyramine acting upon the auricles of the rabbit has a more powerful action at pH 8·0 than at pH 7·0. He obtained an exception however in the case of choline and its esters, in which the depressant action was more marked in the acid solution. This has been confirmed by Clark (1928).

I am fully aware of the shortcomings of these observations so far as our problem is concerned, They show the extent to which the reaction alone can influence surface films, and ceteris paribus the ionizable groups upon an interface their immediate environment. As I have said previously, we yet lack evidence as to how far the influence of such an ionizable group can extend in terms of molecular distance.

Fortunately these observations do not stand alone. The work of others supplies examples of both control of molecules at a distance, and also of the significance of surface as an integrating factor. The work of Ramsden (1919, 1926), for instance (which has now extended over many years), shows that substances can definitely influence a surface film when they are present in a concentration less than one molecule thick. His theory of emulsions is fundamental to problems of the stability of such systems as we have been considering, and to surface problems in general.

Very important too is the recent work of Hardy and Nottage (1928) who have shown that orientation effects may be transmitted from a surface such as that of a metal, capable of influencing a surrounding film for a distance of 7–8μ. Here we have at once a means by which our mosaic

* With bicarbonate and CO_2, most cases have shown a greater response in alkaline solution. One preparation gave the opposite result. Evans and Underhill (1923) observed no change in response to histamine at different pH, probably because they were using rather a larger dose. We have confirmed their finding that spontaneous contraction takes place upon passing from an alkaline to an acid solution. This appears to be quite distinct from the histamine response.

may radiate its effect throughout the cell. It is also perfectly possible to appreciate how a coordinate structure may be maintained even in a medium which is apparently entirely liquid. We have in effect chains of liquid crystals which need not be more than a few molecules thick. Once that we have seen this, it becomes possible to develop some dim conception as to how the miracle occurs that colloid molecules, which have apparently no particular location in the cell, can yet control the synthetic and other reactions which are proceeding. The theory is all that is needed also to enable us to understand how substances can reach a special site in the cell. Between our chains of molecules, fixed by their radiating webs, there will exist paths from the external to the internal surface. Special direction is very likely given to the types of change induced in organic substances by specific enzymes spatially located. One author (Quastel, 1928) has gone so far as to postulate that the enzymes themselves represent points of surface activation, and that specific groups are needed to adsorb specific substances. It is possible that this is true, but the evidence would be more convincing if activation could be demonstrated with some chemical grouping or configuration likely to exist in the wall of the bacterial cell. It is evidently possible to free the enzymes from these walls, as Stephenson (1928) has found. The only attempt which I know to show that protein-like grouping is catalytically active is one by Falk (1917) who investigated the catalytic effect of —CONH— groupings upon the hydrolysis of the lipases. But any such specialized catalytic effects can only influence reactions continuously in a rather indirect way, because inside the cell they could only operate upon such molecules which reach them.

Though I have dwelt in this lecture upon the integrating effect of certain cell surfaces, and its connection with changes in reaction, I do not wish it to be thought that these are the only aspects of importance in regard to cell coordination. Before closing my lecture I should like to deal with conceptions of the energy metabolism, that is with the metabolism of carbohydrates. These have been summarized by Meyerhof in the so-called Meyerhof cycle. With the conception of an organized structure behind us I feel that it is easier to understand how such a system may work in a quite orderly manner. Meyerhof has pictured the changes in the case of a muscle as like those of a cranked motor car; the metabolism proceeds slowly until the need for a sudden outburst of muscular activity. It has been made probable by the researches of Warburg and his associates that a process of this kind is general for most tissues, and they have shown that a special feature of the cancer tissues is its continuous formation of lactic acid even in the presence of oxygen.

Study of the nervous system was made possible by the power of isolating groups of tissues for study, for instance the nicotine methods supplied the means of tracing out the sympathetic system. We have now in our hands at least three excellent agents for investigating the chains of biochemical reaction in the cell, namely cyanide, lack of oxygen and certain anaesthetics. They have been used with signal success by Warburg and others. I think that we should now consciously develop reagents of this type, capable of intra-cellular chemical dissection. Think how valuable to us would be a

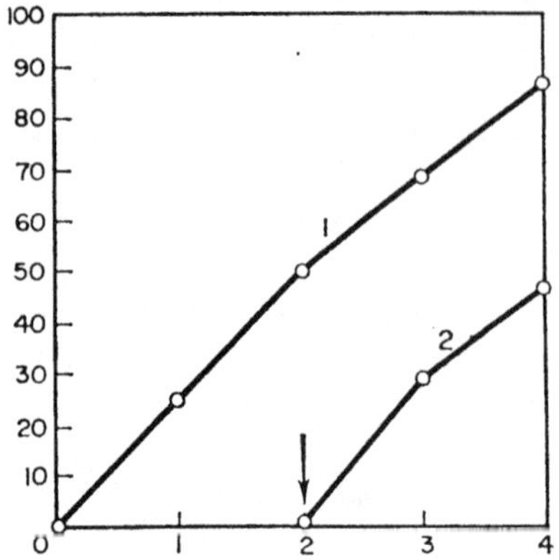

FIG. H6.* The rate of utilization of glucose: Curve 1, Aerobic incubation; Curve 2, Anaerobic and aerobic incubation. At the arrow, the suspension was transferred from an anaerobic to an aerobic tube.

reagent which would stop selectively the process of carbohydrate metabolism inside the cell. In the absence of oxygen, most cells produce lactic acid much faster, though it must be emphasized that the cancer cell does this even when oxygen is present. This rule is so general that exceptions to it are likely to prove of importance. J. T. Irving (1927) working at Oxford found that kidney tissue commences to split down sugar to lactic acid only when oxygen is admitted (Fig. H6). The matter has been confirmed by Case (1929). We have here a variant of the usual "chemical reflex", if you will allow me to use the term. The meaning of this is not yet available, but it is quite probably connected with the functions of the kidney in regulating the reaction of the blood. Oxygen seems to be of especial importance in connection with the nucleus. As illustration of this has come recently to our notice. In the course of some experiments upon the action of insulin upon pigeons to which I shall allude in a subsequent lecture, it was found that the blood sugar (as estimated by Hagedorn and Jensen method) did not fall below 0·06 per cent at the convulsive level. As this was much higher than the usual value obtained for the mammal, it appeared that a large part of the reducing value of birds' blood (totalling about 0·2 per cent) must be due to substances other than glucose.

* By permission of the Editor. Irving (1927), *Biochem. J.* **21,** No. 4, 882.

Dr. Irving suggested that the matter might be proved by making use of the glycolytic power of the red blood corpuscles,* and performed a few experiments. These have been considerably extended by Dr. J. M. Gulland at Oxford, who has proved that much of the reducing value of pigeon's blood is not due to glucose. Experiment has also brought to light the fact that glycolysis in the avian erythrocyte will only take place in the absence of oxygen. The chain of reactions for the avian corpuscles is therefore the normal one for cells so far as the production of lactic acid is concerned. It is the mammalian erythrocytes which are the exception. As the site of glycolysis of the mammalian blood corpuscle lies in the surface layer, the presence of the nucleus seems to alter the reaction of this layer to oxygen. The nature of the chemical influence which it exerts is at present a puzzle.

A useful hint may perhaps be provided by some work of Harrop and Barron (1928), which Dr. Gulland and I have ourselves confirmed. Minute amounts of methylene blue added to mammalian erythrocytes suspended in Ringers solution immediately cause them to take up oxygen as fast as avian corpuscles. If these facts are followed up, they may give some clue to the regions of the cell in which control of glycolysis takes place, a matter of interest in connection with the peculiar metabolism of the cancer cell alone. The site of glycolysis would seem to differ for different cells. Any process, freezing or otherwise, which fractures the external surface of the mammalian erythrocyte stops the action of the glycolytic system. This is also true for liver (Macleod and Simpson, 1928) and for kidney† (Irving), (so far as the glucose is concerned). Mr. Kinnersley and I have also found it to be true for brain. In frozen muscle, however, the glycolytic system becomes highly active when the muscle is thawed. Its association with the tissue is therefore quite different in the latter case. Except in the muscle cell, glycolysis seems to take place at the surface and oxidation in the nucleus.

I can conclude my remarks today by summarizing the gist of discussion. Need has arisen for a conscious development what of may be termed coordinative biochemistry. The theories which have been advanced to account in chemical and physical terms for the fact that processes in the cell are continuously directed are insufficient. The fact that the cell contents are often quite fluid can be reconciled with the fact that the living cell shows a continuous directive power by the view that protein surfaces in the cell constitute a mosaic from which radiate chains of molecules, controlled by their attachment to the central mosaic. These constitute a

* Rüter (1923) has investigated avian erythrocyte glycolysis.

† Irving found that lactic acid was still produced after freezing from precursors in the tissue other than glucose.

fluid anatomy in the cell, and the central mosaic behaves as a kind of central nervous system. The groups responsible for this appear to be terminal groups of the proteins, —COOH and basic groups, and —CONH linkages. By change of reaction and varying adsorption, the activity of such groups can be controlled. Within this matrix, it is possible to picture the dynamic chains of chemical reaction taking place, controlled inside the cell not by ordinary considerations of mass action, but by the exigencies of the matrix.

I am now in the stocks and waiting to be pilloried. May I put forward as an excuse a quotation from A. V. Hill, who has said "It is dangerous to speculate too far, but it is foolish not to speculate at all." I hope at the least that my views will prove sufficiently definite to render them a proper target, and would urge that attempts to follow up the known points of difference in the chemistry of different cells are more fruitful than to dwell upon the tremendous complications that lie ultimately before us. At the least perhaps, these remarks may spur others to make a better attempt to meet a problem not only of supreme biochemical interest, but fundamental to the eugenic control of the public health.

References

Adam (1925), *J. Phys. Chem.* **29,** 87.
Andrus (1924), *J. Physiol.* **57,** 361.
Barcroft, *The Respiratory Function of the Blood,* 1st ed.
Boycott (1928), *Proc. Roy. Soc. Med.* **22,** 55.
Burn and Dale (1922), *Med. Res. Council Rep.*, No. 69.
Case (1929), *Biochem. J.* **23,** 210.
Clark (1928), *J. Physiol.* **64,** 123.
Christiansen, Douglas and Haldane (1913), *J. Physiol.* **48,** 244.
Dakin, *Oxidations and Reductions in the Animal Body* (Monograph).
Dakin and Dale (1919), *Biochem. J.* **13,** 248.
Dale (1929), "Croonian Lectures", *The Lancet* **1,** 1237.
Evans and Underhill (1923), *J. Physiol.* **58,** 1.
Engelkes (1929), *Quar. J. Med.* **22,** 507.
Falk (1917), *J. Biol. Chem.* **31,** 97.
Fletcher and Hopkins (1907), *J. Physiol.* **35,** 247.
Fromageot and Wurmser (1924), *C. R. Acad. Sci. Paris* **179,** 972.
Hardy (1913), Papers in *Proc. Roy. Soc. A* **88,** etc.
Hardy (1926), *Proc. Roy. Soc. A* **112,** 47.
Hardy and Nottage (1928), *Proc. Roy. Soc. A* **209.**
Harkins (1917), *J. Amer. Chem. Soc.* **39,** 354.
Harrop, Barron and Gutzman (1928), *J. Exp. Med.* **48,** 207.
Hartridge and Peters (1922), *Proc. Roy. Soc. A* **101,** 348.
Hill, A. V. (1929), *Mond Lecture Nature Suppl.*, 11 May.
Hill, A. V. (1925), *Proc. Roy. Soc. B* **98,** 506.
Hopkins (1913), Pres. Address Physiol. Section Brit. Association.
Irving, J. T. (1927), *Biochem. J.* **21,** 880.

KEILIN, D. (1929), *Proc. Roy. Soc. B* **104,** 206.
KINNERSLEY and PETERS (1927), *Biochem. J.* **21,** 777.
KOSSEL (1928), The Protamines Monograph.
LLOYD, D. JORDAN (1928), *Biol. Rev.* **3,** 165.
LAIDLAW, (1904), *J. Physiol.* **31,** 465.
LANGMUIR (1917), *J. Amer. Chem. Soc.* **39,** 18.
LEATHES (1923), "Croonian Lectures", *The Lancet.*
LYON (1906), *Arch. Entwick. Mechan.* **23,** 151.
MARSHALL, (1920), *J. Pharm. Exp. Ther.* **16,** 259.
MEYERHOF (1926), *J. Gen. Physiol.* **8,** 533.
McLELAND and PETERS (1919), *J. Physiol.* **53;** *Proc. Physiol. Soc.* **15**
MACLEOD and SIMPSON (1928), *The Fuel of Life.*
MICHAELIS (1925), *The Effects of Ions in Colloidal Systems* (Williams & Wilkins).
MILLER and BARTELL (1922), *J. Amer. Chem. Soc.* **54,** 1866.
MORAN, T. (1926), *Proc. Roy. Soc. A* **112,** 30.
MORGAN (1919), *Physical Basis of Heredity* (Monograph).
MORGAN (1927), *Experimental Embryology* (Columbia University Press).
ONSLOW (1923), *Biochem. J.* **17,** 334 and 564.
PETERS and WALKER (1921), *Biochem. J.* **17,** 160.
PHELPS and PETERS (1929), *Proc. Roy. Soc. A* **124,** 554.
PETERS (1912), *J. Physiol.* **44,** 131.
PHELPS (1929), *J. Chem. Soc.* 1724.
QUASTEL and WHETHAM (1924), *Biochem. J.* **18,** 519.
QUASTEL (1928), *J. Hyg.* **28,** 139.
RAMSDEN (1919), *Trans. Liv. Biol. Soc.* **38,** 3.
RAMSDEN (1926), *Trans. Faraday Soc.* **22,** 2.
RAMSDEN, *Emulsions,* Clayton, ed. 2, App. 1.
REINDERS, van Bemmelin's Festschrift 1910, (quoted from Freundlich).
RÜTER (1923), *Zeit. Ges. Exp. Med.* **37,** 151.
SEIDELL (1916), *United States Public Health Rep.* **31,** 314.
STEPHENSON (1928), *Biochem. J.* **22,** 605.
WALKER (1928), *Biochem. J.* **22,** 292.
WARTBURG (1923), *Collected Papers.* 1909 *et seq.* Also *Jahres. ges. Physiol.* **1,** 136.

Lecture II.* The Ministers of Metabolic Change

EXPERIMENTAL RESEARCHES UPON THE NUMBER, NATURE AND DISTRIBUTION OF THE WATER SOLUBLE VITAMINS

In my last lecture, I tried to envisage in chemical terms the living cell as a working coordinate system, and to review some of the evidence which bears upon this. There are certain chemical substances, hormones such as adrenalin and thyroxin, of which the chemical constitution is known. These act upon the body as a whole, and we know in general their effect. Thyroxin, for instance, raises the metabolic rate. We have still to learn what chemical or other changes these compounds are undergoing in the tissues. The intracellular function of such substances is still obscure. In a rather different class stand the vitamins. Without a supply of these essential substances, termed by Sir Archibald Garrod the "exogenous hormones", the life of an organism comes to an end. Until we can make these compounds in the laboratory and study their actions upon groups of tissues, we cannot hope to understand properly the working system of the cell. They represent links of pre-eminent importance, and among them the B group seem to be concerned with metabolism in some particularly subtle way. They are coordinating agents of a special type, and they minister to metabolic change. Very few living cells appear to be independent of small amounts of ready made chemical substances. Before we discuss what the "B" complex does, we must first know with how many entities we have to deal.

It is curious to reflect now that some eight years ago, the interest in vitamins of the B or water soluble group had flagged. It had all been apparently settled then; water soluble B, the name given by McCollum and Davis (1915) to the complex was growth promoting to rats and cured pigeon and fowl beri-beri. This was the original factor of Eijkmann. But the lull was illusory; today it is almost difficult to keep abreast of the literature in this branch alone, nor is this attention undeserved. Some have been inclined to think that we need not worry much about "B" in an ordinary diet. I think that this conclusion can never be placed upon a firm basis until we have more quantitative data. At least two groups of workers take a different view (Plimmer, 1928; Cramer and Mottram, 1927).

Approach to the problem from different experimental angles has led to a turmoil of apparently conflicting evidence. The problem has become

* Delivered in the Lecture Hall of The Institute, 25 October, 1929.

now not what is vitamin B, but how many factors compose this group. Until we can answer this, we cannot make the best progress towards chemical definition of the complex. I hope to show that many of the apparent contradictions in the literature are not irreconcilable.

In this lecture, I propose to give a connected account of work which has been proceeding in the Department of Biochemistry, at Oxford, and to try to show the points of contact and the differences between our findings and those of others. In this way I hope to convey a picture of the state of the problem at the present time. Upon the work of the pioneers in this field there is no need to dwell. Excellent reviews upon the subject are available.*

My own interest in the problem arose from an attempt to cultivate in pure synthetic solutions the protozoon organism, *Colpidium colpoda* in 1920. At the time when I started the work, I believed that I could grow this upon a medium consisting of salts together with ammonium glycerophosphate as a sole source of phosphorus, carbon and nitrogen. When I tried to apply this to the growth of *Paramoecium caudatum*, I could not succeed in the cultivation without the addition of an alcoholic extract of yeast. These latter experiments aroused my interest in the B group of vitamins, and for several years my colleagues and I have been engaged in a pursuit of the question I now know that my original views about *Colpidum* were incorrect When really free from bacteria, this will not grow without the addition of a yeast extract and either extracted yeast cells or casein (Peters, 1929). Neither seems to suffice alone, so that the nutrition needs appear to be rather complex.

I shall deal in the first place with our evidence that yeast extracts can be obtained which are highly curative in their action upon pigeons suffering from polyneuritis, but which are without the effect of the whole B complex. These we have called torulin, following my predecessor in Oxford, Benjamin Moore (1912);† they represent the factor B_1.‡

When I commenced my work upon the antineuritic substance in yeast, I had to choose between two kinds of tests and two kinds of animals. The test is actually fundamental to the subject, and I must therefore crave indulgence for a short time to consider it. It is probably fair to say that all workers in the vitamin field have been guilty at one time or another of the sin of taking things for granted. Should anyone feel that his besetting sin lies in the taking of tacit assumptions, let him come to the vitamin field for discipline, where he will learn that things which are apparently equal to the same thing are not necessarily equal to one another, and

* See the review by Kruse and McCollum (1928).

† Edie, Evans, Moore, Simpson and Webster (1912).

‡ Throughout this lecture, I propose to adhere to the nomenclature of the Biochemical Society, namely B_1 antineuritic, B_2 anti pellagra, etc.

that things which appear the same are not. It is unsafe for instance to consider that yeast is ever the same substance from the nutritional point of view.

Now the test which was supported with much authority by workers like McCollum was the original rat test, as used by Hopkins (1912) in his classical experiments. This test gave clear cut results with growing rats and was generally easy to manage. The curative test in pigeons had been practically abandoned in favour of a weight maintenance test by Williams, Seidell and by Funk himself. One test involved the adult animal, and the other the growing one. It was held with some fairness that both these tests were better, because the trial was made upon a fit rather than upon a pathological animal. I think that it is a truism that if the process of maintenance and growth demands a daily minimum of each essential constituent, then unless the growing cells obtain this quota from somewhere, the weight must remain stationary or fall, for even the adult animal has growing tissues. These tests are therefore admirable for settling whether a given diet is satisfactory for full health (Plimmer *et al.*). They become dangerous when used to follow up a factor, because they depend absolutely upon the extent to which only that one factor is removed, and can succeed only so long as all other essential factors are supplied in the diet or from stores accumulated by the animal. Failure to remember this fact has led to much of the present confusion in the vitamin field. Tests in which some pathological symptom is selected for cure such as skin lesions have the disadvantage that the animals must be treated as hospital cases; the results are, however, definite.* Upon these general grounds, I decided to return to the original curative test. I hope that this lecture will show that the belief was well founded. Through a fortunate accident I early discovered that many of the previous difficulties arose from a failure to recognize symptoms appearing before true avitaminosis had set in. Cold, for instance, or lack of water may induce pseudo symptoms of B avitaminosis, which can be cured by heat or the giving of water alone. Again the cases produced by a rice feeding are not all in a similar condition, that is to say not all examples of a B_1 deficiency. By careful selection of a standard case, the errors can be reduced, and I think that the test as finally elaborated by Mr. Kinnersley, Dr. Reader, and myself (1927) can be regarded as reasonably reliable. It has the advantage of speed and economy of material.

Birds are standardized by placing upon a laboratory diet for 3 weeks. Symptoms are then induced by feeding upon polished rice at a room temperature of 10°–15°C. Any bird which shows symptoms before the 30th day is transferred to a warmer room (21°C), and there given a dose of distilled water containing glucose. If no cure takes place in 2 to 3 hr, a test dose is given.

* Curative tests are, of course, the rule in vitamins C and D (scurvy and rickets).

Recently we have adopted the practice of giving a standard dose before the test dose, to make sure that the bird is reacting normally. Birds showing no symptoms before the 30th day are rejected.

The amount of torulin present is judged from the time of protection after cure. With this test, I obtained a concentrate in 1923, capable of curing pigeons in doses of 3 mg per diem; a charcoal adsorption stage was employed, as I had found that the active material could be removed from the charcoal by a dilute acid alcoholic solution. This made the method simpler than that of Funk (1911) or Tsukiye (1922). In 1924, I was joined by Mr. Kinnersley and should like to say here how much the progress of the work has depended upon his patient skill in laborious and difficult operations. We succeeded in improving the method sufficiently to obtain small amounts of a fraction curative to pigeons in average doses of 0·08 mg per diem (1925). Owing to the manifest impurity of this, we considered then that the pure substance itself would prove to be curative in doses of not more than 0·001 mg per day. This preparation at that time represented an advance upon others, and convinced us of the reality of the curative substance. It dispelled the impression that symptoms in these birds were not specifically cured by a definite substance.

Some authors had gone so far as to regard the symptoms as accidents of a general condition. That they are not incidental, but that they represent the terminal stages of a progressive state seems to follow from the following considerations. From several angles we have examined the relation between the doses given and the time during which a bird is subsequently protected from a further onset of symptoms. We are quite convinced that there is evidence of a daily usage of vitamin under these conditions. When the vitamin level is sufficiently reduced the symptoms appear. The following results for successive test doses given to the same bird demonstrate this.

TABLE 3*

(*a*) Cases in which the second dose was the same in amount as the first

Bird	Protection for	
	First period	Second period
428 ..	5 days	5 days
649 ..	3 days	3 days
709 ..	3 days	4 days

(*b*) Cases in which the second dose was double in amount

581 ..	2 days	4 days
908 ..	3 days	6 days

* Results selected from Kinnersley, Peters and Reader (1927).

It has been found (Randoin and Lecoq, 1929) that pigeons fed by stuffing become polyneuritic in 20–25 days. If, however, birds are allowed to feed naturally, and so to bring into play all their biochemical adjustments, it is found that individual birds react in rather a constant way to rice feeding. This was first shown by us in 1927, and part of our results are reproduced in Fig. H7, which shows the response of five birds to successive periods of feeding. It will be seen that in the case of bird 180, this constancy was observed nine times, and that the time for symptoms can be predicted to within a day or two, and is not the same for each bird. "Les accidents" are therefore suprisingly constant events, and can be specifically cured by torulin. Now Chick and Roscoe later (1927) have shown that the concentrate prepared by our methods will cure young rats suffering from symptoms of polyneuritis, but until recently there was no proof that the curative method could be used for the adult rat, or even for certain that our most highly purified torulin preparations would cure rats. This proof has been recently supplied by Miss Reader,* who has compared the curative action of three preparations upon adult polyneuritic

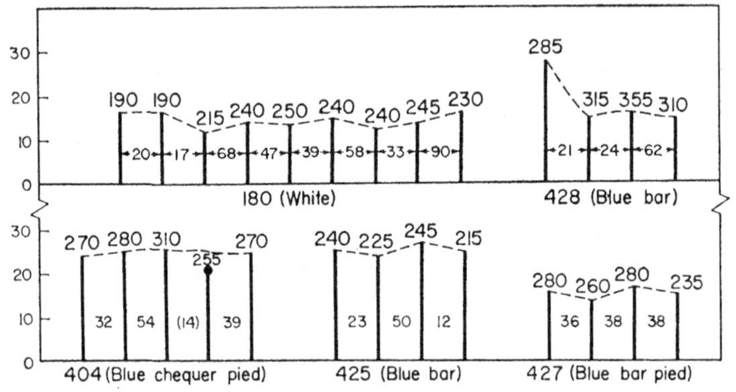

FIG. H7

A. Vertical lines represent the length of time on rice before developing symptoms.
B. Numbers between the vertical lines represent the number of days interval on normal diet between any two experiments.
C. Numbers at the top of vertical lines show the weight of the bird upon the day upon which symptoms developed.

rats with the value obtained by Mr. Kinnersley and myself upon pigeons. She is much to be congratulated upon obtaining these values as not more than some 30 min elapses between onset of acute symptoms and death.

* Unpublished observations.

The preliminary results show that approximately one pigeon day dose is needed by the adult rat per diem. So far as I know this is the first reliable comparison of B_1 activity in terms of a curative test upon a highly purified torulin preparation, and so far as a rat weighing at the time of test 80–100 g requires the same amount of torulin as a pigeon weighing some 200 g, the rat requires relatively more than the pigeon. Calculating from these results and basing the B_1 needs upon the rate of metabolism, it may be concluded that some 50 doses a day are required by a full-grown man, or the amount of B_1 in about $\frac{1}{2}$ oz of marmite.

Two Factors Needed by the Pigeon

It had been found repeatedly in the older literature that curative extracts did not maintain the weight of pigeons feeding upon polished rice; but as yeast preparations could be made which combined a curative with a weight-maintaining action (Seidell, 1922), and further as doubt was cast upon the validity of curative methods, the evidence for two factors had not carried conviction. Now the purified concentrate prepared by us in 1925 was eminently curative in small doses. We were therefore convinced that we were dealing with a genuinely important substance and not with a mere palliative agent. This was supported by the fact that we could follow in a semi-quantitative way the distribution of the curative substance among the various fractions obtained in the course of a preparation. In spite of this, we found that 2-day doses given daily from the 16th day after commencing to feed pigeons with rice did not prevent loss of weight. Further, during the course of our isolation experiments our curative doses never led to resumption of weight. This convinced us that the concentrates described by Seidell must contain more than one factor, in so far as they were capable of maintaining weight. Though there was no proof at the time that this was not some amino acid. it was evident to us at least that those who trusted to a maintenance test were setting themselves the problem of concentrating together at least two factors from the yeast. Our work, of course, established upon a firmer basis much of the older research, in which curative tests have been used. So far as the pigeon was concerned, two factors were present in the B part of the yeast, of which one was the true curative substance.*

Impressed as we were by the reality of the curative substance at that time, we were equally puzzled by some curious features of our feeding experiments, to which I should like now to draw your attention. Figure

* Subsequent work shows that the missing factor must be the one called B_2 below.

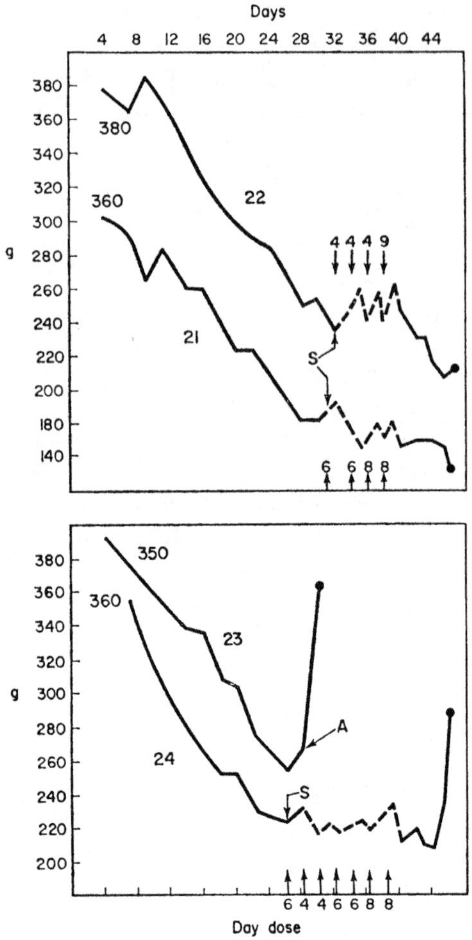

FIG. H8. Chart of pigeon's weight upon polished rice diet.

Nos. 21 and 22, polished rice only. Nos. 23 and 24, polished rice treated with salt.

At S, symptoms of opisthotonus in birds 21, 22 and 24, upon the 31st, 32nd and 26th days respectively.

Dotted line = period of torulin dosing; the number of day doses given, marked with arrows.

At A, No. 23, symptoms of leg weakness; not dosed with torulin.

Weight intervals of 20 g, initial weight of each bird marked on chart.

H8 is a chart of the weight of four birds,* which were fed upon rice until symptoms of opisthotonus appeared and then daily dosed with large amounts of the torulin preparation. These large doses did not suffice to maintain weight in bird 21, though they did doubtfully in birds 22 and 24. It will be noticed that interruption of the torulin dosing led to death within a period of about 8 days. The birds therefore behaved to a preparation which had been carefully standardized upon a curative and protective basis as if the last dose had annulled the storage of the other doses. This contradiction between the preventive and curative activity by the two methods of testing was really an acute form of a difficulty raised in earlier work. Chick and Hume (1917), for instance, had found that the protective and curative doses of pressed yeast were practically identical. Since our experiments were performed with a preparation of torulin purified at least 1000 times as compared with the original yeast, we excluded digestive and absorptive troubles as a factor in the causation of this effect. We were unable to pursue the matter further at the time as our supply of torulin was finished. If we had not so thoroughly established the reality of the curative substance before reaching this stage, we should have been inclined to think that our estimates of curative activity were at fault. Owing to the work which we had done, this explanation was not open to us. I shall return to this important discrepancy between curative and protective activity in my third lecture.

The experiments given in Fig. H8 showed another point, the significance of which I have realized only recently. Birds 23 and 24 were fed with salted rice with the idea of washing out the vitamin by excessive consumption of water. To our surprise the birds died with massive oedema and dropsy. Bird 23 which received no dose of torulin added to its weight over 100 g of water in 2 days preceding death. The bird which was dosed with torulin did not get oedema until the torulin was discontinued. This torulin preparation therefore was an anti-oedema agent.

Further Work Upon B_1

Shortly after our results were published dealing with the pigeon, workers in many countries realized that their results with the rat could be explained only by the view that the so-called vitamin B was a complex.† Prominent among these Goldberger, Wheeler, Lillie and Rogers (1926) showed that a thermostable factor was present known as B_2, the anti-pellagra factor. We need not here go into the question of whether rat pellagra is the same as human pellagra. The factor certainly relieves a

* These four charts were shown to the Physiological Society in 1925. Only one of them has appeared in print, No. 21.

† Smith and Hendrick (1926), Hauge and Carrick (1926), Goldberger, Wheeler and Rogers (1921); see also Sherman (1929).

skin condition in rats. Some of the proofs for two factors were not entirely convincing, as they depended upon the supplementary behaviour of quite different kinds of food stuffs. An experiment of this type may always have the alternative explanation that an inhibitory substance is masking a positive growth effect.

If I may be allowed for a short time to dwell upon this point. Supposing that foodstuff A does not support growth, but will do so when supplemented with B. If A contains in addition to its growth factor a poison (say barium) which neutralizes the effect of the growth factor, and B sulphate, when these two are placed together the barium would be neutralized and the full effect of the growth factor would have its play. Influences of this type must always be operative to a certain extent in normal diets, and may explain anomalous results. Proof can only be considered to be rigid when such effects are excluded.

This is why the experiments of Chick and Roscoe (1927) were of value. Using a torulin concentrate prepared by our methods, and therefore containing only a few milligrams of solid per dose, they showed that growth stopped unless the ration was supplemented with dried yeast, autoclaved as in the experiments of Goldberger *et al.* for 2 hr at 120°C. Thereby they supplied a convincing proof of the necessity of two factors, and of the value of our torulin concentrate to rats. This work reduced the chances of inhibitory effects to a minimum, a chance which is real. Heaton has recently shown (1929) that injection of malt extracts into growing rats stops growth, and Mr. Kinnersley and I* have found that a toxic ether soluble anti-curative substance is occasionally present in our yeast concentrates. It is gratifying to us to find that our methods of torulin preparation are being extensively used, and in fact when material of this activity can be so easily obtained, there is no real excuse for not using it to settle fundamental points. Any small advance in chemical isolation is worth a pile of evidence with supplementary foodstuffs, owing to the gain in experimental clarity.

Further Progress in the Chemistry of B_1

In 1926, Jansen and Donath published an account of a crystalline substance obtained from rice polishings, able to protect a bondol from polyneuritis in doses of 0·002 mg per diem, and a pigeon 0·01 mg.

It is to be noted that a protective substance is not necessarily curative, though it has often been assumed to be so. Supposing, for instance that a substance has a sparing effect upon the utilization of a curative one, then it will be protective, but quite different from the substance which it spares.

* Unpublished experiments.

Describing their technique in brief, it consisted in an adsorption upon Fuller's earth, fractionation with silver baryta, phosphotungstic acid and platinic chloride, all of which had been used by previous workers with only limited success. The fraction so obtained was reached in a yield of 25 per cent of the original amount contained in the rice polishings, and was effective to pigeons at 0·04 mg per diem. From this alcohol and acetone gave a poor yield of the crystals.

The following were some of its properties. It had a formula $C_6H_{10}N_2O$. It was precipitable by mercuric sulphate, and gold chloride in alcoholic solution, and gave an intense red Pauly reaction. Over two years ago, Mr. Kinnersley and I repeated Jansen and Donath's work. One of our experiments is given in Table 4.

TABLE 4. FRACTIONATION OF RICE POLISHINGS BY JANSEN AND DONATH'S METHOD (25 KILOS)*

Successive stages	Curative dose recovered	Approximate activity for recovered pigeon	Jansen and Donath: Yield %	Jansen and Donath: Calculated activity for similar stages	Jansen and Donath: Yield %
(1) Acid alcohol extract	34,000	100 mg		200 mg	
(2) Filtrate from eluted Fuller's earth	7000	5–6 mg	25	1–3 mg	80
(3) Filtrate from first barytaAgNO₃pp	3780	0·9 mg	12	0·4 mg	40
(4) Filtrate from phosphotungstic acid and alcohol	1600	0·06 mg	5	0·15 mg	
(5) ditto platinic chloride	400	0·08 mg	2	0·04 mg	25

Activities for pigeons = weight of day dose.

It will be noticed that though the number of doses which we recovered is much less that that of Jansen and Donath,† the biggest loss occurred at the Fuller's earth stage, at which they recovered 50 per cent more vitamin than we did. They used a special kind of Fuller's earth known as Cheribon. For the rest, the activities of the material which we obtained agrees well with those of the Dutch. Down to this stage, therefore, we

* Kinnersley and Peters, unpublished results.

† In a very recent paper, Jansen (1929) describes further improvements in the method of fractionation.

proved to our satisfaction that their preventive and our curative tests were in agreement. Attempts to obtain the crystalline material however failed, nor did we trouble further, because it is clear that the method used by Jansen and Donath upon rice polishings gives the curative activity stated, at any rate so far as the platinic chloride stage (0·04 mg per day per pigeon).

We have found that their preliminary stages cannot be applied to the concentration of the torulin in yeast. Later in the fractionation, after removal of gum and other substances, it is possible to use their methods with modifications; and in 1927 we separated from yeast in small amounts, fractions of the general activity 0·04 mg, the most active being one of 0·027 mg. Our purer preparations were not stable in alcoholic solution at room temperature, though at earlier stages torulin has remained stable in alcoholic solution for a matter of four years. A sketch of our method of fractionating yeast is given in Table 5.

TABLE 5. GENERAL METHOD OF FRACTIONATION (Kinnersley and Peters).
BAKER'S YEAST

Extracted with hot water

Filtrate Treated with neutral lead acetate	Yeast residue (discarded)
Filtrate Treated with baryta (to remove gum)	Lead acetate precipitate (discarded)
Filtrate, acidified with H_2SO_4 and $BaSO_4$ removed (Hopkins reagent added—Mercuric sulphate)	Baryta precipitate (discarded)
Filtrate. Brought to pH 7·0 and treated with Norite charcoal	Mercuric sulphate precipitate (discarded)
Norite charcoal	Filtrate (discarded)

The Norite charcoal is extracted successively with N/10 HCl and 50 per cent acid alcohol giving extracts of rather different character of which the first is the purest. The material is subsequently fractionated into absolute alcohol after the removal of metals. After this it can be further purified with phosphotungstic acid.

Approximate activity at successive stages, mg/day dose.

Original yeast	500 mg
Water extract	50 mg
Charcoal concentrate	0·3–1 mg
Phosphotungstic acid	0·08 mg
Purest preparation reached	0·01 mg

Quite recently we have separated in very small amounts a preparation which gave a curative activity of 0·01 mg per day dose; this it will be noted is the same curative activity as that of the Dutch. We have no reason to think that this represents a pure compound. Guha and Drummond (1929) also state that they have obtained preparations from wheat germ curative in doses of 0.0025–0·005 mg per day, i.e. considerably more active that the activity of the Dutch crystals. These results suggest that more active curative preparations can be obtained than the crystalline ones from rice polishings.

There are certain differences between the behaviour of our torulin preparations and those described by the Dutch workers. The purer torulin preparations do not keep well in alcoholic solution, do not give intense Pauly* reactions, and are not precipitable by mercuric sulphate; in fact a precipitation with this latter reagent is used by us as a means of removing impurities.

I quote below one experiment out of many, illustrating the instability of the preparations in ether–alcohol mixtures. This is especially marked in the purer concentrates.

Experiment. Illustrating the instability of purified torulin. 800 torulin

Filtrate
Concentrated to dryness on water bath, and finally at room temperature in the dessicator
Extracted with dry absolute alcohol

Precipitate
(inactive)

Filtrate. (400 *doses* still present)
Four volumes ether added

Precipitate

Filtrate
150 *doses*

Precipitate
Extracted with 6 ml dry ether which was added to fil. later.
450 *doses.*

* The above-mentioned preparation does not contain more than 3 per cent of any substance giving a Pauly reaction similar to that of histidine.

doses taken, of approximate activity 0·1 mg. Three volumes of purified ether were added gradually, when a precipitate was formed.

Further fractionation with purified ligroin and resolution in dry alcohol led to gradual disappearance of activity and ultimately only some 150 doses could be found in all the precipitates added together.

We have tried to account for this behaviour in several ways, and have come to the conclusion that the only hypothesis which fits all the facts is that treatment with ether removes a stabilizing agent. When the vitamin therefore is thrown out in relatively anhydrous ether–alcohol mixtures and the precipitated vitamin subsequently extracted with ether, the preparation remaining is now highly unstable; and though doses given immediately are active, those given later are not.

The differences in stability and colour reaction between the yeast and rice preparations cannot be cleared up in the present state of knowledge. Possible explanations are (*a*) that the vitamin forms part of each, but that it is associated with some substance in the rice preparation which makes it stable, or (*b*) that the two curative substances are different though closely allied in structure.

The conclusion that rice polishing preparations are different has also been reached by certain Japanese workers, Kinugasa (1927) found that the anti-neuritic vitamin from rice polishings was precipitable by mercuric sulphate in sulphuric acid solution, thereby confirming the Dutch workers. They have also reached the conclusion that the diazo reaction is not given so intensely by their purer preparation. Their preparations were not very active. Allowing a substantial margin, it seems that the day-dose was of the order of 0·1 mg for pigeons.

The instability of our later preparations is strikingly reminiscent of the behaviour of the purified enzymes (Willstater Euler, etc.). Here also traces of impurities (peptides) to which they are adsorbed suffice to keep them stable, and upon removal the stability disappears. We have always found that treatment of the material extracted from the charcoal, with neutral lead acetate, constantly precipitates some of the vitamin. This we believe to be B_1 in some combined form.

It may not be immediately obvious why a method so successful with one class of material fails with another. This is primarily owing to the fact that where a minimal amount of substance has to be separated from large quantities of impurity, initial steps are in reality always adsorption stages, even if they do not appear to be so. The apparent character of the substance being separated (in this case the vitamin) depends upon its irregular association with impurities. This has been a common experience with the enzymes too.

The Basic Nature of Torulin

From the earliest work upon the curative substance, it has been generally believed that torulin is a base. That it is not a primary one, and probably not a secondary was shown indirectly by McCollum and Simmonds (1918). A yeast concentrate would still support the growth of rats after treatment with nitrous acid. In 1923, I was able to prove clearly that the torulin concentrates were stable to nitrous acid. This has been subsequently confirmed for purer preparations. As in the case of the Dutch preparations, our purest torulin does not contain sulphur. It is probable that the vitamin is a tertiary base. In preliminary experiments, Mr. Kinnersley, Mr. Phelps,* and I, have found that even the better torulin preparations are adsorbed by charcoal at alkaline reactions, approx. pH 9·0. This it will be remembered from my first lecture is a property of bases, so that our adsorption experiments support the view that it is a base.

Sahashi (1928) has recently claimed that 2–6 di-OH quinoline and also 4 (or 5) glyoxaline ethyl methyl carbinol have curative properties. The latter substance is somewhat similar to a possible formula for Jansen and Donath's compound. Dr. J. M. Gulland has recently prepared these compounds and we have tested them upon the pigeon (1929). The compounds unfortunately were not curative and did not behave like torulin.

The Use of Torulin Concentrates in Elucidating the Number of the Water Soluble Vitamins

Let us next consider how far these attempts to concentrate torulin have been of value in settling some outstanding problems in connection with the nutrition of the rat and of the pigeon, and of some of the lower organisms. We may first turn to the rat.

The rat. I have mentioned the fact that Chick and Roscoe demonstrated the necessity of two rat factors by using torulin concentrates. Dr. Vera Reader working in my department has recently proved that a further factor in addition to B_1 and B_2 is needed for full nutrition in the rat.† At the start of her work, she wished to know what was the relation between the dose of torulin for the growing rat and for the pigeon. For this purpose, young rats were fed upon a diet, complete except for the B complex, with the following supplements, B_1 in the form of a torulin preparation made by Mr. Kinnersley by the newer methods of alcohol fractionation, and B_2 in the form of alkaline autclaved marmite. This preparation of B_2 it must be noted was that used by Hassan and Drummond (1927) in their experiments with high protein diets, but was not identical with the "Goldberger" B_2, who used autoclaved dried yeast. It was found that even large supplements of these B_1 and B_2 sources to the diet would not induce growth for any prolonged period.

Figure H9 indicates the type of growth which takes place, a spurt for a matter of 2–3 weeks and then no more. Growth, however, readily takes

* Kinnersley, Peters and Phelps, unpublished experiments.

† A preliminary account of this appeared in *Proc. Biochem. Soc., J. Soc. Chem. Ind.* **47**, 1247, (1929).

place upon including in the diet fresh marmite, in doses of 0·5 g per day. This means that at least two essential factors have been destroyed in the alkaline autoclaving. Both are, therefore, alkali-thermolabile. Even supplements of the Oxford torulin concentrate up to eight pigeon doses per diem are unable to restore growth. At first it was thought that this third factor (named B_3 for convenience) was destroyed by heat at 120°C, but what really appears to happen is a destruction of half the activity, suggesting that the compound responsible for the effect is racemized by the heat like an amino-acid.

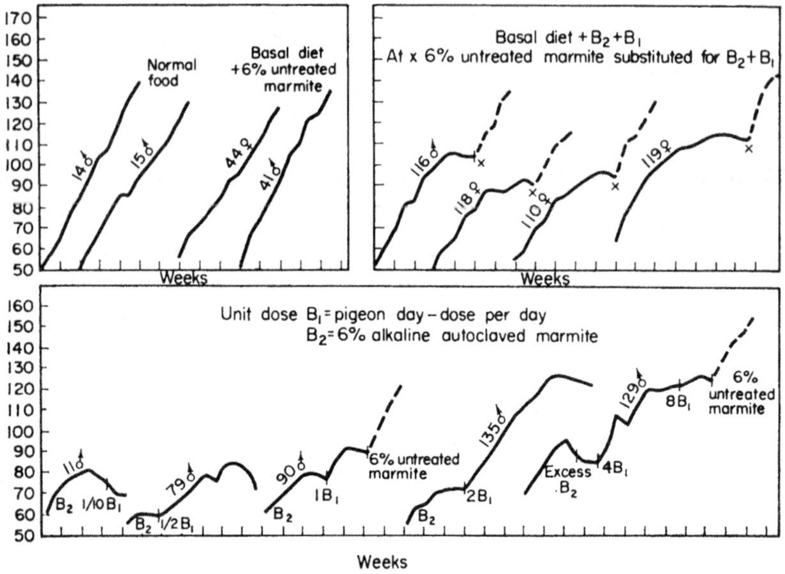

FIG. H9.*

(1) Comparison of growth upon normal food and basal diet + marmite.

(2) Basal diet + B_2 + B_1, showing the effect of substituting untreated marmite.

(3) Effect of excess B_2 and B_1.

With what factor can this alkali-labile B be identified? Our first thoughts turn to the B factor of Williams and Waterman, This is essential according to them for full nutrition in the pigeon, but has not been shown to be necessary for the rat. I reproduce in Fig. H10 some curves which fully

* By permission of the Editor. V. Reader (1929), *Biochem. J.* **23,** 672.

confirm the results of these two authors. They show that marmite and rice is a maintenance diet, and that such birds rapidly gain in weight when placed upon wheat.* The factor present in the wheat is not protein, because a supplement of casein to the marmite diet will not lead to increase in weight. Hence marmite does not contain the Williams and Waterman factor. The alkali-labile factor is therefore different from that of Williams

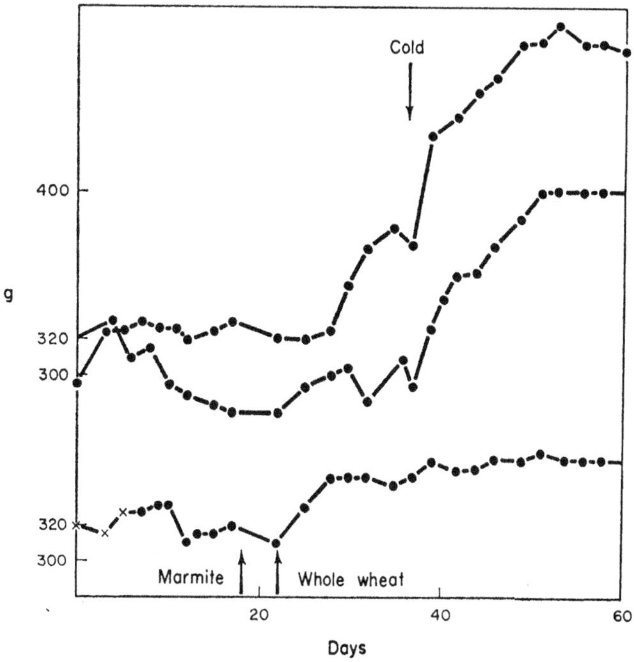

FIG. H10. Chart of pigeon weight showing the "Williams and Waterman" effect.

Ordinates = weight in g. Abscissa = days.
At arrow pigeons transferred from polished rice + marmite (g daily) to whole wheat. At the thick arrow, the birds were subjected to severe cold.

and Waterman. The weight reached by birds upon wheat corresponds to that reached upon our "standardized" diet. It was possible that B_3 was identical with a factor described by Hunt (1928).

* Out of 18 birds so treated, 10 showed a response like the upper curve, 5 like the middle, 3 showed the lower curve. The majority therefore responded intensely to the stimulus of cold which suggests a connection between this factor and the storage of fat.

Hunt has claimed that if autolysed yeast be extracted thoroughly with water, a residue is obtained which has a definite supplementary effect when added to two fractions separated from the filtrate. So he has three factors, one of which is not extracted from an autolysed yeast residue by cold water. This evidence is rather similar to that of Kennedy and Palmer (1928), who could not obtain maximum growth unless they added to the Harris yeast extract of the Osborne and Mendel fraction 2, the autoclaved residue of the yeast. Provisionally it may be suspected that these two sets of workers are dealing with the same phenomenon.

It is unlikely that the new alkali-labile factor is that of Hunt, because it passes quite readily into aqueous solution upon boiling yeast with water.

The provisional conclusion which we have reached at this stage of our discussion is that there are at least two "B" factors for pigeon nutrition, and at least three factors for the rat; there may be one more if we accept the evidence of Hunt. Let us now reconsider the nutrition of the pigeon in the light of the French work, and of certain further work of our own.

The pigeon. Randoin and Lecoq (1929) have recently summarized the results of their experiments upon pigeons living upon adequately balanced synthetic diets, administered by forcible feeding. They distinguished four factors: (1) antinévritique or "équilibre nerveux"; (2) utilization nutritive ou d'entretien; (3) utilization cellulaire, a micro-organism factor, and (4) antipellagra. (1) and (2) are alkali-labile. So the French workers find two alkali-labile factors to be necessary for pigeon nutrition. Are there then two factors necessary for pigeon nutrition other than that of Williams and Waterman? Mr. Carter and I have come to the conclusion that there are.* I have mentioned previously that torulin concentrates do not maintain the weight of pigeons fed upon polished rice. A yeast extract such as marmite, however, when given to such pigeons maintains the weight constant, approximately, at a level of weight below the standardized level (curve A, Fig. H11). This fact forms the basis of Seidell's test. When marmite is autoclaved in alkaline solution containing therefore only B_2, the weight of the pigeons falls (Curve B). The fall in weight is not stayed by the addition of a daily dose of torulin (Curve C).

Hence factors B_1 plus B_2, do not maintain the weight of pigeons, whereas fresh marmite does so. The latter does not contain the factor of Williams and Waterman. The story for the adult pigeon therefore is similar to that for the growing rat. Alkaline autoclaving destroys another factor besides torulin. It seems to us to be probable that the extra alkali-labile factor is identical with the rat B_3, but this conclusion can only be regarded as provisional. It is interesting to note that our experiments upon pigeons feeding naturally upon a polished rice diet, have led us to

* Carter, Kinnersley and Peters, unpublished experiments.

the same conclusions as those of the French workers, who have used synthetic diets and a forcible method of feeding.

It has been stated by Seidell (1929) that the pigeon does not need B_2 (antidermatitis) in its nutrition. Hauge and Carrick (1926) and Hogan and Hunter (1928) on the other hand have claimed that the growing bird needs this. In view of the work to which I have alluded, it is not quite clear that B_2 in their sense was not the B_3. Reference to Fig. H11, curves B and C, shows that Mr. Carter and I have obtained no evidence for the necessity of B_2 for the pigeon. This confirms Seidell.

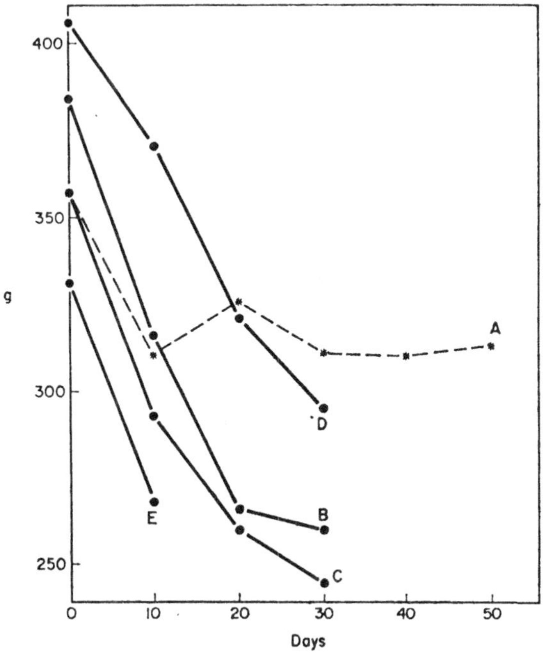

FIG. H11. Curves represent the average fall in weight of groups of pigeons fed from the beginning upon polished rice with the following additions.

A. 1 g untreated marmite daily after 10th day.
B. 1 g alkalized marmite daily throughout.
C. 1 g alkalized marmite + 3-day doses of torulin daily throughout.
D. Three-day doses of torulin.
E. Control with no additions.

Let us now summarize the known factors of the B complex.

THE "B" COMPLEX

	Needed by		
	Growing rat	Adult pigeon	Distinctive property
A. Factors for which evidence is complete			
(1) Antineuritic proper (B_1) Eijkmann[3]	yes	yes	alkali-labile
(2) Antidermatitis (B_2) Goldberger *et al.*	yes	no[1]	alkalistable
(3) Third rat factor (B_3) (Reader)	yes	probably[2] yes	alkali-labile
(4) Third pigeon factor (Williams and Waterman)	no	yes	thermolabile
B. Factor for which evidence is not yet complete			
(5) Fourth rat factor	yes	not known	thermostable

NOTES: (1) When pigeons are living upon an adult diet.
(2) Provisionally identified with the rat factor. Probably the same as Randoin and Lecoq d'utilisation nutritive.
(3) Oryzanin (rice), torulin (yeast).

It can be seen from the above that we have complete evidence for four factors, of which three are needed by the rat and three by the pigeon.

Though the above table represents the factors for which the evidence is clear, there is one other which must be mentioned. Coward, Key and Morgan (1929) have obtained conditions under which rats will grow upon a light white casein, but not upon a Glaxo casein. The results appear to be real and reproducible, and we must await more evidence. It should be mentioned that the light white casein does not contain the B_2 which has been considered above.*

THE WORK OF OTHERS VIEWED IN THE LIGHT OF OUR OWN EXPERIMENTS

Seidell (1929) uses a weight maintenance test in pigeons and must therefore be concentrating on two factors. Levene (1928) in his latest work has obtained highly active preparations for the rat, and recognizes two rat factors. According to our views, his work must be complicated with one unrecognized factor, and it is not clear where this is supplied.

Reconciliation of our work with that from the Lister Institute is at first sight difficult, because these researches have been conducted upon the assumption that only B_1 and B_2 are necessary. Aykroyd and Roscoe (1929) in work

* Reader (unpublished results).

upon the distribution of B_2 have used curative tests as well as tests upon the growing rat. It is therefore beyond doubt that their experimental conditions were such that B_1 plus B_3 was supplied in the growth experiments. In considering this source of supply, I must first point out that there is still confusion as to what is meant by B_2 the antipellagra factor. The original workers, Goldberger *et al.* used yeast dried and autoclaved, at 120°C for 2 hr. Others following Drummond and Hassan (1927) have used alkaline autoclaved yeast extracts. The degree of moisture is different and also the pH of heating. Miss Reader* has found that the alkaline autoclaved marmite still contains a factor which cures dermatitis in rats, and has acted of course as B_2 in her experiments. Since the rat B_3 is not completely destroyed even upon heating at 120°C for 1 hr at pH 6·0, it is very probable that the autoclaved dried yeast as used by Chick and her colleagues as a source of B_2 still contains B_3. This would account for some of the discrepancies. This is borne out by the fact that Chick does not get a complete rat nutrition with a B_2 preparation made from white of egg. Her observation that one of the rat factors is present in white of egg is of much interest as it provides a source of supply in an unheated preparation.

Some recent evidence from Eddy's laboratory by Williams, Waterman and Gurin (1929) demands a reference. They have autoclaved Brewer's yeast at different pH in the presence of water for 6 hr at 15 lb pressure. When these fractions were fed to growing rats, the animals were only protected in their B_2 sense at the more acid pH viz. 1–2. Now we know that this does not hold for B_2 proper, nor for torulin. The results receive a full explanation in terms of B_3.

NOTE: Since writing the above there has come into my hands a report by Williams and Eddy (*Carnegie Institute Year Book*, No. 27, 375, 13 Dec. 1928), in which the authors allude to the pigeon third factor of Williams and Waterman as B_3, and to a further rat factor as B_4. The latter may possibly be identical with the third rat factor of Reader, though its attachment to casein suggests a closer resemblance to that of Coward *et al.*

Meanwhile I have thought it best to adhere to the nomenclature in the *Biochemical Journal*, pending any adjustment of nomenclature by the Biochemical Societies concerned.

THE PART PLAYED BY TORULIN IN THE NUTRITION OF THE LOWER ORGANISMS

Let us now turn to a consideration of the part played by torulin in the life of the lower organisms. Our knowledge of this is more fragmentary, and I cannot do more here than summarize briefly some of our results.

Bios. The earliest growth factor of all to have been described is "Bios". This is not the time and place to consider this in detail; we may, however, allude to the chief points in the problems, see Tanner (1925). The bios question started with Liebig (1871) and Pasteur (1872). The latter said that yeast would grow upon an artificial medium and the former that it would not. The difference (or one of them) between the results depended upon the amount

* Unpublished results.

of the original sowing. Later Wildiers (1901) insisted upon the importance of small traces of an organic substance for the growth of a small seeding. This he partly defined, and his explanation of the original controversy was that Pasteur had used a large sowing.

Much confusion has centred round the bios problem, but the following points must be emphasized. Two methods have been used rather indiscriminately by workers (*a*) the fermentation test in which the amount of CO_2 evolved is the criterion, and (*b*) the growth test. Considering the matter critically with the knowledge which we now possess, it is obvious that the two methods are not identical. Further in the earlier experiments, I do not think that there was any clear recognition of the fact that different yeasts vary in their behaviour. This has been shown in a striking way recently by Eddy, Kerr and Williams (1924).

A "bios" preparation for their yeast was inactive as bios towards another yeast owned by Lucas (1924).* In 1926, when we discussed the whole matter at the British Association, we referred to the tendency for workers to consider all factors capable of promoting growth in the lower organisms as "bios". Such a course could only lead to confusion, as it involves the tacit assumption (*a*) that all yeasts have the same nutritive needs and (*b*) that other micro-organisms have similar requirements to yeast. Our own preliminary work showed that this was not true. At this time Peskett brought forward evidence that allelocatalysis did not occur in yeast, and that some of the torulin concentrates were more active weight for weight than a crystalline preparation of Eddy, Williams and Kerr. Many workers have, of course, tried to establish some connection between the B vitamin and Bios.

Peskett (1928) has recently made a considerable advance in the technique of testing. In this he does parallel tests with two sowings of yeast, a large and a small one. He often finds that an extract which will affect the large sowing will not influence the small one. Hence he distinguishes between two properties of added extracts. Many substances appear to act as stimulants in the sense that they will promote growth when a large sowing of yeast has been made. The true "bios" in Peskett's sense is something which will take the culture up to maximum growth from a small seeding.

Peskett (1929)† has tested our purer torulin concentrates and finds that these have neither bios nor stimulant effect. We must conclude therefore that the growth promoting "bios" is not the same as any of the accepted vitamin factors. I have also found† that, although the less pure torulin extracts promote CO_2 production by yeast, the purest do not do so either by themselves or in the presence of added hexose phosphate.

* See also Tanner and Devereux (1928), Tanner (1925).

† Unpublished observations.

Other micro-organisms. Though torulin is not required by yeast, other micro-organisms are not so independent of it. As Peskett and Hawking (1927) have shown, there is no need for the supply of torulin to yeast as it can synthesize this factor from an inactive medium.

In much of the work done up to the present time upon the growth of micro-organisms, sufficient trouble has not been taken to purify the chemical substances used in the tests, owing to a failure to realize how small were the amounts of growth-promoting factors needed. Before she joined us at Oxford, Miss Reader was studying the nutrition of a strain of *Streptothrix corallinus* in Drummond's department. She found that glucose carefully purified by the method of Funk and Freedman (1913) would not grow the organism, though it was necessary to recrystallize three times from alcohol before this stage was reached. The accessory substance needed was found to be present in a specimen of torulin which Mr. Kinnersley and I supplied to her for test; it acted as a powerful growth-promoting agent. This effect has been followed carefully through the various stages of torulin fractionation (1928).

The distribution and properties of the two substances torulin and the *S. corallinus* growth factor agree closely, there being only one exception. It is possible to inactivate the torulin concentrates for the pigeon by treating with alkali, without affecting their *Streptothrix* activity. The quantitative correspondence is rather remarkable as it holds right down to the purest torulin (0·01 mg/day), some 50,000 times the concentration in the original yeast. How incredibly small is the amount needed may be judged from the fact that of this preparation (which there is no reason to think is pure), only 6×10^{-4} mg per 100 ml* is needed to produce a maximum growth of the organisms (some 400 mg dry weight). It is indeed remarkable that so small an amount should act as a coordinating agent to the metabolism of the whole. Owing to the correspondence found, we are inclined to think that the micro-organism actually needs torulin itself, and that the one instance of discrepancy between the effect upon the pigeon and *S. corallinus* means that the micro-organism can reactivate "alkalized" torulin.

If B_1 is a fundamental factor essential to the life and growth of all cells, it is reasonable to think that organisms which grow in the absence of added B_1, synthesize it in the course of their growth. It has been possible to employ the fact that such small amounts of torulin are needed by *S. corallinus* to test for the presence of the vitamin in solutions too dilute for pigeon tests.

Miss Orr Ewing and Miss Reader have found that a strain of meningococcus is culturable upon a broth from which all B_1 has been removed

* Much smaller amounts produce a noticeable effect.

by charcoal. When this culture fluid is added to a synthetic medium for *S. corallinus* instead of torulin, growth takes place. In this we have an example of synthesis of B_1 by an organism which is independent of it during growth, just like yeast.

It may be mentioned in this connection that *B. coli* and other micro-organisms have been shown to synthesize parts of the B complex, which is probably the explanation of the curious phenomena of refection.

Anomalous result. It is not Nature's fault but ours when we make no discoveries. I wish to speak in conclusion of one that has arisen from a cloud no larger than a man's hand. During the course of some *Streptothrix* tests in the earlier stages of a torulin preparation. Miss Orr Ewing noticed that certain anomalous results appeared. Without entering into details, it may be said that there was a lack of the usual quantitative agreement between the amount of growth for different strengths of the impure vitamin. More vitamin produced relatively more growth. The matter was nearly left, but a lucky instinct led Miss Reader to the discovery that the effect can be imitated specifically by the addition of mannitol to a torulin

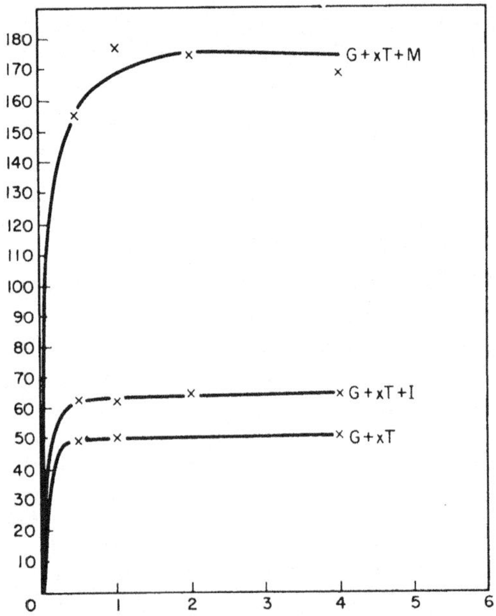

FIG. H12.* Effect of addition of mannitol upon the growth of *Streptothrix corallineus* in presence of torulin weight of organisms in 14 days, i.e. after growth is complete.

M = 0·5 per cent Marmite. G = 0·5 per cent glucose.
I = 0·5 per cent inositol. T = 1/20 unit ion of torulin.

* By permission of the Editor. V. Reader (1929), *Biochem. J.* **23,** 65.

preparation. Mannitol plus glucose gives much more growth than mannitol alone. Inositol and dulcitol will not do as substitutes. The following curves, taken from her paper, illustrate clearly what I mean (Fig. H12).

Here we have an instance of a substance which can act as a co-vitamin and of which the chemical constitution is known. Whether this be the substance in yeast responsible for the effect is not yet certain. Some evidence exists that the results are due to some substance other than mannitol, which produces its effect in lower concentration.* There is no doubt that the matter has opened up an interesting line of work.

I have little doubt that these results for a few micro-organisms are only the beginning of investigation into a field of enquiry which will prove to be fully as large as that of the vitamin itself. It is probable that in addition to new acquaintances, we shall find among the growth promoting factors for various bacteria not a few familiar friends. That a number of these co-ordinating links will be revealed by future research, I do not feel inclined to doubt. It is also significant that Heaton (1926) found that torulin had a growth promoting action upon epithelial tissue cultures. I can now put into a table the present position of our knowledge of the content of our yeast fractions in these various factors under discussion at various stages.

SUMMARY TABLE

Bakers Yeast

Stage	Activ. mg	B_1	B_2	B_3	Yeast stim.	Bios	S. cor. gp.	Epithelin gp.
Aqueous extract	50	+	+	+	+	+	+	+[4]
Charcoal concent.	1·0	+	nil[3]	trace[2]	+[1]	?	+	+[4]
Alcohol fract.	0·2	+	nil[3]	nil	+[1]	?	+	+[4]
Phosphotungstic	0·08	+	nil[3]	nil	nil[1]	nil[1]	+	?
Best fraction	0·01	+	nil[3]	nil	nil[1]	nil[1]	+	?

NOTE: Bios, B_2 (antidermatitis) and *S. corallinus* growth-promoting factor not destroyed by alkali.

[1] Peskett, unpublished results (see 1926).
[2] Reader.
[3] Chick and Roscoe.
[4] Heaton.
Activity (mg) = pigeon day dose.

In this short review, I have had no opportunity to give more than a passing allusion to the work of other laboratories, but I hope that I have been able to present an outlook upon the problem based upon actual experience, which may enable other work to be better understood. It will be

* Reader, unpublished observation.

doubtless realized that a unitarianism is not the modern faith in vitamin work. The vitamins in this follow a fashion in biochemistry. In enzyme chemistry, for instance, Waldschmidt Leitz (a pupil of Willstater) has taught us that trypsinogen and erepsin together with enterokinase constitute a battery of at least four enzymes, with definite specific functions. Why the mammalian, or indeed the lower organism should have lost its synthetic capacity in some cases and retained it in others is of course a mystery, though no more mysterious than its dependence upon a supply of tryptophane. I am tempted to think that along with the evolution of so highly organized a tissue as the brain, has gone a suppression of the power of forming certain ring compounds allied to the alkaloids. The risk of synthesizing poisonous alkaloids might be too high a price to pay for the retention of this degree of independence, and the variations which do not produce them might therefore tend to survive. The risk may be no mean one when we consider that aldehydes and ammonia are both present in tissues.

In this lecture I have dealt with the gradual unravelling of a tangled skein of substances, all of which minister to metabolism in some way or another in animal organisms. This must not blind us to the necessity of salts, or to the need for a supply of essential substances for tissue formation and energy in the body cells. I have tried to show that progress comes from a patient attempt to concentrate some of these factors, and to indicate how such concentrates become a basis for reproducible results free from some errors to which this type of work is liable. In following up these factors, it is safer to rely upon the cure of some particular symptoms rather than upon the restitution of such a complex process as growth, however valuable this latter may be in other connections. I know that in urging this I am setting myself against much of the current opinion so far as the B complex is concerned. We have yet, of course, to learn the function of these factors in human nutrition, though two of them at least, the antineuritic in beri-beri and the antipellagra are already known to be of significance.

Now that the "B" group of factors is reasonably definite, we are in a position to consider the part which some of them are playing in the life of the animal, and it is this question to which I shall devote my next lecture.

My thanks are due to the Medical Research Council for grants in aid of this work, and to those working at Oxford who have kindly given me permission to refer to unpublished results.

References

AYKROYD and ROSCOE (1929), *Biochem. J.* **23,** 483.
CHICK and HUME (1917), *Proc. Roy. Soc. B* **90,** 44.
CHICK and ROSCOE (1927), *Biochem. J.* **21,** 698.
CHICK and ROSCOE (1929), *Biochem. J.* **23,** 498.
COWARD, KEY and MORGAN (1929), *Biochem. J.* **23,** 695.
CRAMER and MOTTRAM (1927), *Lancet*, 19 November.
EDDY, KERR and WILLIAMS (1924), *J. Amer. Soc.* **46,** 2846.
EDIE, EVANS, MOORE, SIMPSON and WEBSTER (1912), *Biochem. J.* **6,** 234.
FUNK (1911), *J. Physiol.* **43,** 395.
FUNK and FREEDMAN (1923), *J. Biol. Chem.* **56,** 851.
GOLDBERGER, WHEELER, LILLIE and ROGERS (1926), U.S. Public Health Reports, 41, 297.
GUHA and DRUMMOND (1929), *J. Soc. Chem. Ind.* **48,** 295.
GULLAND and PETERS *Biochem. J.* **23,** 1929 (in press).
HASSAN and DRUMMOND (1927), *Biochem. J.* **21,** 653.
HAUGE and CARRICK (1926), *J. Biol. Chem.* **69,** 403.
HAWKING (1927), *Biochem. J.* **21,** 728.
HEATON (1926), *J. Path. Bacteriology* **29,** 293.
HEATON (1929), *J. Path. Bacteriology* **32,** 565.
HOGAN and HUNTER (1928), *J. Biol. Chem.* **78,** 433.
HOPKINS (1912), *J. Physiol.* **44,** 425.
HUNT (1928), *J. Biol. Chem.* **78,** 83.
JANSEN and DONATH (1926), *Proc. K. Akad. Wetensch., Amsterdam* **29,** 1390.
JANSEN (1929), *Rec. Trav. Pays. Bas.* **48,** 984.
KINNERSLEY and PETERS (1925), *Biochem. J.* **19,** 820.
KINNERSLEY, PETERS and READER (1928), *Biochem. J.* **22,** 276.
KINUGASA (1927), *Bull. Imp. Hug. Lab. Exp. St. Res.* **57.**
KRUSE and MCCOLLUM (1929), *Physiol. Reviews* **9.**
LEVENE (1928), *J. Biol. Chem.* **79,** 465.
LIEBIG (1871), *Ann. Chem. Phys.* **4,** 235.
LUCAS (1924), *J. Phys. Chem.* **28,** 1180.
MCCOLLUM and DAVIS (1915), *J. Biol. Chem.* **23,** 181
MCCOLLUM and SIMMONDS (1918), *J. Biol. Chem.* **33,** 55.
MCCOLLUM, SIMMONDS and BECKER (1925), *J. Biol. Chem.* **64,** 161.
ORR EWING and READER (1928), *Biochem. J.* **22,** 440.
PALMER and KENNEDY. (1928), *J. Biol. Chem.* **76,** 591.
PASTEUR (1872), *Ann. Chem. Phys.* **4,** 25, 146.
PESKETT, PETERS and READER (1926), *Rep. Brit. Ass.* 297.
PESKETT (1927), *Biochem. J.* **21,** 1102.
PETERS, KINNERSLEY, ORR EWING and READER (1928), *Biochem. J.* **22,** 445.
PETERS (1921), *J. Physiol.* **55,** 1.
PETERS (1923), *Biochem. J.* **18,** 858.
PETERS (1928), *J. Physiol.* **68,** *Proc. Physiol. Soc.* **2.**
PLIMMER (1928), *Public Health*, September.
PLIMMER, RAYMOND and LOWNDES (1929,) *Biochem. J.* **23,** 546.
RANDOIN and LECOQ (1929), *Bull. Soc. Chem. Biol.* **11,** 745.
READER (1928), *Biochem. J.* **22,** 434.
READER (1929), *Biochem. J.* **23,** 689.
SAHASHI, (1928), *Bull. Inst. Phys. Chem. Res. Tokyo* **7,** 1191.
SEIDELL (1922), *J. Amer. Chem. Soc.* **44,** 2042.

SEIDELL (1926), *Bull. Soc. Chem. Biol.* **8,** 746.
SEIDELL (1929), *J. Biol. Chem.* **82,** 633.
SHERMAN, H. C. 1928), *J. Nutrition* **1,** 191. Ed. Review.
SMITH and HENDRICK (1926), U.S. Public Health Reports, **41,** 201.
TANNER and DEVEREUX (1928), *J. Bact.* **14,** 317.
TANNER (1925), *Chem. Rev.* **1,** 397.
TSUKIJE (1922), *Biochem. Z.* **131,** 124.
WALDSCHMIDT-LEITZ (1928), *The Enzymes.*
WILDIERS (1901), *La Cellule* **18,** 313.
WILLIAMS and WATERMAN (1928), *J. Biol. Chem.* **78,** 311.
WILLIAMS, WATERMAN and GURIN (1929), *J. Biol. Chem.* **83,** 321.

Lecture III.* Tissue Anarchy

AS ILLUSTRATED BY OUR KNOWLEDGE OF THE CHANGES PRODUCED IN THE HIGHER ANIMAL BY A LACK OF THE VITAMIN B COMPLEX

It has been said that the living organism differs from an ordinary machine in that it is able to put itself right when it is out of order. This is, of course, true only in a very limited sense. There comes to any living organism a certain temperature beyond which life cannot exist, a certain degree of anoxaemia which proves fatal. So unless a diet is adequate in quantity as well as in quality, certain progressive changes ensue which lead ultimately to tissue anarchy.

In my first two lectures I have given my views as to how biochemical coordination in the cell is effected, and have examined in detail evidence for the existence of a group of coordinative factors representing the dietary complex formerly known as vitamin B. I now wish to consider more especially the part played by some members of this group in intermediary metabolism. Though much information is to be obtained from the normal organism, it is an axiom of our subject that the experiments of Nature have provided valuable clues to normal processes. No better illustration of this could be found than "Inborn errors of metabolism" and no better exponent than Sir Archibald Garrod. Had it not been for scurvy, rickets or beri-beri, the world would be ignorant today of certain important aspects of nutrition. It is equally a truism that until diseases pass into the experimental stage, it is almost impossible to gain accurate information upon what is happening. This is so with beri-beri. Much of the earlier pathology of experimental beri-beri has been obscured as to its clear issues by the simultaneous presence in the same animal of more than one vitamin deficiency. My previous lecture showed that vitamin B consisted of at least four factors. Even yet we cannot be certain upon many points, but I feel that the time is ripe for some attempt to appraise the present position. In this lecture therefore I shall consider some abnormalities in metabolism arising from lack of vitamin B factors in the diet. As in the case of my last lecture, I cannot deal adequately in an hour with the diverse literature which clusters around the pathology of vitamin B deficiency; nor is this necessary. The earlier workers Eijkmann (1897), Schaumann (1910), Vedder (1913), and McCarrison (1921), have dealt so thoroughly with the clinical pictures in beri-beri of all types, that further

* Delivered in the Lecture Hall of The Institute, 1 November 1929.

effort in the matter is superfluous. The histopathology of B_1 and B_2 deficiency also has been recently investigated by Findlay (1928), and Stern and Findlay (1929), I shall therefore confine myself to the contact of our work at Oxford with some of the major problems, which have been raised. It must be clear, I think, that relation to the disease in the human subject can only be made definite, when we know accurately the possible number and actual functions of the various B factors absent in a polished rice diet. Our own observations have been largely incidental to work upon the isolation of B_1 (torulin) and so deal mainly with the action of this factor.

Inanition and avitaminosis B. I shall first deal briefly with the relation between inanition and avitaminosis B. The matter was early raised and has been contentious. Are the symptoms due to avitaminosis or to starvation? The question is really whether they result from a balanced general malnutrition or from a one-sided inanition due to lack of a special factor. Let us consider the matter from a general standpoint. It is a familiar fact to most of us that there are people of our acquaintance who are tremendous eaters, but who never get fat. Others add to their corpulence the slightest excesses in their diet. Most of us have pondered upon the meaning of this. Why is it that some individuals absorb food without storing it? Now I feel that one of the really striking things about vitamin B is that an animal starved of vitamin B is unable to use the food passing through it. Before appetite disappears such food can be shown to be absorbed.

This matter was demonstrated by Hopkins (1912) in his classical experiments carried out from 1906–12. It has been recently studied by Bickel and his colleagues (1924) Working with the dog, they found that the nitrogen balance becomes negative in the absence of vitamin B, even when the animal is upon a diet otherwise adequate. The food is being absorbed, but not being used. The analogy between such a type of avitaminosis and the ways of our "Cassius-like" man is so marked that I find it hard not to believe that the types of metabolism have something in common. Starvation in this case is internal. Loss of appetite is secondary to the internal starvation, and a reflection from tissue cells which have lost their rightful powers of assimilation. The point is fundmental to the understanding of the pathology of the disease and to matters of diet in general, and it bears upon other forms of cachexia. We are apt to think of human nutrition in terms of digestion and absorption. This is important of course; in rickets one of the functions of the vitamin would appear to be the facilitation of absorption, but it is not all. The food after entry into the circulation may be of no avail to the animal. Nutrition is ultimately a question of the individual tissue cells. If in the normal individual we introduce glucose from the gut into the blood stream, this is stored or burnt rapidly and in a matter of 2 hr the blood sugar is again normal in amount; the tissue cells are again bathed by their normal concentration of sugar. Glucose, however, in the blood stream of an animal starved of vitamin B is apt to remain there and so to cause a hyperglycaemia. The animal is sugar intolerant and in that sense behaves as a diabetic. It must be remembered that other forms of sugar intolerance are produced by thyroid and pituitary troubles. Avitaminosis must therefore always involve starvation, though the primary event be one missing unit.*

Much time and energy has been spent upon the question as to what features of avitaminous inanition are special to lack of vitamin B and what are general to all forms of deprivation of food. There is firstly the question of the relation of avitaminous symptoms to loss of weight.

The experimental facts of importance are that animals when starved and given water, die early without showing any symptoms; if given vitamin they occasionally do (Eijkmann and van Hoogenhuize, 1916; Drummond and Marrian, 1927). In the last stages of avitaminosis they lose appetite; if given the vitamin they immediately recover it. By forcible feeding (Braddon and Cooper, 1914; Randoin and Lecoq, 1929), the birds get symptoms quicker with much less loss of weight than those allowed to feed naturally. This happens when care has been taken to render the diet complete in every respect except the vitamin.

It is generally agreed that the loss of weight *per se* is not the cause of the illness. A diet given *par gavage* is not in itself harmful, because if the food is forcibly fed together with vitamin a bird remains in good health.

It is not, however, immediately obvious to me that the cause of death is the same in the forcibly fed bird as in that of the bird allowed to feed naturally, owing to the hyperglycaemia constantly induced by "stuffing". Tissue cells must be working at a wrong concentration of glucose. Loss of weight is indeed stemmed, but much of this may be protective against a final dissolution, owing to the liberation of essential factors from tissue stores. The sacrifice of one tissue for another is one of the most striking points of this type of cachexia. Loss of appetite is actually protective whether or not this happens accidentally. Kon and Drummond (1927) have controlled the question of inanition by feeding control birds with an amount of food similar to that actually consumed by the avitaminous animals. In this way they have shown that loss of weight in birds feeding naturally is due to the diminished food intake.

Inanition in avitaminosis has often been attributed to lack of appetite alone, because it is an undoubted fact that vitamin B extracts stimulate appetite (Cowgill, Deuel, Smith, 1925). In his classical experiments, Hopkins showed that loss of appetite was not the primary happening. In this connection, I think that our own work proves that the lack of appetite is a tissue adjustment. As I pointed out in my previous lecture, many rice-fed birds exhibit a constant time for the induction of symptoms under our conditions (Kinnersley, Peters and Reader, 1928). Now a bird

* See E. G. Grey (1928) for a general discussion upon constitutional factors and Marrian (1928) for the effect of B deficiency upon the suprarenals.

fed by stuffing will usually get symptoms sooner. Hence the delay in the birds feeding naturally is due to loss of appetite; and since this occurs in a constant way for a given bird, it proves that the loss of appetite is a tissue adjustment. One of the most striking features of feeding a group of birds upon a rice diet, as McCarrison (1928) has observed and as we can confirm from an extensive experience, is the idiosyncrasy shown in the rate at which symptoms develop.

We have recently handled an astonishing case of this. Ten birds were placed on a diet of polished rice, daily doses of torulin and Oxford tap water. The majority died before the 50th day. One of them survived at a low level of vitality for 120 days. This may be taken to represent 10 years in the life of a man (see Fig. H15).

Facts of this nature suggest an answer to the controversy as to whether we have given enough vitamin B in our normal diets or not. This too must be a matter of idiosyncrasy. Most individuals probably have enough and to spare, but there will be a few for whom an ordinary level of "B" is totally insufficient.

It must be remembered in this connection that there is direct evidence in the case of man that individuals living upon polished rice succumb to beri-beri at different time intervals. I have recently been in contact with a case of this nature in a child of seven years, who had been living upon a diet which was apparently adequate for the rest of the family. Yet the child had not only not gained in weight, but was becoming easily fatigued and listless at work with poor memory. It was decided to increase all sources of the B complex in the diet with the surprising result that a gain of 1½ lb took place in a fortnight, and a noticeable improvement in the more psychological symptoms. We may conclude that in this case we are dealing with an individual who is extravagant for some reason in the use of some parts of the B complex.

A clear picture of the relation of inanition to avitaminous symptoms cannot be reached without realizing that the symptoms are not specific for lack of B_1 in pigeons, to which matter I shall refer later in this lecture. Meanwhile, it may be said that so far as the action of torulin is concerned, provided that symptoms appear before the 30th day of rice feeding, true B_1 opisthotonus symptoms can be cured quantitatively by torulin at any stage of inanition at which they appear. Very often the cure heralds a further continued fall in weight, suggesting that failure of vitamin has jammed the tissue mechanisms so that further adjustments cannot be carried out, until more vitamin is present.

Cramer, Drew and Mottram (1922) considered that lack of vitamin B specifically affects the lymphoid tissue, and that the poor nutrition is really due to the failure in the normal functioning of the lymphocyte. However this may be, it is most striking to see the complete disappearance of all tissue fat and atrophy of such organs as the testis, which occur in the preliminary stages before tissue anarchy in some form or the other has set in. The general loss of weight and sacrifice of one tissue for another are common features of all kinds of malnutrition, though accentuated in B avitaminosis. It surprises me to find how little interest this curious problem has excited. What determines the changes in adsorptive power of the tissues so that fragments set free by autolysis of the testis (say) are caught by the heart and the brain? The matter suggests a careful study of the differential adsorptive powers of the various organs concerned.

Accepting the relation between B_1 dosing and the disappearance of B_1 induced "opisthotonus", we may now consider the function of torulin.

Pharmacological action of vitamin B_1. It might be considered that one method of determining the function of B_1 was to investigate its pharmacological behaviour. This has been tried.

Uhlmann (1918) found effects similar to those of pilocarpine. Voegtlin and Myers (1919) supported a secretin-like action. Anrep and Drummond (1921) and Cowgill (1921) could not confirm the conclusions. Leszcynski (1926) has claimed to have found a parasympathetically active substance in yeast fractions which remains after heating to 150°C, a procedure which destroys the vitamin. The presence of histamine and choline in extracts is likely to explain these effects.

As Mr. Kinnersley has prepared sufficient amounts of purified torulin lately, Mr. Fisher* and I have re-investigated the question using these extracts. We have tried to obtain responses with doses of from 10–40 pigeon units of torulin upon the virgin guinea-pig's uterus, the frog's heart, the rabbit's heart, and finally upon an anaesthetized pigeon preparation. Our experiments have not been extensive but have been negative, and we have not considered that further use of the valuable preparations was justified. Our provisional conclusion therefore supports those of others that torulin cannot be considered to have a pharmacological action in the ordinary sense. Such a view fits in with much that we know about it. Investigation is more likely to be fruitful by research upon its function in intermediary metabolism.

Relation of B_1 to the metabolism of carbohydrates. We may therefore ask whether the amount of vitamin has to be adjusted to the particular needs of the food stuff metabolized. Here we find a considerable divergence of opinion. It is held by some that the amount of the vitamin B complex has to be adjusted to the needs of the metabolism as a whole.† By others the view expressed early by Braddon and Cooper is held with equal tenacity, that vitamin B has to be adjusted to the amount of carbohydrate metabolized. If pigeons are forcibly fed with carbohydrates, it has been found that symptoms appear sooner for some of these than for others.

The "stuffing" experiments have been criticized upon the grounds that often large amounts of a hypertonic solution of the carbohydrate in question must be introduced into the crop (Kon and Drummond, 1927). This would not apply to all the experiments. It has been shown for instance that the giving of glucose can be made to increase the glycogen in the liver of an avitaminous

* Fisher and Peters, unpublished experiments.

† Plimmer *et al.* (1927)

bird (Collazo, 1924, and Randoin and Lelesz, 1926). This means that the sugar is absorbed, but Collazo claims that it rapidly disappears from the liver; the analogy to diabetes is indeed close.

Another means of deciding whether the vitamin is concerned with carbohydrates is by research upon the sugar metabolism. It was early shown by Funk and v. Schönborn (1914) that there was a hyperglycaemia in "stuffed" rice fed birds. Collazo in birds concluded that there was intolerance for sugar as I have mentioned earlier, and Evans and Lepkowsky (1929) have also found this in rats. We can confirm it for birds, and it suggests a deranged metabolism of carbohydrate. Further Bickel and his colleagues* (1924 and following) make much of the alteration in the ratio of C/N occurring in the urine of dogs in avitaminosis as a concomitant of loss of weight; the ratio increases, and the incomplete oxidation of carbohydrate is called by them "dysoxidative carbonuria". Collazo and Morelli (1925) have found increased lactic acid in the blood of avitaminous birds, and Rosenwald (1926) increased lactic acid in the urine of the dog. The amounts of such urinary lactic acid is not enough to explain the increases of C, and lately Collazo and Morelli (1927) find that the dog develops lipuria, which may account for much of the change. In a careful study Stanislaw Kon (1929) has reached the conclusions that changes in the ratio C/N in rat's urine are due in especial to deficiency of B_2.

Our own work supplies a convincing proof of the connection between deficiency of B_1 and changes in the intermediary metabolism of carbohydrate from a rather different angle. My relation to this problem in especial commenced with the modest attempt to find out whether our so called "glucose" cures were due to a lowered blood "sugar". Examination of the blood sugar of birds selected at random gave me no satisfactory answer, and I was therefore led to inject insulin into the birds. I found as others have done, especially Riddle and Honeywell (1923) that ordinary doses of insulin had no effect upon the birds. Large doses generally gave convulsions (10–20 units), though the latent period is a long one, up to 24 hr. To my surprise the convulsions exactly resembled those of avitaminosis, except that their duration was shorter (see photograph, Fig. H14). So far as I know this has not been previously pointed out. In view of this fact, Mr. Kinnersley and I were led to examine systematically the various features of the carbohydrate "cycle" in the brains of avitaminous birds. In this connection we first examined the glycolytic power of pigeons' brains, that is to say the appearance of lactic acid from glucose placed in contact with brain tissue. We soon found that there was no marked difference in glycolytic activity between the brain of the avitaminous bird and the normal; but during the course of this work

* Shiff and Zilva (*Biochem. J.* **22**), have criticized stringently the experimental work of Bickel *et al.*

some hints were obtained of an increased amount of lactic acid in the birds showing symptoms. This was therefore investigated. To cut a long story short, applying to the brains of pigeons a guillotine technique (McGinty and Gesell, 1925) in which two knives were so arranged that they severed the head and split it in two halves at once, we have frozen the brains in liquid air as soon as possible after death. It was a little time before we realized how quickly changes were taking place in these brains. The major part of the change in lactic acid occurs in 40 sec after death. Figure H13 summarizes the experiments. Points marked as squares in the lower curve represent values obtained upon normal birds lightly etherized. Above this

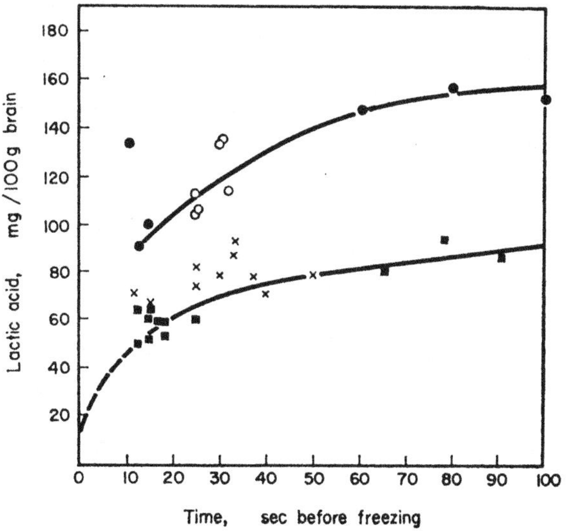

FIG. H13.* Rate of appearance of lactic acid after death in pigeon brains.

● = Anaesthetized opisthotonus. ○ = Stunned opisthotonus.
■ = Anaesthetized normal. × = Stunned normal.
Upper curve anaesthetized opisthotonus. Lower curve anaesthetized normal.

curve will be noticed crosses obtained upon birds killed by stunning. Such birds gave higher values. As this has occurred too often to be accidental, we think that injury leads to rapid production of lactic acid. The curve for the normal has been extrapolated to 15–20 mg per cent upon the assumption that the amount of lactic acid in an intact brain is identical

* By permission of the Editor. Kinnersley and Peters (1929), *Biochem. J.* **23**, No. 5. Lactic acid estimation was performed by the method of Friedemann, Cotonio and Shaffer (1927).

with that in the blood. Whether this be correct or not, it is evident that evolution of lactic acid is rapid, a not unexpected result in view of the findings of MacLeod and his co-workers (1928), and of Davenport and Davenport (1928) upon the rate of evolution of lactic acid in muscle. It is to be noticed that the values obtained for the avitaminous animals are of a different order. Now this has no significance in the present connection unless it can be shown either that there is no more sugar present initially in the abnormal brain or that the abnormal brain has not a more active glycolytic system. In some of the experiments, we have not found more sugar present than in the normal bird. This disposes of the first objection.

A more subtle objection to the truth of this view has been investigated by Dr. J. M. Gulland* in work to which I have already alluded briefly. The true blood sugar in the bird is some 60–80 mg less than the apparent blood sugar. The non-glucose reducing substances are uric acid, —SH compounds and ergothioneine in part. Supposing that in the avitaminous bird there was less "pseudo" sugar, then there would be more true sugar available for glycolysis. That that is not so, is shown by the fact that analyses of the non-glucose reducing substances in the avitaminous bird are not widely different from the normal.

The question as to whether the glycolytic system itself is more active is more difficult to settle. If we take the results most in favour of such a view, the glycolytic system would have to be at least twice as active, and in the case of one of the points, it would have to be four times as active. This we feel to be unlikely, as we found no marked difference *in vitro*. We have therefore reached the conclusion that there is more lactic acid present in the brain of the avitaminous bird at the instant of death.

TABLE 6. EFFECT UPON LACTIC ACID OF ALLOWING THE BRAIN TISSUE TO THAW

Bird	Time 1st half	Lactic acid	Time 2nd half	Lactic acid	Treatment
77	5 sec	60 mg	65 sec	92 mg	Both halves frozen at times stated, and then allowed to thaw for 1 hr at room temperature before extraction with trichlor. acid.
174	15 sec	60 mg	90 sec	88 mg	First half allowed to stand in trichlor. acid at room temperature for — min before smashing. Second half placed at time stated in ice cold trichlor. acid and immediately ground.

NOTE: Lactic acid mg per 100 g *tissue*

* Gulland and Peters, unpublished experiments.

Following up this work Mr. Kinnersley and I* have recently obtained some evidence as to the part of the brain especially concerned. By chance we found that frozen brain tissue does not change in lactic acid content appreciably upon thawing, contrary to muscle tissue. The experiments in Table 6 demonstrate this for two normal birds. It will be seen that the figures obtained lie closely upon the normal curves. Hence it is possible to allow the frozen tissue to thaw sufficiently so that the parts of the brain can be separately picked out. In practice, we have always removed the separate parts as soon as this was possible and before the tissue was completely thawed, in the event of any unforeseen error induced by glycolysis.

TABLE 7. THE LACTIC ACID IN DIFFERENT PARTS OF THE BRAIN IN NORMAL AND B. AVITAMINOUS BIRDS

Results in mg per 100 g tissue.

(1) *Normal.* (*a*) and (*b*) represent the average results for groups of five birds respectively. The average times for immersion in liquid air for the first halves was 11 sec, and for the second halves 86 sec after death.

		Brain					
		Cerebrum		Cerebellum		Remainder	
		LA	Sugar	LA	Sugar	LA	Sugar
First half	(*a*)	56	138	46	124	55	122
	(*b*)	52	159	53	129	58	119
Second half	(*a*)	96	96	85	115	92	100
	(*b*)	87	113	94	109	85	106

(2) *Avitaminous.* (In opisthotonus). The two halves of the brain immersed in liquid air within 6 sec and the parts of the brain separately picked out as soon as possible.

Bird	Brain						Blood	
	Cerebrum		Cerebellum		Remainder			
	LA	Sugar	LA	Sugar	LA	Sugar	LA	Sugar
762	89	145	94	119	124	122	170	164
147	65	197	64	166	108	189	54	357
86	113	224	80	233	157	—	—	—

Normal averages for the three different parts of the brain were obtained by analysis of the brains of two groups of five pigeons, the three parts

* Kinnersley and Peters, unpublished experiments.

chosen being the cerebrum, cerebellum and the rest. Table 7 shows the results. It will be seen that the lactic acid is the same in amount in each part of the brain in the normal animals, and that the results lie upon the curve in Fig. H13.

We have found, however, that in the avitaminous bird conditions are quite different. In Table 7, it will be noticed that there is always much more lactic acid present in the lower part of the brain than in the upper. In Bird 147, the values for the cerebrum and cerebellum are not far from normal whereas those for the remainder are high. Further, it is clear that the condition of the brain is not a mere reflection of the blood lactic acid. This appears especially in bird 762. The biochemical change is therefore located especially in the lower part of the brain.

In a paper by Iwata (1928) which I have only seen in abstract, the author believes that injury to the medulla is the cause of symptoms in B avitaminosis, and that the causative agent is furfural, arising by incomplete combustion of carbohydrates. Our work points to a medullary origin of the symptoms perhaps, but makes the toxic product an excess of the normal metabolite, lactic acid. Walshe (1920) supported a "toxic" origin of the nervous symptoms.

In a proposal to connect these symptoms with actual deprivation of the vitamin in the lower part of the brain, one other possibility must be seriously considered. McGinty (1929) has claimed recently that interference with the oxygen supply to the brain leads to an immediate increase in the amount of the lactic acid in the blood coming from the brain. Though the blood in the avitaminous bird is usually bright red in colour, it was possible that circulation was so much disturbed that the increased lactic acid found arose in this way. The results upon localized lactic acidosis make this improbable, but Mr. Fisher* and I thought that some direct evidence should be obtained if possible. Records of the average blood pressures of normal and avitaminous birds were therefore taken under light ether anaesthesia. The averages were as follows, for the normal 50 mm, and for those in opisthotonus 48, 80, 90, 70 mm. These figures taken with the general visual impression gained during the experiments of the excellent state of the blood supply to the head, made it unlikely that the increased lactic acid is merely due to lack of adequate circulation.

If the increased acid is directly connected with the lack of B_1, it should disappear upon the cure of the symptoms, and should not be present until symptoms are established or are threatening. We have been able to show that some few hours after cure of the symptoms the brain lactic acid is normal again, and that at any rate over a large part of the time of feeding, the lactic acid found in the brain is not increased.

* Unpublished experiments. Subsequent work suggests that these low pressures may have been due to over-ventilation.

A final test upon the matter seems to lie in introducing torulin into the site of the metabolic lesion. I have found that doses of torulin introduced into the cranial cavity have a more rapid effect upon the symptoms than torulin introduced elsewhere. Complete cures may be effected in this way in less than 30 min.

I feel that the evidence here presented is strongly in favour of the view that vitamin B_1 is specifically connected with the processes involved in the removal of lactic acid. It is probable that the symptoms originate in the lower part of the brain owing to a local deprivation of torulin in the tissue. These new facts establish in my opinion an undoubted connection between vitamin B_1 and carbohydrate metabolism, and connect it with a definite part of the carbohydrate cycle.*

If B_1 as suggested here, is concerned with lactic acid removal, then it is possible to understand why increased feeding tends to earlier appearance of symptoms. It presumably leads to an increased production of lactic acid.

The next question which arises is whether the increased production of lactic acid takes place only in the lower part of the brain, and is not found in the rest of the body. Work now proceeding upon this by Mr. Fisher and myself shows that there may be little change in the rest of the body, so far as lactic acid is concerned†. If these results are confirmed by further work, they will explain why certain observers have been unable to obtain changes in oxygen uptake with tissues from the avitaminous animal.

Relation Between the Symptoms of Opisthotonus and Avitaminosis

Let us now ask a further question. Supposing that we connect this lactic "acidosis" in the brain with the avitaminous symptoms, are such symptoms always associated with such an acidosis? This question is of fundamental interest in connection with the nervous system, but is of interest too from the narrower standpoint of these lectures. I have indicated previously that the cases of opisthotonus occurring during a rice feeding in pigeons are not all cases of B_1 deficiency. A number of instances have collected in our own work, and that of others of opisthotonus-like symptoms in pigeons. Most of these we have seen ourselves. Confining the problem to "metabolic" changes (as distinct from the effects of drugs, etc.), the following conditions produce typical "opisthotonus":

1. Avitaminosis.
2. Starvation.
3. Anoxaemia.

* These results seem to be the link needed to relate the work of Holmes and Holmes upon the carbohydrate metabolism of isolated brain tissue with its living activity. They suggest possible applications to epileptic and other such conditions.

† However, the lactic acid content of the heart may be as high as 140 mg per 100 g tissue, and pectoral muscle 90 mg.

4. Insulin overdose.
5. Light anaesthesia.
6. Certain disturbances in birds sensitized by rice feeding.

So far as we know* (1) is associated with increased lactic acid: (4) with decreased lactic acid in the brain. In (5) there appears to be no change. The state of the brain in the other conditions has not yet been investigated.

The evidence bearing upon this statement is as follows:

(1) *Avitaminosis.*† The symptoms are prolonged, and it is not certain in view of the two types of rice disease in pigeons (McCarrison, 1928) that the opisthotonus is always due to lack of the same vitamin. For a detailed account, see Kinnersley, Peters and Reader (1928).

(2) *Starvation,* with or without flushing with water; stated by Eijkmann and van Hoogenhuize (1916) to be due to washing out of vitamin, and to be cured by giving vitamin; Marrian, Drummond, Baker and Woollard (1927) think that starvation of itself produces it, and attach importance to the fact that the appearance of symptoms is associated with a pre-mortal fall in the blood sugar. Probably our "glucose" cures must be placed in this category.

(3) *Anoxaemia.* Symptoms appear during the early stages of CO poisoning as a temporary stage. Findlay has shown that nitrite produces them and that they can be cured by oxygen. Hess (1921) found that the giving of sub-lethal doses of cyanide to normal birds induced the condition, and that birds threatening symptoms were more easily affected by small doses of cyanide than normal birds. These facts have been confirmed. Hess uses them to support his view of the relation between vitamin B and tissue oxidation.

(4) *Insulin.* In convulsions produced by insulin, the brain lactic acid has been found to be low,‡ in confirmation of Holmes and Holmes (1925).

(5) *Light anaesthesia.* Marrian (1928) has recorded a case at the beginning of ether anaesthesia. We have observed it occasionally for ether, but many times in the case of chloroform, where it may be quite prolonged. (See photograph, Fig. H15).

(6) *Various disturbances.* Under this heading must be placed the symptoms precipitated by cold and muscular exercise, which are possibly circulatory in origin, but about which not much is yet known.

It is clear that lactic "acidosis" cannot be held to be the only concomitant of the symptoms. Can we then relate the various effects in one complete scheme, to act as a working basis? If we take a broad view of all the above

* Kinnersley and Peters, unpublished results.

† I have assumed throughout these lectures that symptoms of "opisthotonus" not alleviated by dosing with B_1 are not due to uncomplicated B_1 deficiency.

‡ Kinnersley and Peters, unpublished results.

FIG. H14. Insulinized pigeon showing "opisthotonus".

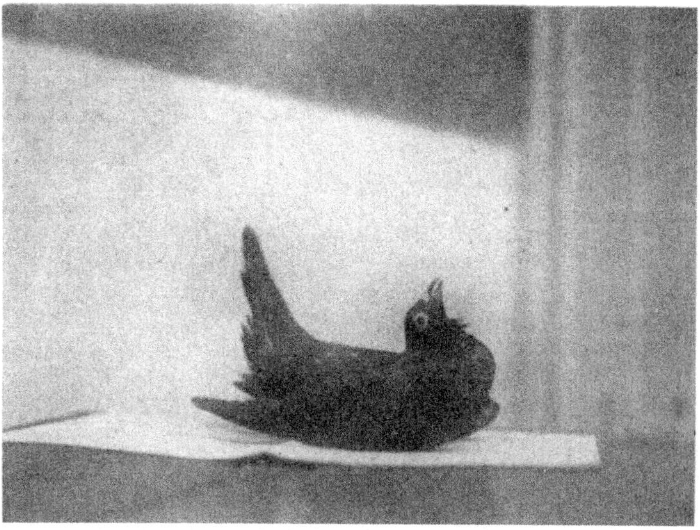

FIG. H15. Chloroformed pigeon showing "opisthotonus".

(except 5) they are conditions of metabolic extremity, to which we should therefore expect the central nervous system to respond. Supposing that we believe that in all these cases some change in the medulla gives rise to the symptoms—in the case of asphyxia, the medulla has always been believed to be the seat of change, and there is some evidence of its connection with the symptoms of avitaminosis.* Interference with the normal cycle of change may be conceived to produce the symptoms of opisthotonus. Picturing the state of affairs as in the diagram (making use of Meyerhof's carbohydrate cycle), we see that failure of the normal state of dynamic change may occur in several places. Upon this view, asphyxia stops the cycle at A, insulin at I, and cyanide at C. A failure of circulation may either lower the glucose supply or the oxygen supply or slow the removal of waste products. The vitamin would appear to act at V. Opisthotonus according to this view in the pigeon is the first warning of disability in this cycle, and the vitamin is a specific coordinating agent. These views cannot be pressed too far, but they serve to give a coordinate picture of apparently unrelated happenings.

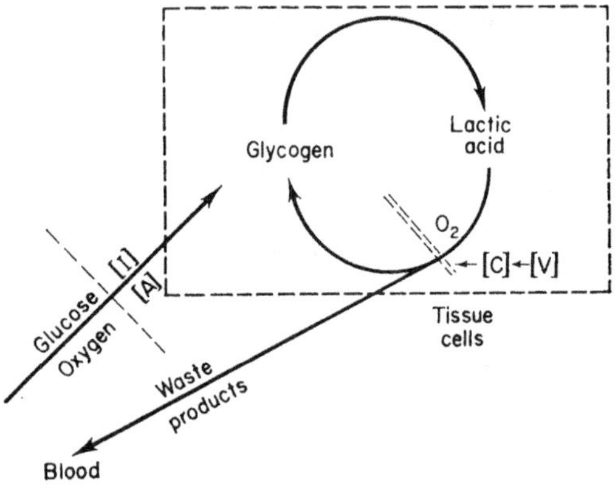

FIG. H16.

Olmsted and Logan (1923) and Kermack, Lambie and Slater (1929) have reached very much the same position in their investigation of insulin convulsions. They have stressed the similarity between insulin and asphyxial effects. It is interesting in this connection to note that polyneuritic symptoms in the rat resemble insulin symptoms in the same animal.

The "after effects" of insulin overdoses in pigeons are rather surprising. In many cases the birds cannot see to feed themselves, in one case a bird had to be hand fed for a fortnight after a dose of insulin. The effect is surprisingly like that of a temporary anoxaemia.

* Above experiments.

To summarize the position which I have reached, opisthotonus in the pigeon arises from any cause which interferes with the "reflex cycle" of carbohydrate change. In B_1 deficiency this is specifically concerned with the removal of lactic acid. Now that we can allocate a function to torulin in the intermediary metabolism, we are in a position to consider the action of some other members of the B complex.

Wet and dry beri-beri. In this connection I wish to draw your attention first to some experiments of ours, which not only throw light upon other work but help to provide a partial solution to some of our own difficulties. I have mentioned in my last lecture the puzzle presented by the finding that an undoubted curative extract became less efficient during a prolonged rice feeding, and that a second alkali-labile factor is needed to maintain weight in pigeons. Now two types of "rice" disease have been long recognized clinically and together contribute to the mixed symptoms so often occurring in human deficiency diseases. Many years ago now, Vedder and Williams (1913) thought that there were two diseases, one associated with neuritic symptoms, and the other with cardiac failure. The idea was two vitamins and two diseases. A similar view was later put forward by McCarrison (1921) to explain nutritional oedema.

Kohman (1920) has produced nutritional oedemas in rats by "carrot" feeding, and this has been confirmed to a certain extent by Frisch, Mendel and Peters (1929).

In the last few years, McCarrison has repeatedly insisted upon the existence of two types of rice disease in pigeons, called by him "polyneuritis columbarum" and "beri-beri columbarum".

The symptoms upon which he relies for his diagnosis of the latter are (1) enlargement and degeneration of the heart; (2) the presence of oedema or serous effusions or both; (3) a degeneration of the muscles; (4) chronic passive congestion of the abdominal viscera; and (5) congestion and ecchymoses of the upper intestinal tract. The production of the condition is held to be due to the presence of an insufficient amount of vitamin B in the diet, not to complete lack of this. In explanation, he has returned to the older view of Eijkmann that the ultimate cause of beri-beri is a positive and toxic agent produced in the course of a disordered metabolism, arising out of insufficiency of vitamin B in the food.

A particular point to notice is that his two types of cases are not distinguished clinically, as they both show symptoms of head retraction. Most of the beri-beri cases occurred after the thirtieth day of rice feeding, and would therefore be excluded by our tests; but the few occurring before the thirtieth day might possibly have been accounting for our so-called abnormal cases. Fortunately we find that the cases which get oedema before the thirtieth day are few in number, and do not appreciably affect our tests. In order to test the broad view of McCarrison, we have given doses of torulin (3-day doses) daily with a polished rice diet. I have mentioned previously that an active preparation obtained in 1925 did not protect pigeons against loss of weight upon a rice diet. More interesting than this, however, it is not protective against polyneuritic symptoms after some 30 days. The curves in Fig. H17 and the data in Table 8 show this. They have been obtained recently be Mr. Carter, Mr. Kinnersley and myself, and are the individual data upon which the composite curve B of Fig. H11 (Lecture II) are based. It will be noticed that at various intervals after the thirtieth day of feeding, some birds fell sick with symptoms and that some showed fluid at autopsy. This occurred even though a daily dose of torulin was being given far in excess of the amount required to cure and protect in the case of a bird suffering from symptoms before the thirtieth day.

We may reckon matters in this manner. The normal bird gets opisthotonus in 20–30 days, and therefore may be said to have a store of 20–30-day doses. We give the bird 3-day doses per diem; it loses weight and gets symptoms in 40–50 days. Such birds have been given by the mouth in this period 120 doses, to which may be added the initial store of 20 doses, out of which 40 should have been used. There should therefore still be excess of 100 doses. This gives a substantial margin, even if we assume that some of the doses were not properly absorbed.

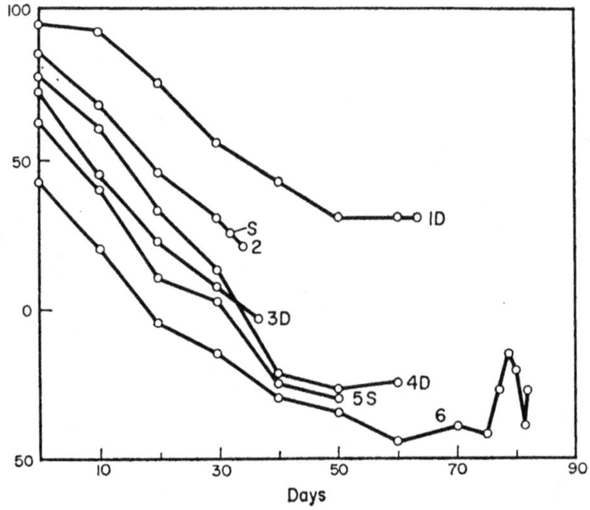

FIG. H17. Chart of pigeons weight

Pigeons 1–6 from Table 8. S = Symptoms of opisthotonus.
D = Died without symptoms.

(For details see Table 8)

Now at first I thought that such birds could not be suffering from a B_1 deficiency, and that we had proved this because torulin administered by the mouth did not always cure. However, we find that many of the cases can be cured by increasing the amount of torulin given by the mouth, and if not in this way by injection. These results confirm the earlier ones quoted previously, and show that torulin is less efficient in these later stages. Further it would seem that our newer torulin preparations still have

TABLE 8.* TEN PIGEONS PLACED UPON POLISHED RICE AND THREE-DAY DOSES TORULIN DAILY

Details of Experiment. On *68th day*, the surviving birds, 6, 7, 9, 10, given 10 g salted rice daily (forcibly fed) + 3-day doses torulin daily.

On *77th day*, same birds given in addition 0·5 g salt in 5 ml water daily, introduced into the crop to test for latent oedema. This led to death of 10 on 78th day, with a rise in weight of 60 g, and of 9 on 80th day with a rise of 40 g.

On *80th day*, the survivors, 6, 7, were given three extra doses torulin by mouth and six by intramuscular injection. This led to a temporary remission of the oedema in 6 (Fig. H14), but the extra doses did not protect for long as it died with rise of weight on 82nd day.

For the subsequent history of pigeon 7 (see Fig. H18).

No.	Weight g Initial	Weight g Final	Day of death	Symptoms	Fluid[3]	Weight g heart	Beri-beri[1]	Remarks
1	425	290	64th	–	+	4·2	+	Symptoms 36th day
2	390	260	37th	+	–	3·2	–	Killed 37th day
3	450	244	37th	–	–	3·0	–	
4	450	250	63rd	(paralytic)	–	3·0	–	Not cleared up by extra dose of torulin by mouth
5	425	250	53rd	+	–	3·4	+	Not cleared up by extra dose of torulin by mouth
6	384	205	82nd	–	+	2·2	–	
7	390	220	123rd	+	+	4·6	+	
8	340	245	37th	–	–	3·0	(border)	
9	380	195/235[2]	80th	–	+	4·4	+	
10	440	240/318[2]	78th	(wing drop)	+	3·9	+	Not cleared up by extra dose of torulin

[1] As diagnosed by ratio heart weight, final body weight (after Mr. Carrison).
[2] Higher figure due to oedema.
[3] Fluid in pericardial and abdominal cavity.

* Unpublished results, Carter, Kinnersley and Peters.

an anti-oedema effect.* Figure H18 represents the weight of one pigeon of the group No. 7. This case did not get oedema until the 115th day of feeding, even with extra salt. At this stage, the daily torulin given by the mouth proved to be ineffective, whereas extra torulin given by injection

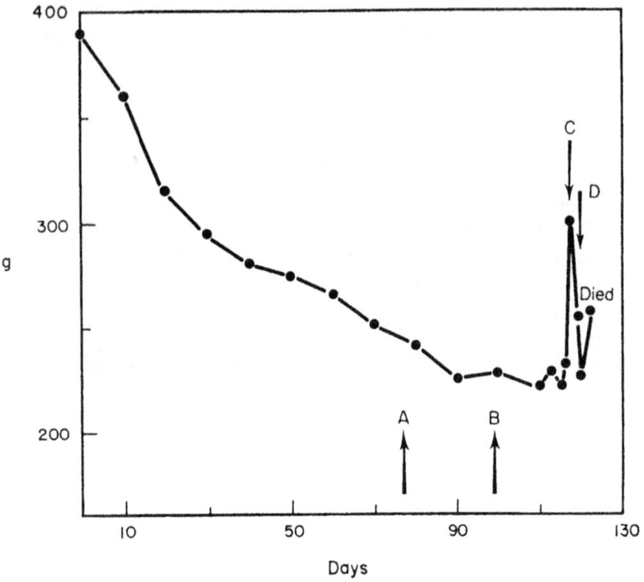

FIG. H18. Chart of pigeon 7 (Table 8)

For diet see Table 8.
At A. Given 0·5 g salt (NaCl) in 5 ml water daily in addition to 10 g salted rice (forcibly fed) and 3-day doses torulin.
B. Torulin preparation changed, four doses given daily of a purer ($N/_{10}HCl$) preparation.
C. 12-day doses torulin injected.
D. 8-day doses torulin injected.
Bird found dead with extensive oedema on 122nd day, after the usual daily dose of torulin. Note the rise in weight.

controlled the oedema of some 80 mg temporarily.† The bird was unfortunately lost later overnight owing to a failure to realize how much torulin such birds need. It will be noticed also by reference to the table that 5 of these 10 birds had heart weights which would be classified as "beri-beri" by McCarrison.

* Such preparations have been shown by Dr. Reader to contain no "B_3".
† The clinical significance of this nutritional "tissue anarchy" in water hardly requires emphasis!

It must be emphasized that a similar dose of torulin given by the mouth in the form of marmite protects against loss of weight and also against oedema. We feel that it is safe to conclude that in the absence of some third factor which is destroyed by an alkaline heating of marmite (?B_3), torulin is utilized with much less efficiency than in its presence. Further the effect of this factor is in some way concerned with the mobilization of water. Mr. Carter and I have found that the best way to show up an incipient oedema in these birds is to administer salt in doses of 0·5 g in water or to give salted rice. Birds with a tendency to oedema rise in weight. Birds fed with additions of alkaline autoclaved marmite to the polished rice tend to get oedema in our experience, but we are not certain that this is not due to the salt present in the marmite preparation.

These experiments must be still regarded as preliminary, but I think that they provide some more facts in reference to the puzzle set in my previous lecture, and to the question as to whether B_1 is protective as well as curative. This depends upon the presence of some other alkali-labile factor (? B_3), present in marmite, which is anti-oedema in the sense that it makes effective the action of the torulin. Upon this view torulin plus anti-oedema is really equivalent to the anti-beri-beri vitamin. Until it is possible to supply the third factor in the absence of the other two, we must regard this as provisional. It does, however, fit in with certain other facts. Birds which have been standardized upon our laboratory diet give many more cases of uncomplicated head retraction within the month, possibly owing to the storage of this third factor. Our work has brought us very near to Vedder's old hypothesis of the necessity of two vitamins for protecting against beri-beri. Since the giving of large amounts of B_1 does not prevent the appearance of beri-beri, McCarrison's view that the condition arises owing to a diminished amount of B does not hold *sensu stricto*. Some other factor is wanted to explain the results and this we believe to be concerned in rendering the torulin efficient.*

We are also convinced that the presence of oedema may in many cases merely be masked by the absence of salt. Perhaps this accounts for the fact that McCarrison did not find a strict correlation between the oedema and beri-beri.

The heart. The heart undergoes more than one change in a rice feeding. There is the "beri-beri" change, about which I do not wish to speak further today. There is, however, another condition of a special interest, which seems to differ from changes in the heart muscle, *per se.*† Wishing to obtain a clinical test for distinguishing hearts with pericardial effusions,

* This is a different conception from that of Rosedale (1929), which we do not feel to be supported by our facts.

† Aalsmeer and Wenckenbach, 1928.

I happened by chance to take the stethoscope to the birds, and to my surprise found a marked irregularity of rhythm—a slow pulse which became enormously accelerated upon exertion. Thereby I had rediscovered a fact described by Schaumann, who does not appear to have followed it. This interesting condition has been investigated by Mr. Carter of Oxford, and Dr. A. N. Drury (1929) of Cambridge. They find that it is due to a heart block, which is apparently central in origin, because it is abolished by cutting the vagus nerve. In fact it is the first definite symptom to appear in a bird apart from the fall in weight. It is quite remarkable to find how slowly these hearts may beat, perhaps 50 a minute, instead of the normal 250.

With regard to the origin of the condition, it appears in some 10 days in birds receiving daily doses of torulin, and is therefore not due to lack of B_1. Is it due to lack of any other known factors? Mr. Carter* has found that the giving of marmite only cures the condition extremely slowly, so that the factor responsible is only present in traces in this yeast extract. Cod liver oil effects no improvement, but whole wheat effects a rapid and complete cure. If anything, the factor appears to run parallel to the Williams and Waterman factor, but until further work has been done it would not be wise to lay any stress upon this. What then is the function of this mysterious Williams and Waterman factor? From my own experiments, there is a suggestion that it is associated with the storage of fat (see Fig. H10), Lecture II). Admittedly the evidence for this is at present slender, but it is interesting to note that Evans and Lepkowski (1929) have found a sparing effect of fat upon vitamin B (alcoholic extract of rice polishings). The two sets of observation would appear to be connected. Whether we are really to include this factor in the B complex must be left for future work to decide.

The evidence, which I have outlined here, points strongly to the presence of one B factor associated especially with carbohydrate B_1, one indirectly with the metabolization of water, and there is slender evidence that another is concerned with fat. It is tempting to ask, therefore, whether there is evidence that the other great group of foodstuffs, the proteins has its B factor too. This matter has not been investigated by us, but has been raised elsewhere. So far as I am aware, a connection between "B" and protein was first raised by Hartwell (1922) working in Mottram's laboratory. She found that there was a definite relation between the amount of vitamin B and of protein in the diet of the lactating rat. The matter was demonstrated from another point of view by Reader and Drummond (1926) who found that high protein diets would suffice for the growth of rats if the amount of vitamin B in the ration was raised. Later, Hassan

* Carter, unpublished experiments.

and Drummond (1927) used the finding that alkalized marmite was effective in this sense as an argument in favour of two factors in the B complex. So there is evidence in support of a B factor associated with protein, present in a yeast extract containing the antipellagra or antidermatitis factor B_2.

Barnet Sure and his co-workers (1928) have obtained additional evidence lately in favour of these views.

Therefore I think that the battles which have been fought over the question as to whether B is concerned with metabolism as a whole, or with protein or carbohydrate alone are to be resolved in the "Sir Roger de Coverley" way in which so many biochemical problems have been settled. Each is true in a limited sense. Much more evidence will be needed before this proposition can be accepted without reserve, and if it be completely true, it will leave behind it food for thought and speculation as to how—and perhaps why—the animal organism must take these factors in small amounts to assist the metabolism of each class of foodstuff.

We are now in a position to review the influence which this knowledge must have upon our general conception of biochemistry. The work has brought us to the portals of intermediary metabolism, and from those portals we gaze upon fresh vistas of experimental work in which we may be able to investigate each of the cardinal types of metabolism in isolation from the other inside the cell. During a rice feeding, for instance, a bird cannot utilize its carbohydrate property or store fat; so far as its carbohydrate metabolism is out of gear, it is living upon its fat and protein. This is borne out by the respiratory quotient which investigators have obtained (Magne and Simonnet, 1922). It is extremely interesting that such animals are said to have no ketosis. In my first lecture I pointed out how valuable it would be to have a reagent capable of disconnecting carbohydrate metabolism in the cell. I think that it is not too fanciful to believe that in the operation of B_1, we have such a means of dissection. Future work upon the animal, I think, will have to take far more account of the nutrition of the tissues previous to experiment. By taking this into our ken, we may at last reach an adequate integrated conception of the chemical reflexes in cells, and estimate their relative importance. Then at last shall we obtain a picture of the coordinative working of a system, the marvellous intricacy of which we can but dimly guess. For a rigid utilitarianism is not a final aim. The scientific artist portrays the harmony of nature. Poincaré expresses this when he speaks of the search for this special beauty, the sense of the harmony of the world. Only those who have studied them can appreciate the fugues of Bach, which may justly mirror the concurrent and harmonious actions of the living cell. Biochemistry is a study of Nature's counterpoint. Many have not seen its beauties, but they are there for those that will.

(I wish to express my gratitude to those who have very kindly allowed me to refer to their unpublished work.)

References

AALSMEER and WENCKENBACH (1928), *Weiner. Archiv. f. Inn. Med.* **16,** 193.
ANREP and DRUMMOND (1921), *J. Physiol.* **54,** 349.
BICKEL (1924), *Biochem. Z.* **146,** 493.
BICKEL (1925), *Biochem. Z.* **166,** 251.
BRADDON and COOPER (1914), *J. Hyg.* **14,** 331.
CARTER and DRURY (1929), *J. Physiol.* **68**; *Proc. Physiol. Soc.* **1.**
COLLAZO (1922), *Biochem. Z.* **134,** 194.
COLLAZO (1924), *Biochem. Z.* **140,** 256.
COLLAZO and MORELLI (1925), *J. Physiol. Path. Gén.* **24,** 77.
COLLAZO and MUNILLA (1929), *C.R. Soc. Biol.* **99,** 1448.
COWGILL (1921), *Proc. Soc. Exp. Biol. Med.* **18,** 290.
COWGILL, DEUEL and SMITH (1925), *Amer. J. Physiol.* **73,** 106, etc.
CRAMER, DREW and MOTTRAM (1921), *Lancet,* 1202. Dec.
DAVENPORT and DAVENPORT (1928), *J. Biol. Chem.* **76,** 651.
EIJKMANN (1897), *Arch. f. Path. Anat.* **149,** 147.
EIJKMANN and VAN HOOGENHUIZE (1916), *Proc. K. Akad. Wetensch. Amster.* **18** (2), 1467.
EVANS and LEPOWSKY (1929), Proc. Physiol. Congress, *Amer. J. Physiol.* **90,** 340.
FINDLAY, (1928), *J. Path. Bact.* **31,** 355.
FRIEDEMANN, COTONIO and SHAFFER (1927) *J. Biol. Chem.* **73,** 335.
FRISCH, MENDEL and PETERS (1929), *J. Biol., Chem.* **84,** 167.
FUNK and SCHÖNBORN (1914), *J. Physiol.* **48,** 328.
GARROD, A. E. (1923), *Inborn Errors of Metabolism.* 2nd ed. Ed. Frowde.
GREY, E. C. (1928), *J. Hyg.* **27,** 268.
HARTWELL (1922), *Biochem. J.* **16,** 78.
HASSAN and DRUMMOND (1927), *Biochem. J.* **21,** 653.
HARTWELL (1928), *Biochem. J.* **22,** 1212.
HESS (1921), *Z. Physiol. Chem.* **117,** 284.
HOLMES and HOLMES (1925), *Biochem. J.* **19,** 492 and 836.
HOLMES and HOLMES (1926), *Biochem. J.* **20,** 1196.
HOLMES and HOLMES (1927), *Biochem. J.* **21,** 412.
HOPKINS F. G. (1912), *J. Physiol.* **44,** 425.
IWATA (1928). Quoted from *Chem. Abstr.* **22,** 3909.
KERMACK, LAMBIE and SLATER. (1929), *Biochem. J.* **23,** 414.
KINNERSLEY, PETERS and READER (1928), *Biochem. J.* **22,** 276.
KOHMAN (1920), *Amer. J. Physiol.* **51,** 378.
KON and DRUMMOND (1927), *Biochem. J.* **21,** 632.
KON, S. (1929), *J. Nutrition* **1,** 467.
LESZCYNSKI (1926), *Polka gaz. lek.* **5,** 909. Quoted from *Abstr.*
MCCARRISON (1921), *Studies in Deficiency Disease.* Oxford.
MCCARRISON (1928), *J. Ind. Med. Res. Memoirs,* No. 10 (and other papers in this Journal).
MCGINTY and GESELL (1925), *Amer. J. Physiol.* **75,** 70.
MCGINTY (1929), *Amer. J. Physiol.* **88,** 312.
MACLEOD and SIMPSON (1928), *Fuel of Life.* Oxford University Press.

MAGNE and SIMMONNET (1922), *Bull. Soc. Chim. Biol.* **4,** 41.
MARRIAN (1928), *Biochem. J.* **22,** 843.
MARRIAN and DRUMMOND (1926), *Biochem. J.* **20,** 1229.
MARRIAN, DRUMMOND, BAKER and WOOLLARD (1927), *Biochem. J.* **21,** 1336.
MEYERHOF (1926), *J. Gen. Physiol.* **8,** 533.
OLMSTED and Logan. (1923), *Amer. J. Physiol.* **66,** 437.
PLIMMER, ROSEDALE and RAYMOND (1927), *Biochem. J.* **21,** 913.
RANDOIN and LELESZ (1926), *Bull. Soc. Chim. Biol.* **8,** 15.
RANDOIN and LECOQ (1929), *Bull. Soc. Chim. Biol.* **11,** 745.
READER and DRUMMOND (1926), *Biochem. J.* **20,** 1256.
RIDDLE (1923), *Proc. Soc. Exp. Biol. Med.* **20,** 244.
RIDDLE and HONEYWELL (1923), *Proc. Soc. Exp. Biol. Med.* **20.** 248.
ROCHE (1929), *C.R. Soc. Biol.* **99,** 671.
ROSEDALE (1929), *J. Ind. Med. Res.* **17,** 217.
ROSEWALD (1926), *Biochem. Z.* **168,** 324.
SCHAUMANN (1910), *Archiv. Schiffs. Trop. Hyg.* **14,** 217.
STERN and FINDLAY (1929), *J. Path. Bact.* **32,** 63.
SURE, BARNET (1928), *J. Biol. Chem.* **76,** 659, etc.
UHLMANN (1918), *Z. Biol.* **68,** 457.
VEDDER (1913), *Beri-Beri.* Bale Sons and Danielson.
VEDDER and WILLIAMS (1913), *Phillipp. J. Sci.* **8,** B.
VOEGTLIN and MYERS (1919), *J. Pharm. Exp. Ther.* **13,** 301.
WALSHE, F. M. R. (1920), *Med. Sci. Abstr. II,* 411.
WILLIAMS and WATERMAN (1928), *J. Biol. Chem.* **78,** 311.

NOTES ON THE HARBEN LECTURES

LECTURE I

page 219. The differences between the attack made upon —SH compounds by mono-substituted (primary) and di-substituted (secondary) arsenical tervalent compounds was only cleared up in recent years (see Dohme Lectures, *Bulletin John Hopkins Hospital* **97,** 1–42, 1954).

page 223. 0·2 μ was the approximate limit of the optical microscope using 1/12 in immersion. The phages and viruses are well below this limit.

page 224. It was not possible then to pursue the nucleic acids further, as there were only two nucleic acids then recognized.

page 225. The groups given here were only a beginning. The list would now be extended with present knowledge of protein —SH, protein —S.S.— reactions. There is also hydrogen bonding, the conception introduced by Pauling.

page 228. Danielli (*Biochem. J.* **35,** 470, 1949) has calculated an interfacial pH using the Donnan equilibrium. It seems to be safer to call this apparent pH, as with so few ionized molecules involved, the conception of pH based upon statistical considerations may possibly not apply.

page 234. The conception of Hardy and Nottage was used here as a possible way in which influences could spread from a surface. It must be recognized now that the problem as to how far orientation or other effects can spread from the surface to molecules adjacent to it is still a matter of doubt. On the one hand there are statements that surface effects can spread from molecule to molecule for some distance. The experiments of Hardy and Nottage have been criticized (Bowden and Bastow, 1935), and more recent work suggests that such influences do not spread further than 1 to 2 molecules from the surface. The direct influence upon the crystallization of fatty acids of steel and copper (Hardy and Nottage) has been criticized upon the grounds that one cannot produce an even fracture. It seems likely, however, that some influence can extend for a few molecules from an interface. In this connection experiments by Miss McLoughlin upon the influence of epidermal layers of skin upon underlying structures are surely of significance.

Water. The majority of cells contain about 80 per cent of water, and this is enough to give a definite colour with a pH indicator like phenol-red, as was found by Chambers. It is a question, however, how much of this water is freely moving and how much is "bound" by proteins especially. The amount which is bound has been determined by finding the shrinkage in volume on adding together, for instance, gelatin and water.

Lecture II

This lecture is a summary of knowledge at that time upon the number and the distribution of the water soluble "B" vitamins, with some observations on the chemistry of thiamine (vitamin B_1).

page 241. Further work on *Colpidium*, now known as *Tetrahymena geleei* has been done by Kidder and Dewey, who have proved that this small organism requires a battery of substances to make it grow. Most were undiscovered 30 years ago.

pages 242 *and* 243. The conditions are described for the successful use of the pigeon test for thiamine. These are still not known to all workers (for example see R. R. Williams).

pages 244 *and* 245. The matter of idiosyncrasy is fundamental and was rigidly proved for the pigeons and found to occur in rats by Miss Reader (now Mrs. Smith).

page 247. These cases of oedema are interesting, and I have not seen other work upon them. In thiamine deficiency, they are produced by feeding with salt solution. A dose of salts having no effect on the normal had a massive effect in the animal deficient animal. The matter would repay further study.

page 251. Our pure thiamine does not give the usual Pauly azo reaction. We developed a special test, the formaldehyde azo test which had a high degree of specificity (Kinnersley and Peters).

pages 245 *and later*. The essential conclusion to which I came was that there was evidence then for 4 B factors, 3 needed by the rat, and 3 by the pigeon. This conclusion was not contradicted in subsequent work as there is evidence for many factors in the original vitamin B group, viz. thiamine, niacin, riboflavin, pyridoxine, pantothenic acid and cyanocobalamine. Our work on maintenance nutrition in pigeons has been published as also that upon the heart factor needed by pigeons (Carter, Kinnersley and Peters, 1930, and Carter, 1930). It has been stated in the literature that some of the factors here described do not exist, a statement which is not correct.

The studies started by Mr. Kinnersley, Mr. Carter and myself were continued by Mr. Carter and Mr. O'Brien, for some years. It was difficult to reach decisions as to the explanation of the phenomena owing firstly to the fact that the concentrate of B factors fed were not sufficiently pure; but also to the curious circumstance, which was well-attested, that the results upon the weight of the birds of giving supplements to the food depended largely upon the degree of depletion. Some of the factors studied were retained tenaciously by the animal at a time when it would have run out of its stores of thiamine completely. The degree of depletion was difficult to judge, and made the results for individual birds variable. The curves shown in Fig. A1 give a good idea of what happens. It was concluded in their final paper (1939) that 4 factors were needed for the complete nutrition of the pigeon; thiamine, riboflavin, a factor B_3 and a factor B_2. It was probable that the activity of B_5 was due mainly to its content of vitamin B_6 (pyridoxin), though it would also have contained nicotinic acid.

It seems to me that it would be worth returning to the investigation of the missing factor, though it requires a type of research which many are not prepared to face at the present time. One thing is clear; so far as the experiments go, the results are real and require interpretation. I mention this because of certain statements which have been made in the literature, suggesting that factors B_3, B_4 and B_5 do not exist in any form. From what we know now, it is likely (as stated above) that B_5 was a mixture of pyridoxine and nicotinic acid; the Williams and Waterman B_3 could have been an early indication of the need for pantothenic acid. The phenomena may well become important again in any study of the nutrition of birds.

In parallel with the studies of thiamine and the biochemical lesion due to its deficiency, prolonged studies were made of the factors needed to keep a pigeon in full nutrition, by studying the factors concerned in maintenance nutrition. There was a hope at one time that it would be possible to devise a diet for the pigeon, which would be complete in all respects with the exception of thiamine, so that the test for thiamine would involve only a study of a loss and gain in weight. To this end, we studied for a year the behaviour of a specimen of wheat, which could be treated with heat so that only the thiamine was destroyed. The results of using this for thiamine tests were excellent, and we were just going to publish the method, when we thought it would be advisable to study another specimen of wheat. The tests were never published. because the second sample of wheat behaved quite differently from the first. It was not possible to destroy the thiamine present even by a much more prolonged treatment than that formerly applied to the first specimen. It is an interesting fact that thiamine can become so stable in some wheats, but it was very disappointing that a year's work had to be foregone.

Dietary Heart Block

Allusion is made to this condition in the Harben lectures (p. 286). When fed upon an incomplete diet for about 13 days, pigeons were found by C. W. Carter to develop a heart block which was investigated with Drury and found to be central in origin, because it disappeared when the vagus nerve was cut. The block rapidly disappeared upon exercise to appear again when the bird was resting. If a normal bird is considered to have a heart rate of 170, these birds might have heart rates of about 120 at rest. Carter proved that this was not due to thiamine deficiency, as it could not be cured by large does of thiamine. It was a factor present in whole wheat, which was heat labile. So far as I am aware there has been no further work upon this.

Lecture III

page 268. It is doubtful whether the immediate causes of loss of appetite are yet understood.

page 275. This is the start of our observations upon the biochemical lesion in the brain of the avitaminous pigeon.

page 284. The point is registered that in the later stages thiamine could not be absorbed from the gut, but needed injection, an important clinical matter.

page 287. It may be recalled that these speculations preceded the finding of nicotinamide, riboflavin and pyridoxine co-factors.

APPENDIX 2

SURFACE STRUCTURE IN THE INTEGRATION OF CELL ACTIVITY*

THE importance of surface structure and of adsorption processes in the phenomena displayed by living matter has been long recognized, and has formed the theme of many writings and discussions.† This might seem to preclude any necessity for considering the problem further. It is because I feel that the exact significance of surface in its intervention in cell activities is not yet sufficiently understood, that I venture to put forward again a summary of views, presented last year in greater detail (Peters, 1929). To avoid any misconception, I wish to make it clear at the outset that much of this report must be regarded as still highly speculative and lacking in experimental support; but I propose to be dogmatic partly in the hope of providing a more concrete case for discussion, but more especially because I have felt, no doubt in common with other biochemists, the need for some coherent working hypothesis of the biochemistry of living cells.

Two phrases have been much used in the past in connection with life, the normal architecture of the cell and dynamic equilibrium (Hopkins, 1912). The view which is presented here differs from most others in the stress which is laid upon architecture. Its keynote is the complete (or nearly complete) structural organization of the cell. I believe this to be organized not only in respect to its grosser parts such as the nucleus, but also in regard to the actual chemical molecules of which it consists. Owing to the microheterogeneous nature of the system, surface effects take precedence over ordinary statistical, mass action relationships, and become in the ultimate limit responsible for the integration of the whole and therefore the direction of activities. It is believed that the directing effect of the internal surfaces is displayed predominantly by an organized network of protein molecules, forming a three-dimensional mosaic extending throughout the cell. The enzymes would form part of this structure, their activity being largely controlled by the mosaic. This conception of living matter reconciles some difficulties; for instance independent chemical reactions could proceed simultaneously in various parts of the cell, but at the same time the mosaic could react as a whole to stimuli transmitted from the cell surface. It obeys one of the tests of a useful hypothesis, in that it suggests lines of experimental attack. I hope to show that

* Reprinted from *Trans. Faraday Soc.* **26**, No. 115, Part 12, December 1930.

† See especially Warburg in relation to cell oxidations (1923), also Gortner, 1929.

it is already possible to reach some conclusions as to the actual groupings concerned and the nature of their action.

The case for this working hypothesis can be developed best by starting with concrete facts. The most surprising manifestation of living matter is its power of reproduction. With this is intimately connected the problem of what may be termed, "directed chemical change" in the cell. It has to be remembered that the influence of heredity extends beyond the species to the individual, and is not confined to gross anatomical structure in the ordinary sense. Heredity may even show itself in variations in the blood sugar level as between one species and another, as Riddle and Honeywell showed in pigeons (1923). Realisation of all that this involves in the way of constant dynamic adjustments in the metabolism of carbohydrate will define the extent of the problem.* These facts are well known; my excuse for drawing attention to them here is to indicate that the problem of greatest significance is that presented by this orderly, continuous and coordinated direction of chemical processes.

During the pioneer developments of biochemistry, it has been usual to consider living matter in rather vague terms as a heterogeneous colloidal system in dynamic equilibrium, or to fall back upon conceptions of specific enzymes or of selective permeability. Investigation of these systems *in vitro* by chemical method must clearly form an essential step in our progress towards knowledge of the living stuff. But however valuable they may be in their own sphere, I feel that they no longer serve a useful purpose, when applied to the cell *in vivo* as general conceptions. I think even that they almost conceal some confusion of thought. Enzymes are essentially organic catalysts, capable by their very definition of accelerating chemical change. Specific enzymes may guide chemical change in the sense that an enzyme may determine that, out of a number of possible reactions, this one will take place rather than that; but enzymes are servants. They cannot determine what gets to them, nor can they be responsible for maintaining the integrity of structure which must exist; nor do we obtain much help from saying that the cell is in colloidal equilibrium. When liberated from the cell, by destroying the cell organization, reactions run to completion; inside the cell they may not. They are under some kind of control.†

When we inspect the problem more closely, its difficulties increase. Though living matter is often a gel, it may equally well be a sol; and is therefore capable of exhibiting its activities within a fluid matrix. Lyon (1906), for instance, showed long ago that the nuclei of cells could be centrifuged to one side without interfering appreciably with their function.

* At rest, a man consumes an amount of fuel equivalent to the total blood sugar in about 20 minutes, and in exercise much more. If this were not being continually replaced, he would become immediately unconscious.

† See also Hardy, VI., *Coll. Symp. Mono.*, for emphasis of this point.

Extreme order has therefore to be reconciled with a fluid anatomy. Reflection upon this matter for many years has convinced me that this cannot be done adequately without borrowing conceptions from physiology, and especially from neurology. The cell must be considered as a reflex entity, structurally organized so far as even its chemistry is concerned, with chains of chemical substances acting as it were as reflex arcs. The surfaces presented by the more highly organized molecules and particularly the proteins I regard as a kind of central nervous system. I have here purposely avoided reference to the organized structure of cells which is visible to the microscope, because attention to this tends to obscure the real issue relating to the parts below the limits of microscopic vision. Significant as the lipoids may be as surfaces of separation between the ultimate units of living matter, in the sense to which Leathes (1923) has drawn our attention, I believe that the notion that the patterns of a cell surface are connected somehow with the dynamical behaviour of the cell itself is still more important. While a pupil of Professor Barcroft, in 1912, I was able to establish definitely that oxygen entered into true chemical combination with the colloid haemoglobin. Rather later, in work upon the toxic war gases with which I was connected, it was demonstrated that organic chemical structure as such had to be considered in relation to the toxic effect of arsenical and other compounds in inducing lung oedema, essentially a physical change (Walker, 1928).

So long, however, as it was not legitimate to think of organized structure at an interface, the problem before us presented itself as incapable of any experimental attack whatever. It is to the genius of W. B. Hardy that we owe the first experimental evidence that surface structure in the chemical sense exists. This is a necessary consequence of the theory of orientation of molecules at an interface, which has been amply proved by Harkins and Langmuir in independent work, and by the fascinating researches of Adam. Further allusion to such well-established work is unnecessary. It is now legitimate to postulate rigid structure in surfaces of even mono-molecular thickness, and to believe that they must be present in cells. Here we have at once the key to our problem of directional control. With the idea of such surface mosaics behind us, can we hazard any guesses as to the types of structure which exercise this control? It is likely that we must look to substances of the protein type. Though complex carbohydrates have been shown to be capable of modifying antigenic activity (Heidelberger and Avery, 1923), the fundamental antigens appear to be protein in nature. Clearly we must look to the compounds which differ from one cell to another rather than to those which are alike. There is much to support the idea that the proteins are the truly different entities. The view that the proteins of the individual carry the stamp of individuality together with its implications as to the origin of the so-called medical "Diathesis" was early advanced by Garrod (1909). Implicit perhaps in this is the suggestion that it must be the proteins which are the responsible directive agencies; but so far as I know this was first definitely put forward by J. Lloyd (1928).

Reflection upon the properties of proteins in general does not tell us how this effect is exercised. To reach any plausible conclusions about this, it is necessary to consider the process of cell division Taking the nucleus as the most important part of a cell and the "gene" theory as a guide (Morgan, 1919), it appears to be a definite fact that a particular piece of chromatin actually exerts its influence upon the subsequent history of the individual. It is therefore pertinent to our enquiry that so far as we know, the chromatin is nucleo-protein in nature, a compound of nucleic acid and of protamines. Competent authorities consider that the nucleic acids are identical in the animal and vegetable kingdom respectively. This focuses attention upon the protamines for any differences that we find. Considering their importance, we ought to be in possession of many more facts about the protamines—they appear to constitute a vast field for enterprising organic chemists. We know that protamines are in the main compounds of the more complex amino acids, histidine, arginine, and lysine (Kossel). If we isolate in thought the instant in the reproductive cycle at which the "gene" first commences to exert its directive effect, we reach the conclusion that the surface structure which must intervene and guide the dynamic activity of subsequent events consists essentially of —CONH—, and terminal —NH_2 and —COOH groups united variously into a surface mosaic and exhibiting optical activity.* It is of course likely that many of the —NH_2 groups will be combined with nucleic acids.

I am aware that this represents a simplification which to many minds must appear not only irrelevant but shocking. I shall be told of the films of lipoids, the traces of metals, and the units of nucleic acids, with which the protamines must form considerable edifices. But it should be remembered that the lipoids and metals are not the essentially different entities, however important their catalytic and other effects may be. Since the nucleic acid-protamine edifices will create surface in three dimensions, their presence would appear only to transfer the problem to 3 dimensions instead of 2; and if we concentrate upon the points of difference where surely must lie hid the key to our problem, the essential difference would appear to lie in a pattern of just the groups above mentioned. Though it may savour of "washing in Jordan", I therefore suggest that the way to unravel this complexity is to investigate properly the behaviour of —NH_2, —COOH, and CONH— groups in interfaces. Capable of infinite variation, the structures I feel to be in essence simple. We have to

* To this should be added the special glyoxaline and guanidine grouping peculiar to histidine and arginine.

deal with their effects, catalytic and otherwise, upon compounds with which they may be brought into contact, with the possible effect of ionization, and of modification of action of the mosaic by the selective adsorption of other ions. As a guide to research upon the catalytic effects, we have now the admirable work of Hinshelwood and his colleagues (1930) and of others upon heterogeneous reactions. Not much is known yet of the directive influence of optically active substances upon chemical reactions, though Erlenmeyer has investigated this in the case of mandelic acid (see H. and E. Erlenmeyer, 1922). A considerable body of information is now available

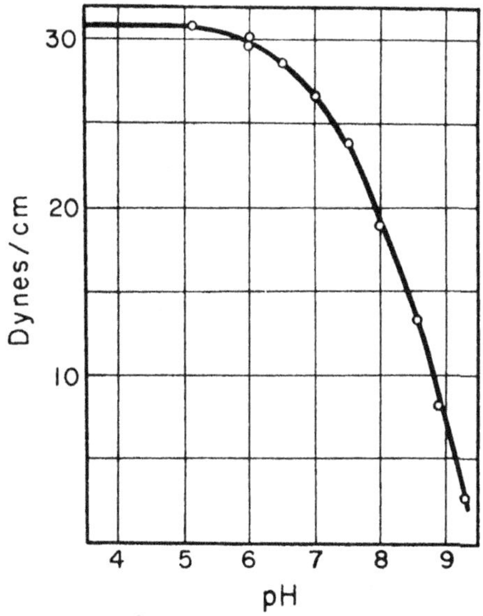

Fig. 1. Interfacial tension. Palmitic acid in benzene—aqueous phosphate solution of varying pH (Concentration of palmitic acid 0·393 gm 100 g benzene).

upon the state of the proteins. It is hardly necessary to refer to the early work of Hardy or to the extensive researches of Sørensen, Pauli, Svedberg, and their colleagues. From my present standpoint, the state of the proteins will only influence the "master" activities of the cell in so far as it may alter the nature of the active surfaces present.

One essential part of the above story is related to the ionizable groups. Our own interests in this connection have been directed mainly to an attempt to advance our knowledge as to the role of hydrogen ion concentration in adsorption effects at interfaces. Estimations by Kerridge and Winton (1929) suggest that the mean H ion concentration of living matter in the guinea pig's uterus is approximately pH 7·0.* Consideration of the ionization curves of simple carboxylic acids and simple amines shows that at this reaction, these will be present almost entirely as salts. If therefore the $—NH_2$ and —COOH groups under consideration ionize at the interface as do the simpler compounds in solution, there would be no chance for any effects due to small changes in ionization to occur at the reaction of living matter. But there are experimental grounds for the belief that these groups when set in an interface ionize at a different reaction to that which they would adopt when in true solution. Reinders (1910) appears to have been the first to draw attention to the relation between H ion concentration and interfacial tension in the case of a fatty acid dissolved in oil. In 1920–22 Hartridge and I found that dilute benzene solutions of some pure substances had a different interfacial tension against alkaline and acid buffered solutions (phosphate) only when the terminal group of the substance dissolved in the benzene possessed an ionizable polar group. This might be either free or combined to form a triglyceride. That this was not due to the method of measurement was proved by the use of a modified Rayleigh ripple method designed by my colleague.

In investigations which are not yet complete, some further points have been elucidated. Figure 1 shows the fall in interfacial tension for palmitic acid in benzene solutions against aqueous phosphate solution of different H ion concentration. The fall in interfacial tension starts rather abruptly at about pH 5·8, soon becomes linear when plotted against the pH, and continues practically linear to pH 9·5, at which point it is practically impossible to measure the tension. These changes are therefore occurring immediately over the mean range of C_H of the cell. That we are studying an effect due essentially to the —COOH head of the acid is shown by the fact that capric acid with a much smaller length of carbon chain begins to show a fall in interfacial tension also at approximately pH 6·0. The total range in pH units, over which these changes occur, is approximately pH 6·0–10·0 or some 4·0 logarithmic units. Contrasted with this, the total range over which the ionization of a simple fatty acid in solution, extends, such as proprionic acid, when plotted against the pH is about 4·0. The correspondence between these two can hardly be coincidence, and taken together with the finding that palmitic and capric acids begin to show their fall in interfacial tension when exposed to the same H ion concentration approximately, makes it very probable that the effects to which I have drawn attention are the direct consequence of ionization at the interface.

The experiments therefore indicate that ionization of the group —COOH

* The Sørensen logarithmic rotation is here used throughout. Other determinations suggest pH 7·3.

takes place at a more alkaline reaction at the interface than in true solution, or expressed in another way there is an apparent shift in the dissociation constant of the acid some 2·0–3·0 pH units to the alkaline side.* The correspondence of this curve with that of the ionization of some proteins upon the alkaline side of their isoelectric point can hardly be coincidence, though I do not wish to digress by entering into it here. With a cell at a mean H ion concentration of pH 7·0, —COOH groups ionizing at surfaces exist at a reaction at which their ionization effects may be most felt.

That the problem is more complex than this is of course obvious; groups adjacent to the ionizable ones will modify their activity, as is well known for the case of an —OH group in (say) lactic acid. I have found that the fall with erucic acid appears to start at a more acid reaction. It is interesting to obtain some concrete ideas as to the numerical values involved. At a pH 7·0, calculating from the bulk concentration, there is one H ion present in an area of which the face accommodates some 260,000 carboxyl heads. If we change the pH to 8·0, the C_H diminishing thereby 10 times, the number of —COOH groups in contact with one H ion increases some 4·6 times to about 1,200,000. Such a change makes a difference in interfacial tension of about 8 dynes, in the system considered in Fig. 1. Over this range the change in the surface ionization of the palmitic acid is some 25 per cent. Hence in theory the introduction of acid at such an interface in amounts sufficient to cover about 25 per cent of the surface should produce quite considerable changes in tension. Even allowing for buffer effects in the aqueous phase, there seems to be ample margin for the conclusion that an amount of acid much less than that required to cover the surface would produce measurable changes in tension. This argument is of importance in considering part of the objections raised by Hill (1925) to the surface tension theory of muscular contraction. I do not, however, wish to suggest that H ions are responsible for muscular motion, as this view appears to have been excluded upon other grounds by recent work (Meyerhof, 1930).

Closely associated with this problem is that of the influence of C_H upon adsorption and precipitation processes. My attention to this was directed by the finding that adsorption of the antineuritic vitamin from yeast extract was much influenced at a certain stage by the C_H of the solution from which adsorption took place, Kinnersley and Peters (1927). This was curious in view of the comparatively neutral character of the charcoal surface, at any rate so far as ionization is concerned. Fromageot and Wurmser (1924) had found that the adsorption of certain acids increased with

* The apparent $pK[\log(1/K)]$ is approximately 7·8. This constitutes a measure of interfacial ionization.

increasing C_H of the adsorbate, though there was no simple relation between the degree of dissociation and the adsorption. My colleague, Mr. H. J. Phelps, and I were able to show that upon a highly purified charcoal (Bartell and Miller, 1922) the adsorption of propionic and caproic acids followed the dissociation curves closely. In each case the undissociated molecule was the most adsorbed (Phelps and Peters, 1929). The acid extracted charcoal used in our experiments gave results in general agreement for the bases, though the changes in adsorption did not follow the ionization curves at all closely. Recently, working with a highly purified cellulose charcoal, Mr. Phelps in the Biochemical Laboratory, Oxford, has obtained a charcoal upon which the adsorption of weak acids and bases follows the ionization curves very closely, the undissociated molecule being selectively adsorbed (Phelps, 1930). With this charcoal, adsorption of propionic acid falls from a maximum value to nil under the conditions of the experiments between pH 3·2 and 6·0. Similarly the adsorption of propylamine rises from nil at pH 7·0–8·0 to a maximum value in solutions more alkaline than pH 11·0. These experiments leave no doubt in the mind that adsorption upon unionized surfaces follows closely the ionization curves, and indicate that at the mean reaction of living matter the simple acids and bases will not be appreciably adsorbed by neutral surfaces. It is interesting to note that mono-amino mono-carboxylic acids are but little adsorbed by charcoal at any pH as Michaelis originally showed. This is perhaps connected with the presence of a *Zwitterion*. It was interesting also to us to find that histamine and histidine showed change in the degree of adsorption with change in pH over the range shown by the vitamin, namely pH 5·0–8·0.

The above results raise points germane to the behaviour of proteins in very acid and very alkaline solutions, which need not detain us here. In the present connection they show that simple acids and bases would not be expected to be adsorbed upon uncharged surfaces at the mean C_H of the cell, but that adsorption of more complex substances, such as histamine, must take place and may be expected to influence profoundly the directional mosaic at the interfaces.

In addition to simple adsorption of the above type, we shall be likely to have replacement effects occurring together with the adsorption. From this point of view, the case of Fuller's earth, which Mr. Phelps has also studied, provides data of value. Fuller's earth adsorbs bases in some degree over the range of pH 2·0–12·0, approx. Alkaline solution adsorption is taking place upon the same lines as adsorption of bases upon cellulose charcoal; this he has proved by showing that the curve for adsorption of *n*-butylamine upon silica gel follows closely that upon charcoal between pH 7·0 and 11·0. As the acidity is increased, the effect of calcium substitution (first pointed out by Seidell) begins to be noticed, and we have true adsorption and replacement taking place side by side. It seems to me that further investigation of the behaviour of these simple substances is not only of practical interest in the use of adsorbents as chemical reagents, but is also likely to provide us with much information of value in the interpretation of living phenomena. The presence of a constant amount of histamine, for instance, might be expected to alter by adsorption the surface mosaic, which has been pictured, and so to disturb permanently the direction of cell reactions, and we have shown histamine to be adsorbed by charcoal over the range pH 6·0–8·0.

Before leaving the question of C_H, I wish to draw your attention to some phenomena in connection with the precipitation of certain basic substances by phosphotungstic acid which I have had reason to investigate lately. If a solution of 10 per cent phosphotungstic acid is neutralized to pH 6·0 and added in a concentration of about 5 mg per ml to a dilute solutions of a histamine salt (0·1 per cent) previously neutralized to pH 8·5, a precipitate will not appear. Addition of dilute H_2SO_4 drop by drop causes a precipitate to appear quite sharply at about pH 7·8. I have found that the C_H at which these precipitates appear depends largely upon the constitution of the base, as Fig. 2 will show; and the observation has been used by Mr. Kinnersley and myself as a means of separating impurities from vitamin B_1. Guanidine is peculiar in that it exhibits a precipitate in alkaline solution which disappears in solutions of pH 6·0 to reappear again in strongly acid solution. This property of precipitation in alkaline solution seems to be largely lost by substitution in the $—NH_2$ group, e.g. di-methyl guanidine. Such observations are largely empirical, but they emphasize if this be needed the large effects produced by small changes in reaction; and it is interesting to note that the glyoxaline and guanidine structures are common to histidine and arginine, the basic amino acids present in nucleoproteins.

These mosaics with a protein basis, which I have pictured, act as a kind of central nervous system in the cell, and form a stable background for chemical change, even when the whole is in the liquid state. There is little doubt that the outside surfaces of cells (such as bacteria) must form structures into which will be built enzyme systems, capable of catalysing highly specific chemical change. Washed bacteria have been much studied by Quastel and his colleagues (Quastel, 1928), who have made it clear that ordinary conditions of chemical equilibria apply to them. I think, however, that it is a definite mistake to believe that such considerations apply blindly within the cell. The forces which operate from surfaces here override in large part other rules. A *reductio ad absurdum* I admit, but I cannot conceive how the living cell can behave as it does, unless practically all the molecules inside it are under some kind of directed activity. One of the main functions of the surface enzymes must be to tranform the molecules presented to them, preparatory to their entry into the cell so as to simplify the compounds reaching the interior.

A calculation will indicate why the conditions under which chemical processes occur must be considerably different inside cells, quite apart from the probability that equilibrium as such is non-existent. An organism which lies upon the limit of microscopic vision is approximately some $0{\cdot}2\mu$ in diameter. According to the laws of mass action as applied to the dissociation of water, we are accustomed to write the equation of water $C_H \times C_{OH} = K_W$, and to believe that approximately one water molecule in some 555,000,000 molecules is dissociated at the ordinary temperature. Now the volume occupied by this amount of water is a cube with a side of some $0{\cdot}25\mu$, in length. Therefore, such a small organism, upon our ordinary conceptions, will contain on an average only one H ion at the mean pH 7·0. The number of molecules of other substances present in such a small volume is much smaller than the number of water molecules as may be learnt by studying the figures of Cameron (1929), for the red blood corpuscle, a much larger structure. These considerations apply with even greater force to ultra-microscopic organisms, and I feel that there is a very high degree of probability that isolated compartments of this nature occur in the larger cells. The fact is that in these small organisms we have reached a point at which averages are likely to give a poor picture of the state from moment to moment, and at which it is therefore dangerous to apply thermodynamic arguments directly. This might conceivably be of no importance in a system which possessed no structure. Upon kinetic considerations alone, I am assured by a competent authority that if a reaction required for its continuance the presence of one H ion, the effect of wide temporary fluctuations in the C_H, would mean merely that the reaction stopped during the temporary absence of H ion. If one H ion were present for half the time, the reaction would go half as fast. But in cells we have structure and large molecules. If one reaction proceeds when an H ion is present, and another can only proceed during temporary absence of H ion, then the failure of averages is likely to become of great significance. It is possible to conceive as an instance of some semipermeable membrane, in which the presence or absence of an H ion, determined whether some substance entered the cell. The entry of this substance might completely alter the character of the reaction which normally required the presence of one H ion at least for its accomplishment. Temporary fluctuations in micro-heterogeneous systems in unstable equilibrium may become of the greatest consequence—I would almost say must.* The fact that living processes exhibit a high order of accuracy in spite of these possibilities is itself an argument in favour of a constant directional control

* See also E. C. Guye, *Physico-Chemical Evolution*, 1925.

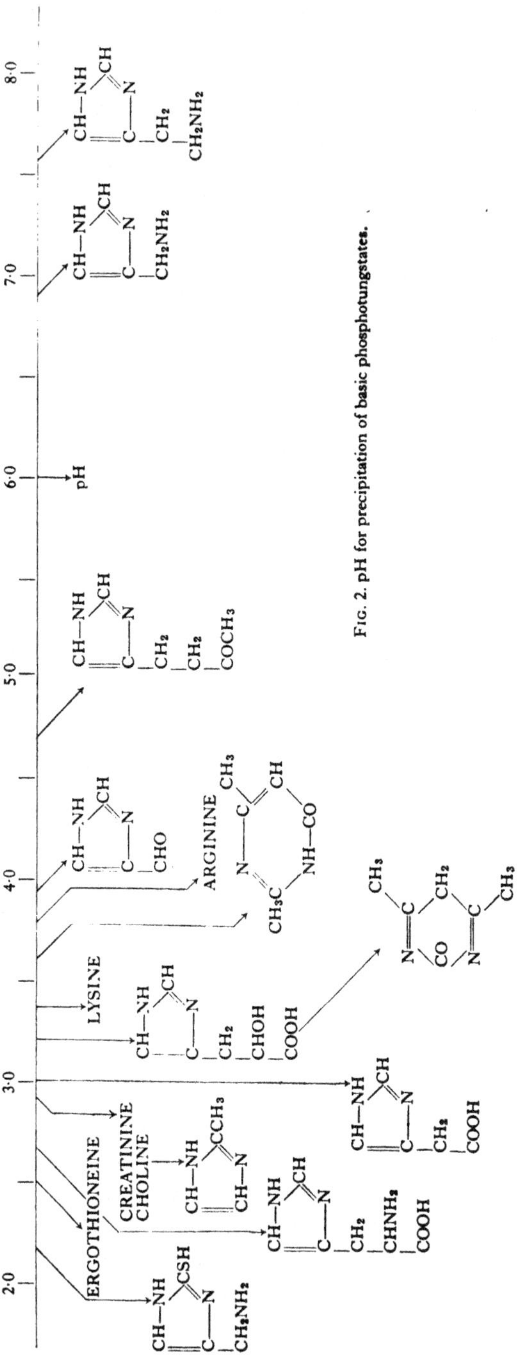

Fig. 2. pH for precipitation of basic phosphotungstates.

in cells. Though the mean C_H is some pH 7·0, large variations from this must be possible locally at interfaces, quite capable of producing optimum conditions for the action of Robison's bone phosphatase for instance, which in a bulk concentration *in vitro* shows an optimum action at pH 9·0. It must be admitted that we need to know how far it is possible to localize such pH effects at an interface, but that these ideas are not fantastic can be gathered from studying the action of acids upon the vinegar eel. I have seen this surviving with apparent comfort in a solution of pH 2·0 which is also the approximate reaction of our gastric juice. It is inconceivable that the inside properties of this animal differ so profoundly from all other organisms as to be capable of having a mean pH approximating to this acidity.

In the ultimate structural units of the cell, we have reached the limit for the application of ordinary statistical considerations, and must substitute some more anatomical view, based upon control by surface. There is work already showing the control of molecules at a distance. Ramsden (1926) has found that substances definitely influence surface films when present in concentrations less than one molecule thick. Most important of all, Hardy and Nottage (1928), in their adhesion experiments, have found that orientation effects may be transmitted from a surface such as that of a metal, capable of influencing the surrounding molecules for a distance of 7–8μ. Quoting from my previous discussion.

"Here we have a means by which our mosaic may radiate its effect throughout the cell. It is also perfectly possible to appreciate how a co-ordinate structure may be maintained in a medium which is apparently liquid. We have in effect chains of liquid crystals which need not be more than a few molecules thick. Once that we have seen this, it becomes possible to develop some dim conception as to how the miracle occurs that colloid molecules, which apparently have no particular location in the cell, can yet control the synthetic and other reactions which are proceeding. This theory is all that is needed to enable us to understand how substances can reach a special site in the cell. Between our chains of molecules, fixed by their radiating webs, there will exist paths from the external to the internal surface, the *capillaries* of the cell."

With this conception behind us as a working hypothesis, we may with some confidence develop what I have elsewhere termed coordinative biochemistry. I think that reflection upon it will give the schemes of Meyerhof (1926) for the carbohydrate cycle and of Keilin and Dixon for oxidation systems a newer meaning. Whatever may be felt about this, it is most important to understand that when we consider the effects of substances in relation to living things, we have to operate in a structurally organized system, where in consequence relations to surface dwarf other considerations. In this way it comes about that the surface mosaics are responsible for the integration of the activities of the whole.

I hope that I have been able to indicate clearly why a hypothesis of control by surface mosaics is necessary, and to show that it provides a fruitful basis for experimental attack upon a problem, which must surely rank among the most interesting with which science has to deal.

I wish to acknowledge the considerable help which I have gained by the opportunity of discussing these problems with Mr. Hinshelwood and Professor Milne, and also my gratitude to my colleague Mr. H. J. Phelps for his criticisms and for his permission to quote from his recent work. I need hardly say that I would not commit them thereby to agreement with these speculations.

Note added subsequently. There has been brought to my notice the interesting book by du Nouy on *Surface Equilibria of Colloids*, 1926. I regret that reference to it has not been made in the text, but would refer the reader in connection with the above to p. 84 especially, in which the author envisages monomolecular layers so arranged that small disturbances may be perceived far beyond the actual seat of reaction.

REFERENCES

ADAM (1927), *Science Progress*, April.
CAMERON (1929), *Trans. Roy. Soc., Canada*, **23**, 151.
DIXON, M. (1929), *Biol. Rev.* **4**, 359.
ERLENMEYER, E. and H. (1922), *Biochem. Z.* **133**, 52.
FROMAGEOT and WURMSER, (1924), *C. R. Acad. Sci. Paris* **179**, 972.
GARROD, A. E. (1909), *Inborn Errors of Metabolism*, 1st ed., Frowde; (1927), "*Diathesis*" *Brit. Met. J.*, p. 967.
GORTNER, (1929), *Outlines of Biochemistry*, Wiley.
HARDY, W. B. (1913), *Proc. Roy. Soc.* **A. 88**, etc.
HARDY and NOTTAGE, (1928), *Proc. Roy. Soc.* **A. 209.**
HARKINS, (1917), *J. Amer. Chem. Soc.* **39**, 354.
HARTRIDGE and PETERS, (1922), *Proc. Roy. Soc.* **A.101**, 348.
HEIDELBERGER, M. and AVERY, (1923), *J. Exper. Med.* **37**, 275.
HILL, A. V. (1925), *Proc. Roy. Soc.* **B. 98**, 406.
HINSHELWOOD, C. W. (1930), *Kinetics of Chemical Changes in Gaseous Systems*, Oxf. Univ. Press,
HOPKINS, F. G. (1912), *Pres. Address (Physiology), Brit. Ass. Rep.*
KEILIN, (1929), *Proc. Roy. Soc.* **B. 104**, 206.
KERRIDGE and WINTON, (1929), *J. Physiol.* **67**, 66.
KINNERSLEY and PETERS (1927), *Biochem. J.* **21**, 777.
KOSSEL (1928), *The Protamines*, Longman.
LANGMUIR (1917), *J. Amer. Chem. Soc.* **39**, 18.
LEATHES (1923), "*Croonian Lectures*", *Lancet.*
LLOYD, D. JORDAN (1928), *Biol. Rev.* **3**, 165.
LYON (1906), *Atch. Entwiss. Mech.* **23**, 151.
MEYERHOF, O. (1926), *J. Gen. Physiol.* **8**, 533; (1930), *Die Chemischen Vorgänge im Muskel*, 304, Springer.
MILLER and BARTELL (1922), *J. Amer. Chem. Soc.* **54**, 1866.

MORGAN (1919), *Physical Basis of Heredity.*
PAULI, W., *The Proteins.*
PETERS (1912), *J. Physiol.* **44,** 131; (1929), *Harben Lectures, J. State Medicine*, 38.
PHELPS and PETERS (1929), *Proc. Roy. Soc.* **A. 124,** 554.
PHELPS (1930), *J. Soc. Chem. Ind.*, June, *Proc. Biochem. Soc.*
QUASTEL (1928), *J. Hyg.* **28,** 139.
RAMSDEN (1926), *Trans. Faraday Soc.* **22,** 2.
REINDERS, *van Bemmelin Festschrift* (quoted from Freundlich, *Capillary Chemistry*, p. 89, Trans., 3rd ed.).
RIDDLE and HONEYWELL (1923), *Amer. Nat.* **52,** 412.
SÖRENSEN, Papers in *Compt. Rend. Lab. Carlsberg.*
SVEDBERG, Papers in *J. Amer. Chem. Soc.*
WALKER (1928), *Biochem. J.* **22,** 292.
WARBURG, O. (1923), *Jahres. Gen. Physiol.* **1,** 136, Collected Papers.

Author Index

The figures in italic type refer to lists of references

Subject Index

Printed in Dunstable, United Kingdom

91338446R00188